Lecture Notes in Computer Science 12485

More information about this subseries at http://www.springer.com/series/7151

James F. Peters · Andrzej Skowron (Eds.)

Transactions on
Rough Sets XXII

Springer

Editors-in-Chief
James F. Peters 🆔
University of Manitoba
Winnipeg, MB, Canada

Andrzej Skowron 🆔
Systems Research Institute,
Polish Academy of Sciences
Warsaw, Poland

CTN UKSW
Warsaw, Poland

ISSN 0302-9743   ISSN 1611-3349 (electronic)
Lecture Notes in Computer Science
ISSN 1861-2059   ISSN 1861-2067 (electronic)
Transactions on Rough Sets
ISBN 978-3-662-62797-6   ISBN 978-3-662-62798-3 (eBook)
https://doi.org/10.1007/978-3-662-62798-3

Preface

Volume XXII of the *Transactions on Rough Sets* (TRS) is a continuation of a number of research streams that have grown out of the seminal work of Zdzisław Pawlak[1] during the first decade of the 21st century.

The paper by Mohammad Azad investigates bi-objective optimization of classification and regression (CART)-like decision trees for data analysis. The results show that at least one of the considered techniques (L -19) can be useful for the extraction of knowledge from medium-sized decision tables and for its representation by decision trees. The paper by Nicholas Baltzer and Jan Komorowski presents the design and implementation of a multi-core execution process in ROSETTA, which is optimized for speed and modular extension. The parallel ‖ – ROSETTA software was tested using four datasets of different sizes for computational speed and memory usage. This tool is a successor to the Rosetta system developed in 1994 to exploit rough set paradigms in machine learning. The paper by Stefania Boffa and Brunella Gerla introduces sequences of orthopairs generated by tolerance relations as special sequences of rough sets. These special sequences of coverings represent situations where new information is gradually provided on smaller and smaller sets of objects. The paper also investigates several operations between sequences of orthopairs to provide concrete representations for certain classes of many-valued structures. The paper by Arun Kumar contributes to the algebraic and logical developments of rough set theory. The study leads to new semantics for the logics corresponding to the classes of Stone algebras, dual Stone algebras, and regular double Stone algebras. Representations of distributive lattices (with operators) and Heyting algebras (with operators) are proved. Moreover, it is shown that various negations appear in Dunn's Kite negations in this generalized rough set theory. The paper by Dávid Nagy is on similarity-based (tolerance) rough sets, where the system of base sets is generated by correlation clustering. The space generated by the clustering mechanism, on the one hand, represents the interpreted similarity properly and on the other hand, reduces the number of base sets to a manageable size. This work deals with the properties and applicability of this space, presenting advantages that can be gained from correlation clustering. Finally, an errata for a reference which appeared in *Transactions on Rough Sets*, vol. XVII, LNCS 8375, pages 197–173 is included.

The editors would like to express their gratitude to the authors of all submitted papers. Special thanks are due to the following reviewers: Davide Ciucci, Richard Jensen, Mikhail Moshkov, Lech Polkowski, and Dominik Ślęzak.

---

[1] See, *e.g.*, Pawlak, Z., A Treatise on Rough Sets, *Transactions on Rough Sets* IV, (2006), 1–17. See, also, Pawlak, Z., Skowron, A.: Rudiments of rough sets, *Information Sciences* 177 (2007) 3–27; Pawlak, Z., Skowron, A.: Rough sets: Some extensions, *Information Sciences* 177 (2007) 28–40; Pawlak, Z., Skowron, A.: Rough sets and Boolean reasoning, *Information Sciences* 177 (2007) 41–73.

The editors and authors of this volume extend their gratitude to Alfred Hofmann, Christine Reiss, and the LNCS staff at Springer for their support in making this volume of TRS possible.

October 2020
James F. Peters
Andrzej Skowron

LNCS Transactions on Rough Sets

The *Transactions on Rough Sets* series has as its principal aim the fostering of professional exchanges between scientists and practitioners who are interested in the foundations and applications of rough sets. Topics include foundations and applications of rough sets as well as foundations and applications of hybrid methods combining rough sets with other approaches important for the development of intelligent systems. The journal includes high-quality research articles accepted for publication on the basis of thorough peer reviews. Dissertations and monographs up to 250 pages that include new research results can also be considered as regular papers. Extended and revised versions of selected papers from conferences can also be included in regular or special issues of the journal.

Errata

In the paper

Henryk Rybiński and Mariusz Podsiadło: *Rough Sets in Economy and Finance*. Transactions on Rough Sets XVII, pp. 109–173, Springer (2014) https://link.springer.com/chapter/10.1007/978-3-642-54756-0_6, https://doi.org/10.1007/978-3-642-54756-0_6

the correct reference to the paper by J.K. Baltzersen should be the following:

7. Baltzersen, J.K.: An attempt to predict stock market data: a rough sets approach. Diploma Thesis, Knowledge Systems Group, Department of Computer Systems and Telematics, The Norwegian Institute of Technology, University of Trondheim, 1996.

Contents

Decision Trees with at Most 19 Vertices for Knowledge Representation

Mohammad Azad$^{(\boxtimes)}$

College of Computer and Information Science, Jouf University,
Sakaka 72441, Saudi Arabia
mmazad@ju.edu.sa

Abstract. We study decision trees as a means of representation of knowledge. To this end, we design two techniques for the creation of CART (Classification and Regression Tree)-like decision trees that are based on bi-objective optimization algorithms. We investigate three parameters of the decision trees constructed by these techniques: number of vertices, global misclassification rate, and local misclassification rate.

Keywords: Knowledge representation · Decision trees · Dynamic programming · Bi-objective optimization

1 Introduction

Decision trees are used to a large degree as classifiers [5,6,10], as a means of representation of knowledge [4,7], and as a kind of algorithms [20,25]. We investigate here decision trees as a means of representation of knowledge.

Let us consider a decision tree Γ for a decision table D. We investigate three parameters of Γ:

- $N(\Gamma)$ – the number of vertices in Γ.
- $G(D, \Gamma)$ – the global misclassification rate [7], which is equal to the number of misclassifications of Γ divided by the number of rows in D.
- $L(D, \Gamma)$ – the local misclassification rate [7], which is the maximum fraction of misclassifications among all leaves of Γ. One can show that $G(D, \Gamma)$ is at most $L(D, \Gamma)$.

The decision tree Γ should have a reasonable number of vertices to be understandable. To express properly knowledge from the decision table D, this tree should have an acceptable accuracy. In [7], we mentioned that the consideration of only the global misclassification rate may be insufficient: the misclassifications may be unevenly distributed and, for some leaves, the fraction of misclassifications can be high. To deal with this situation, we should consider also the local misclassification rate.

The optimization of the parameters of decision tree has been studied by many researchers [9,11–13,16–19,24,26]. One of the directions of the research is the bi-objective optimization [1–8]. In [7], we proposed three techniques for the building

© Springer-Verlag GmbH Germany, part of Springer Nature 2020
J. F. Peters and A. Skowron (Eds.): TRS XXII, LNCS 12485, pp. 1–7, 2020.
https://doi.org/10.1007/978-3-662-62798-3_1

of decision trees based on the bi-objective optimization of trees and studied the parameters N, G, and L of the constructed decision trees. Unfortunately, these techniques are applicable to medium-sized decision tables with categorical features only and, sometimes, the number of vertices in the trees is too high. In particular, the decision tree Γ_1 with the minimum number of vertices constructed by these techniques for the decision table D NURSERY from the UCI Machine Learning Repository [15] has the following parameters: $N(\Gamma_1) = 70$, $G(D, \Gamma_1) = 0.10$, and $L(D, \Gamma_1) = 0.23$.

In this paper, instead of conventional decision trees, we study CART-like (CART-L) decision trees introduced in the books [1,2]. As the standard CART [10] trees, CART-L trees use binary splits instead of the initial features. The standard CART tree uses in each internal vertex the best split among all features. A CART-L tree can use in each internal vertex the best split for an arbitrary feature. It extends essentially the set of decision trees under consideration. In [1,2], we applied Gini index to define the notion of the best split. In this paper, we use another parameter abs [2].

We design two techniques that build decision trees for medium-sized tables (at most $10,000$ rows and at most 20 features) containing both categorical and numerical features. These techniques are based on bi-objective optimization of CART-L decision trees for parameters N and G [1], and for parameters N and L. Both techniques construct decision trees with at most 19 vertices (at most 10 leaves and at most nine internal vertices). The choice of 19 is not random. We consider enough understandable trees with small number of non-terminal vertices which can be useful from the point of view of knowledge representation. This choice is supported by some experimental results published in [1]. One technique (G-19 technique) was proposed in [1]. Another one (L-19 technique) is completely new. We apply the considered techniques to 14 data sets from the UCI Machine Learning Repository [15], and study three parameters N, G, and L of the constructed trees. For example, for the decision table D NURSERY, L-19 technique constructs a decision tree Γ_2 with $N(\Gamma_2) = 17$, $G(D, \Gamma_2) = 0.12$, and $L(D, \Gamma_2) = 0.22$.

The obtained results show that at least one of the considered techniques (L-19 technique) can be useful for the extraction of knowledge from medium-sized decision tables and for its representation by decision trees. This technique can be used in different areas of data analysis including rough set theory [14,21–23,27]. In rough set, the decision rules are used extensively. We can easily derive decision rules from the constructed decision trees and use them in rough set applications.

We arrange the remaining of the manuscript as follows. Two techniques for decision tree building are explained in Sect. 2. The output of the experiments is in Sect. 3. Finally, Sect. 4 contains brief conclusion.

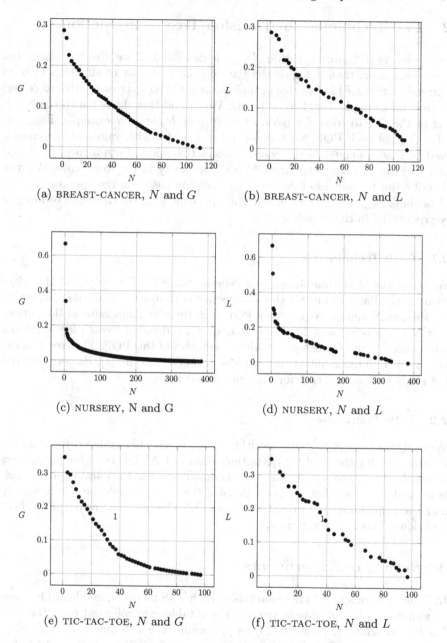

Fig. 1. Sets of Pareto optimal points for tables BREAST-CANCER, NURSERY, and TIC-TAC-TOE for pairs of parameters N, G and N, L

2   Two Techniques for Decision Tree Construction

In the books [1,2], an algorithm $\mathcal{A}_{POPs}$ is described. If we give this algorithm a decision table, then it builds the Pareto front – the set of all POPs (Pareto optimal points) for bi-objective optimization of CART-L trees relative to N and G (see, for example, Fig. 1(a), (c), (e)). We extend this algorithm to the building of the Pareto front for parameters N and L (see, for example, Fig. 1(b), (d), (f)). For each POP, we can get a decision tree with values of the considered parameters equal to the coordinates of this point. Both algorithm $\mathcal{A}_{POPs}$ and its extension have exponential time complexity in the worst case. We now describe two techniques of decision tree building based on the operation of the algorithm $\mathcal{A}_{POPs}$ and its extension. The time complexity of these two techniques is exponential in the worst case.

2.1   G-19 Technique

We apply the algorithm $\mathcal{A}_{POPs}$ to a decision table D. The output of this algorithm is the Pareto front for the bi-objective optimization of CART-L trees for parameters N and G. We choose a POP with the maximum value of the parameter N which is at most 19. After that, we get a decision tree Γ, for which the parameters N and G are equal to the coordinates of this POP. The tree Γ is the output of G-19 technique. This technique was described in [1]. However, we did not study the parameter L for the constructed trees.

2.2   L-19 Technique

We apply the extension of the algorithm $\mathcal{A}_{POPs}$ to a decision table D to create the Pareto front for the bi-objective optimization of CART-L trees for parameters N and L. We choose a POP with the maximum value of the parameter N which is at most 19. After that, we get a decision tree Γ, for which the parameters N and L are equal to the coordinates of this POP. The tree Γ is the output of L-19 technique. This is a new technique.

3   Results of Experiments

In Table 1, we describe 14 decision tables, each with its name, number of features as well as number of objects (rows). These tables are collected from the UCI Machine Learning Repository [15] for performing the experiments.

We applied G-19 and L-19 techniques to each of these tables and found values of the parameters N, G, and L for the constructed decision trees. Table 2 describes the experimental results.

The obtained results show that the use of L-19 technique in comparison with G-19 technique allows us to decrease the parameter L on average from 0.16 to 0.11 at the cost of a slight increase in the parameter G on average from 0.06 to 0.07.

Table 1. Decision tables which are collected for performing the experiments

Decision table	#Features	#Objects (rows)
BALANCE-SCALE	5	625
BREAST-CANCER	10	266
CARS	7	1728
HAYES-ROTH-DATA	5	69
HOUSE-VOTES-84	17	279
IRIS	5	150
LENSES	5	10
LYMPHOGRAPHY	19	148
NURSERY	9	12960
SHUTTLE-LANDING	7	15
SOYBEAN-SMALL	36	47
SPECT-TEST	23	169
TIC-TAC-TOE	10	958
ZOO-DATA	17	59

Table 2. Results of experiments

Decision table	G-19 technique			L-19 technique		
	N	G	L	N	G	L
BALANCE-SCALE	19	0.19	0.38	11	0.20	0.32
BREAST-CANCER	19	0.16	0.26	19	0.18	0.20
CARS	19	0.10	0.39	19	0.12	0.23
HAYES-ROTH-DATA	17	0.06	0.17	17	0.06	0.17
HOUSE-VOTES-84	19	0.02	0.02	19	0.02	0.02
IRIS	17	0.00	0.00	17	0.00	0.00
LENSES	7	0.00	0.00	7	0.00	0.00
LYMPHOGRAPHY	19	0.06	0.20	19	0.10	0.11
NURSERY	19	0.11	0.34	17	0.12	0.22
SHUTTLE-LANDING	13	0.00	0.00	13	0.00	0.00
SOYBEAN-SMALL	7	0.00	0.00	7	0.00	0.00
SPECT-TEST	17	0.02	0.09	19	0.02	0.02
TIC-TAC-TOE	19	0.18	0.39	19	0.21	0.25
ZOO-DATA	17	0.00	0.00	17	0.00	0.00
Average	16.29	0.06	0.16	15.71	0.07	0.11

4   Conclusions

We proposed to evaluate the accuracy of decision trees not only by the global misclassification rate G but also by the local misclassification rate L, and designed new L-19 technique. This technique constructs decision trees having at most 19 vertices and acceptable values of the parameters G and L. Later we are planning to extend this technique to multi-label decision tables using bi-objective optimization algorithms described in [2,3]. Also, our goal is to make more experiments with other numbers of vertices like 13, 15, 17, 21, 23, etc. Another direction of future research is to design some heuristics to overcome the problem of working with larger data set.

Acknowledgments. The author expresses his gratitude to Jouf University for supporting this research. The author would like to give thanks to Igor Chikalov and Mikhail Moshkov for valuable comments. The author also would like to give thanks to anonymous reviewers for their suggestions.

References

1. AbouEisha, H., Amin, T., Chikalov, I., Hussain, S., Moshkov, M.: Extensions of Dynamic Programming for Combinatorial Optimization and Data Mining. Intelligent Systems Reference Library, vol. 146. Springer, Heidelberg (2019). https://doi.org/10.1007/978-3-319-91839-6
2. Alsolami, F., Azad, M., Chikalov, I., Moshkov, M.: Decision and Inhibitory Trees and Rules for Decision Tables with Many-Valued Decisions. Intelligent Systems Reference Library, vol. 156. Springer, Heidelberg (2020). https://doi.org/10.1007/978-3-030-12854-8
3. Azad, M.: Decision and inhibitory trees for decision tables with many-valued decisions. Ph.D. thesis, King Abdullah University of Science & Technology, Thuwal, Saudi Arabia (2018). http://hdl.handle.net/10754/628023
4. Azad, M.: Knowledge representation using decision trees constructed based on binary splits. KSII Trans. Internet Inf. Syst. **14**(10), 4007–4024 (2020). https://doi.org/10.3837/tiis.2020.10.005
5. Azad, M., Chikalov, I., Hussain, S., Moshkov, M.: Restricted multi-pruning of decision trees. In: 13th International FLINS Conference on Data Science and Knowledge Engineering for Sensing Decision Support, FLINS, pp. 371–378 (2018). https://doi.org/10.1142/9789813273238_0049
6. Azad, M., Chikalov, I., Hussain, S., Moshkov, M.: Multi-pruning of decision trees for knowledge representation and classification. In: 3rd IAPR Asian Conference on Pattern Recognition, ACPR 2015, Kuala Lumpur, Malaysia, 3–6 November 2015, pp. 604–608. IEEE (2015). https://doi.org/10.1109/ACPR.2015.7486574
7. Azad, M., Chikalov, I., Moshkov, M.: Decision trees for knowledge representation. In: Ropiak, K., Polkowski, L., Artiemjew, P. (eds.) 28th International Workshop on Concurrency, Specification and Programming, CS&P 2019, Olsztyn, Poland, 24–26 September 2019. CEUR Workshop Proceedings, vol. 2571. CEUR-WS.org (2019). http://ceur-ws.org/Vol-2571/CSP2019_paper_1.pdf
8. Azad, M., Chikalov, I., Moshkovc, M.: Representation of knowledge by decision trees for decision tables with multiple decisions. Proc. Comput. Sci. **176**, 653–659 (2020). https://doi.org/10.1016/j.procs.2020.09.037

9. Boryczka, U., Kozak, J.: New algorithms for generation decision trees - ant-miner and its modifications. In: Abraham, A., Hassanien, A.E., de Leon Ferreira de Carvalho, A.C.P., Snásel, V. (eds.) Foundations of Computational Intelligence - Volume 6: Data Mining, vol. 206, pp. 229–262. Springer, Heidelberg (2009). https://doi.org/10.1007/978-3-642-01091-0_11

10. Breiman, L., Friedman, J.H., Olshen, R.A., Stone, C.J.: Classification and Regression Trees. Wadsworth and Brooks, Monterey (1984)

11. Breitbart, Y., Reiter, A.: A branch-and-bound algorithm to obtain an optimal evaluation tree for monotonic boolean functions. Acta Inf. **4**, 311–319 (1975)

12. Chai, B., Zhuang, X., Zhao, Y., Sklansky, J.: Binary linear decision tree with genetic algorithm. In: 13th International Conference on Pattern Recognition, ICPR 1996, Vienna, Austria, 25–19 August 1996, vol. 4, pp. 530–534. IEEE (1996)

13. Chikalov, I., Hussain, S., Moshkov, M.: Totally optimal decision trees for Boolean functions. Discrete Appl. Math. **215**, 1–13 (2016)

14. Delimata, P., Moshkov, M., Skowron, Á., Suraj, Z.: Inhibitory Rules in Data Analysis: A Rough Set Approach, Studies in Computational Intelligence, vol. 163. Springer, Heidelberg (2009). https://doi.org/10.1007/978-3-540-85638-2

15. Dua, D., Graff, C.: UCI machine learning repository. University of California, Irvine, School of Information and Computer Sciences (2017). http://archive.ics.uci.edu/ml

16. Garey, M.R.: Optimal binary identification procedures. SIAM J. Appl. Math. **23**, 173–186 (1972)

17. Heath, D.G., Kasif, S., Salzberg, S.: Induction of oblique decision trees. In: Bajcsy, R. (ed.) 13th International Joint Conference on Artificial Intelligence, IJCAI 1993, Chambéry, France, 28 August–3 September 1993, pp. 1002–1007. Morgan Kaufmann (1993)

18. Hyafil, L., Rivest, R.L.: Constructing optimal binary decision trees is NP-complete. Inf. Process. Lett. **5**(1), 15–17 (1976)

19. Martelli, A., Montanari, U.: Optimizing decision trees through heuristically guided search. Commun. ACM **21**(12), 1025–1039 (1978)

20. Moshkov, M.J.: Time complexity of decision trees. In: Peters, J.F., Skowron, A. (eds.) Transactions on Rough Sets III. LNCS, vol. 3400, pp. 244–459. Springer, Heidelberg (2005). https://doi.org/10.1007/11427834_12

21. Moshkov, M., Zielosko, B.: Combinatorial Machine Learning - A Rough Set Approach, Studies in Computational Intelligence, vol. 360. Springer, Heidelberg (2011). https://doi.org/10.1007/978-3-642-20995-6

22. Pawlak, Z.: Rough Sets - Theoretical Aspect of Reasoning About Data. Kluwer Academic Publishers, Dordrecht (1991)

23. Pawlak, Z., Skowron, A.: Rudiments of rough sets. Inf. Sci. **177**(1), 3–27 (2007)

24. Riddle, P., Segal, R., Etzioni, O.: Representation design and brute-force induction in a Boeing manufacturing domain. Appl. Artif. Intell. **8**, 125–147 (1994)

25. Rokach, L., Maimon, O.: Data Mining with Decision Trees: Theory and Applications. World Scientific Publishing, River Edge (2008)

26. Schumacher, H., Sevcik, K.C.: The synthetic approach to decision table conversion. Commun. ACM **19**(6), 343–351 (1976)

27. Skowron, A., Rauszer, C.: The discernibility matrices and functions in information systems. In: Słowinski, R. (ed.) Intelligent Decision Support. Handbook of Applications and Advances of the Rough Set Theory, pp. 331–362. Kluwer Academic Publishers, Dordrecht (1992)

||-ROSETTA

Nicholas Baltzer[1,2(✉)] [iD] and Jan Komorowski[1,3] [iD]

[1] Department of Cell and Molecular Biology, Uppsala University, Uppsala, Sweden
{nicholas.baltzer,jan.komorowski}@icm.uu.se
[2] Department of Medical Epidemiology and Biostatistics, Karolinska Institute, Stockholm, Sweden
[3] Polish Academy of Sciences, Warsaw, Poland

Abstract. Technology improves every day. In order for an established theory to maintain relevance, implementations of such theory must be updated to take advantage of the new improvements. ROSETTA, a framework based on Rough Set theory, was developed in 1994 to exploit Rough Set paradigms in Machine Learning. Since then, much has happened in the field of Computer Technology, and to fully exploit these benefits ROSETTA needed to evolve. We designed and implemented a multi-core execution process in ROSETTA, optimized for speed and modular extension. The program was tested using four datasets of different sizes for computational speed and memory usage, the factors considered the primary limitations of classification and Machine Learning. The results show an increase in computation speed consistent with expected gains. The scaling per thread of memory usage was less than linear after five threads with increases in memory based primarily on the number of objects in the dataset. The number of features in the data increased the base memory needed but did not significantly impact the memory scaling by threads. The multi-core implementation was successful, and ||-ROSETTA (pronounced Parallel-ROSETTA) is capable of fully exploiting modern hardware solutions.

Keywords: Bioinformatics · Rough set theory · Parallel computing

1 Introduction

1.1 Background

The application of Rough Set theory to practical matters revolves around the availability and practicality of modern tools and implementations. As the performance and usability of these tools improve, their adoption widens, and the algorithms they apply become more standard for future research.

ROSETTA [14] is a tool, a framework, for Rough Set algorithms. It handles a wide range of functions in the form of a pipeline applied to a dataset. It will commonly include completion, discretization, reduct computation, classification, validation, and publishing.

© Springer-Verlag GmbH Germany, part of Springer Nature 2020
J. F. Peters and A. Skowron (Eds.): TRS XXII, LNCS 12485, pp. 8–25, 2020.
https://doi.org/10.1007/978-3-662-62798-3_2

ROSETTA was built in 1994 and came long before the switch to multi-core processors and hyper-threading, hence it did not have support for parallel computing. The use of threading in computer-assisted science has increased massively in later years and made possible computational loads of exponential magnitude; thus updating ROSETTA to use threaded computations is a step in maintaining the practicality and application of Rough Sets in Machine Learning, an approach that has already proven successful in application [1]. Retaining Rough Set theory implementations is necessary as a wide swath of the computer-assisted science community is turning more towards deep learning, an approach that prohibits any reasonable interpretation of the underlying classifier and thus gives up the discovery of any explanatory path from premise to antecedent.

The design of ||-ROSETTA is an attempt to be as effective as possible in favor of being as efficient as possible. Thread independence and speed are both maximized while retaining the modular simplicity of adding new algorithms. This has the benefit of greatly limiting inter-core communication, but the focus on efficacy maximizes memory usage. The parallelization of ||-ROSETTA is primarily handled at the level of experiments and not within the algorithms themselves. It is possible to run multiple discretization algorithms in parallel, but it is not possible to run a single instance of a discretization algorithm using multiple threads. The only exception to this is a reduct computation that allows for parallel execution within the algorithm itself. Adding parallel execution possibilities within the algorithms themselves would require a unique per-algorithm implementation to ensure that no memory inconsistencies were introduced, and would only work for the algorithms that could compute datapoints independently of each other. All algorithms that relied on computing results sequentially, such as reading data from a file or discretizing data based on previous discretization cuts, would be unable to benefit from parallel execution.

An early effort to provide rough set computations in parallel was the design of PSL [4] – a language for distributed computations – which subsequently was applied to distributing computation of algorithms implemented in the RSES library [3]. Distributed computations differ significantly, however, from multi-core execution.

1.2   ROSETTA

ROSETTA was designed in collaboration between the laboratories of Jan Komorowski and Andrzej Skowron and implemented by Aleksander Øhrn. It included a number of algorithms and modules developed within the RSES framework [3]. It was created in C++, and has seen multiple extensions as new algorithms and techniques have been added to its repertoire [10].

Functionality. The core functionality revolves around a pipeline of algorithms that can be applied in sequence to a dataset for the purpose of analysis and classification. It has been used in many different fields with authors and users of the current package concentrating on bioinformatics applications. The algorithms deal with exporting data, completing data, discretizing data, reduct generation, data filtering, batch classification, data approximation, rule generation, rule filtering, statistical analysis, dataset generation, and dataset partitioning. There are multiple algorithms available that can fulfil each of these roles within the pipeline, and the pipeline itself can be run in several configurations, such as training-testing splits or k-fold cross-validation where algorithms are applied separately to each part of the dataset ahead of the testing phase.

The output also takes multiple forms. Discretization cuts and classification rules can serve as both input and output, and there are several statistical reports available for the classifier (confusion matrix, ROC curve, Accuracies). Recently, ROSETTA also became available as an R package [8].

There have been multiple updates to ROSETTA over the years. These additions have focused on the availability of new algorithms rather than any updates to the core functionality or computational process.

Design. ROSETTA is constructed with a hierarchy design. All structures and algorithms inherit from the persistent base class in a strict hierarchy. For instance, the discretization algorithms inherit from the discretization base class, which inherits from the algorithm base class, which inherits from the persistent base class. All base classes include a set of virtual methods to ensure that additions can function within the established framework of execution and manipulation. In addition to this hierarchy of classes, there is also a set of independent helper classes such as smart pointers, handles, and static I/O functions amongst others.

There are four distinct major categories that perform all actions in the ROSETTA framework; structure, transformer, classifier, and executor. All classes in the structure category fall under the structure base class, while all classes in the transformer-, classifier-, and executor- categories fall under the algorithm base class.

Inheriting from the structure base class are all data objects, such as datasets and rules. These classes contain no transformation methods, only access and manipulation methods. To ensure that structures can be easily manipulated and destroyed, they all require reset and duplication methods via pure virtual functions in the structure base class.

The transformer category contains the processes that transform or extract information from data objects, such as discretization or reduct computations. Transformers may apply any transformation to the data and return any type of data object, with the only requirement (from pure virtual functions in the algorithm base class) being that it must take a structure as argument and return a structure handle. For instance, a discretization algorithm will take a dataset, transform the values, and return the transformed dataset, while a rule generator will take a reduct set and return a set of rules. Transformer is the largest category, and the one that has seen the most expansion over the years. The transformer category base class is algorithm.

The classifier category contains all forms of classification. Inheriting from the classifier base class are the different classifier schemas as well as different classification methods like batching. Classifiers print various statistics such as confusion matrices, accuracies, and ROCs, given a dataset and a set of rules. The classification algorithm is usually the last step in a pipeline.

Executors detail the execution of an algorithm onto a structure. These largely revolve around the continuity of input and output data-structures as well as any error handling that might arise when the input structure of one algorithm does not correspond to anything the algorithm expects or can handle. Execution also comes in different variants, such as cross-validation, where each conformation of the dataset is computed and validated independently of the other conformations.

Each executor pipeline starts with a dataset. This dataset is used as the input to the first algorithm, which in turn can have any structure as its output. To preserve all information from the different algorithms, each output structure is appended to a pipeline results list in order. This means that the algorithms in the pipeline do not need to match inputs and outputs impeccably as long as there is an appropriate input available in the results list. Such a construct allows a pipeline to reuse previous output structures and insert algorithms that do not produce a meaningful output structure. Reusing previous structures means that a pipeline consisting of a reduct computation followed by a rule generation does not need a second reduct computation in order to run a second, different, rule generator. It also means that the pipeline algorithms do not necessarily need to work towards a singular purpose but can generate side-effects such as data and rule exporting or multiple classifiers for comparison purposes in the same pipeline. The results list is ordered and the structures within will be attempted by the algorithm from most recent to last recent.

1.3   Open-MP

Open Multi-Processing (OpenMP) is an application programming interface for multi-threaded operations in a shared memory computation environment. It supports Fortran, C, and C++. It consists of a set of compiler directives, libraries, and run-time variables. OpenMP supports most operating platforms with the three largest being Windows, macOS, and Linux. It allows for a comparatively simple approach to threading computer instructions in both shared and independent memory environments.

2   Methods

2.1   Parallel Implementation

There are two areas where parallel computations are implemented: the cross-validation/batch executor and the reduct computation algorithm. These are both areas that involve independent data requiring significant computation time. The executor is used for cross-validations and batch training/testing validations, and is where most pipelines can be parallelized. The reduct computation algorithm is where single output pipelines that do not use independent execution can be parallelized. In the case of the cross-validation/batch pipeline, executing a reduct computation within a split will only add additional threading if there are available cores to work with. All pointers and static accesses are protected by thread mutexes as defined by OpenMP, and all master counters are volatile and atomic. Data independence was favored over memory frugality where possible to ensure that threads could run with as little inter-communication as possible.

The executor function is the master method of the pipeline. Its purpose is to run the user-specified algorithms on the input dataset until all commands have been processed. It takes a data structure, a set of algorithms, and a set of parameters. The algorithms and parameters come in one big set and first need to be divided into two groups, one group for the algorithms to be applied to the training data and one group for those to be applied to the testing data. After the algorithms and parameters have been divided, a shared random number generator (RNG) is initialized in this non-threaded section to ensure that all threads have the same generator. This is important for validation and retesting purposes.

After the algorithms have been prepared, the parallel section starts. The initialized RNG is copied out to the private memory of every thread to ensure that execution is deterministic for any given seed regardless of thread racing. Giving each thread its own but same RNG that the other threads cannot access means that it doesn't matter which thread finishes first and reuses the RNG for the next iteration, the results will always be the same when the program is run again. The threads are given a static schedule to avoid randomness in succeeding executions. The static schedule means that the workload of every thread is specified from the start and cannot be changed. Even if a thread should finish its workload with iterations left to be assigned from the loop it cannot take on the extra work. This simple scheduling approach removes the need for any oversight of the threads, saving computation and memory resources, and it is highly unlikely that any thread would complete its work fast enough for any dynamic scheduling to improve the execution speed.

Each threaded iteration uses a different split from the dataset as its testing set. In a ten-fold cross-validation, the dataset is split into ten parts, and each iteration will use one of these parts as its testing set while the remaining nine are used for the training. This methodology is well suited to parallel execution as all algorithm sets can be run independently of one another. The training algorithms are executed serially on the input data, which is initially the training dataset. The output of previous algorithms is used as the input for the next algorithm, necessitating a functional link in the pipeline lest an algorithm be given an input list it cannot handle, like a reduct set being used as input for a rule tuner computation. This chain of training algorithms must at some point emit a set of rules for classification to be possible.

After the training is completed, the testing data is prepared in the same way using the second set of algorithms. Usually, the second set of algorithms include a discretization function that simply loads the discretization cuts computed on the training set.

Once the rules have been calculated and the testing data prepared accordingly, a classifier is executed with the rules and testing data. The classification generates accuracies, a confusion matrix, a list of ROC values as well as the AUC and accompanying statistics. All these thread-specific statistics are collated in a structure shared between threads, and after the parallel section has finished, the averages of all thread results are used as the final determinant of the classifier's performance.

The reduct computation is a transformer algorithm. It takes a structure, performs basic sanity checks on this structure to ensure that it is compatible with reduct computations, and then starts the parallel computations. The return structure, a set of reducts, is shared between threads. Failure flags are also shared so that a failure in any one thread can abort the entire process. The parallel section starts by checking the thread team size to determine if locks should be used or not, consistent with checking if threading is used specifically for computing the reducts or if threading is used in a larger context such as cross-validation. If threading is specifically used for the reduct algorithm then appending reducts to the return structure requires locks, a way to ensure that only one thread can run the locked code section at a time. This it to avoid overwriting or corrupting the shared data. If threading is from the cross-validation the reducts are computed serially within each cross-validation thread and do not write to the same data structure. Contrary to the executor function, the reduct computations are assigned to threads dynamically as the ordering of these does not have an impact on the outcome.

The parallel batch execution algorithm is specified in Listing 1.

The implementation was focused on increasing speed while retaining the modularity of the underlying framework. ROSETTA was built with a strict hierarchy in mind. This hierarchy makes it easy to implement new algorithms without having to alter base code such as executor functions or the data structures, as can be seen by the pipeline structure taking a list of algorithms whatever they may be (Listing 1). Instead, new algorithms can simply inherit from the already existing virtual classes and expand from there as needed. Adding multi-threading on a coarse level with separated memory for the threads, such as the executor for cross-validations, means that any new algorithm added can benefit from the multi-threading without having to make any concessions or changes in the algorithm code. The separated memory keeps additions simple while also reducing the need for locks and other inhibiting functions that slow down the computations by forcing threads to wait their turn. This is the easiest way to maintain ||-ROSETTA as a collaborative work and it ensures that additions need not understand the underlying framework to make full use of it.

2.2 Testing Hardware

The computations were performed on a dual socket Hewlett-Packard mainboard with two Intel Xeon X5650 6-core processors running at 2.66 GHz with a turbo

---

Algorithm 1: Pseudocode for the batch classification parallel executor.

> **Input** : An input dataset *ds*, a set of algorithms *algs*, and a set of parameters *params* for those algorithms
> **Output**: A set of statistical results for the classifier

1 **List**<Algorithm> alg-train, alg-test;
2 **List**<Parameter> param-train, param-test;
3 **Structure** table ← *Duplicate*(ds);
4 *//Get the algorithms and their parameters split for the training and testing parts of the cross-validation*;
5 (alg-train, param-train) ← *GetTrainingCommands*(algs, params);
6 (alg-test, param-test) ← *GetTestingCommands*(algs, params);
7 *//The random number generator should be initialized in the serial section so that every thread starts with the same numbers to prevent irreproducible results.*;
8 **RNG** rng ← *GetSeed*();
9 **List**<float> accuracies, rocareas, rocstderrors;
10 **ConfusionMatrix** confusion-matrix;
11 *//This is the beginning of the parallel section. From here, everything is executed multiple times by different threads. firstprivate is a pragma for copying the variables specified into the private memory of each thread that is initialized. Every thread will have its own version of these variables. shared is a pragma for making the following variables accessible to all threads.*;
12 **Parallel** *firstprivate*(rng, alg-train, alg-test, param-train, param-test) *shared*(table, confusion-matrix, accuracies, rocareas, rocstderrors);
13 *//This is the variable for how many parts the dataset should be divided into.*;
14 **int** splits ← *GetNumberOfSplits*();
15 *//This is the pipeline loop. It is executed once for every split of the dataset. Schedule determines how to allocate threads, nowait means that threads should keep going until the entire workload is done, and ordered means that in some places the threads will have to run as if they were executed serially*;
16 **for** *i* ∈ *splits; schedule(static, 1) nowait ordered* **do**
17   *//Build the training and testing tables for this thread*;
18   training-table ← *GetTrainingSample*(rng, table, i);
19   testing-table ← *GetTestingSample*(rng, table, i);
20   **Ruleset** rules;
21   **Structure** parent;
22   **Pipeline** pipe-train, pipe-test;
23   **Batchclassification** results;
24   *//The resulting structure from running all the training algorithms on the training dataset*;
25   parent ← pipe-train.*ExecuteCommands*(training-table, alg-train, param-train);
26   *//Take the rules from the training pipeline and add them to the testing pipeline*;
27   rules ← *GetRuleset*(parent);
28   pipe-test.*SetRules*(rules);
29   *//Evaluate the testing pipeline*;
30   results ← pipe-test.*ExecuteCommands*(testing-table, alg-test, param-test);
31   *//Gather the results from all the threads*;
32   accuracies[i] ← *GetAccuracy*(results);
33   confusion-matrix[i] ← *GetConfusionMatrix*(results);
34   rocareas[i] ← *GetROCArea*(results);
35   rocstderrors[i] ← *GetROCStandardError*(results);
36 **Output**(accuracies, confusion-matrix, rocareas, rocstderrors);

---

boost (version 1.0) of 3.06 GHz and three memory channels. The node had 96 GB of DDR3 memory running at 1333 MHz and CAS latency 9. The software platform was CentOS 7, with Linux kernel 3.10.0-693.21.1.el7.x86_64. The program was compiled with OpenMP 3.1 and GCC 4.8.5-36, and used level three standard compiler optimizations. Scripts used Perl v5.16.3 and took the time values directly from the system.

2.3   Testing Pipeline

The testing pipeline is the set of algorithms to be run on the validation data, using the discretization and the ruleset from the training pipeline. In a 10CV the validation data is one tenth of the whole dataset, where the other nine parts are used in the training phase. The datasets were discretized using Entropy scaling with a termination based on minimum description length [6]. After discretization, the reducts were computed using a greedy algorithm biased towards single prime implicants of minimal length [9]. The reducts were computed modulo decision and approximated with a hitting fraction of 98%. Rules were generated from the reducts, tuned [10], and used with an object-tracking schema at a 0.8 fraction for classification. No datasets had missing values hence did not require completion. The tuned rules were also compared between classifications of the same dataset with different levels of threading used to see if the results would differ between the executions. The rules were compared on accuracy, support, coverage, odds ratios, and risk ratios. While differences were not expected to impact the time of computations in any meaningful way, consistent and similar results were important for replication of experiments.

2.4   Tests

||-ROSETTA was run using four different datasets. The first set was using data gathered from a study on Systemic Lupus Erythematosus (SLE), consisting of 33,007 features and 628 objects [2]. All the values were of type float. The second set was truncated from a study on histone modifications [7], consisting of 115 features and 7,258 objects. The values were of type string. The third set was truncated from a study on the probability of credit card clients defaulting on payment [13], gathered from the UCI Machine Learning Repository [5], and consisted of 24 features and 20,000 objects, all of type integer. The last dataset was a binary table with feature correlations generated using Cholesky decomposition [11], containing 10,000 objects and 1,000 features. These datasets were used to test the scaling of ||-ROSETTA with regards to features and objects separately. The test was run with a Perl script that measured the time taken using the on-board clock. The overhead for the script was estimated at less than a second. The test itself was a ten-fold cross-validation where the training pipe consisted of Entropy-scaled discretization, Johnson algorithm reduct computations, rule generation, and a rule exporter. The testing pipe loaded the discretization from the training pipe and ran an object tracking classification.

3   Results

The results were consistent with the expected scaling for the data used. The increased threading was efficient, though overhead also increased with the number of threads used. For the Histone Modifications dataset, at two threads the overhead was 7%, but at five it had increased to 29% (Table 1). This is likely due to the increased demands on memory access channels and increased number of locks required for additional threads. The comparative total time taken estimates how fast the pipeline is executed compared to a single thread. With ten iterations for the cross-validation, the time taken is not linear to the number of threads but rather to the number of iterations needed to complete all ten validations. The lack of difference in time taken between five and six threads shows this effect. The time taken for two, five, and ten, threads shows the maximal efficiency of the parallel execution as these will complete the entire cross-validation using the full thread-potential.

Table 1. Completion times for the histone modifications dataset.*

Histone modifications

Threads	Time taken (seconds)	Comparative total time taken	Theoretical time per iteration	Comparative time taken
1	11,606	100.0%	1,160 (10)	100.0%
2	6,212	107.0%	1,242 (5)	53.5%
3	5,165	133.5%	1,291 (4)	44.5%
4	4,117	141.9%	1,372 (3)	35.5%
5	3,000	129.2%	1,500 (2)	25.8%
6	2,903	150.1%	1,451 (2)	25.0%
10	1,836	158.2%	1,836 (1)	15.8%

*Threads = the number of threads used for 10CV. Time Taken (seconds) = the system time needed to finish the computations, including the execution script. Comparative Total Time taken = the total time needed as the percentage of the execution time for one thread only. Theoretical Time per Iteration = the time needed per thread to finish one iteration. This accounts for the decreased efficiency of the computing system when using all resources simultaneously. Theoretical Time per Iteration = The amount of time spent on each iteration. This is an average value as iterations can vary in efficiency. Comparative Time Taken = the real time needed as a percentage of the time for using one thread. This is the practical outcome of using more threads for the computation.

The SLE dataset had only 628 objects but 33,007 features. This type of data scales poorly with ||-ROSETTA in a relative sense, and adds significant overhead just from building the dataset, a process that cannot be efficiently threaded. While the estimated time per iteration for the first dataset differed by a factor 1.58 between one and ten threads, the second dataset differs by a factor 2.96, suggesting considerable overhead (Table 2).

The payment default dataset had 20,000 objects and 24 features. The results are similar to the histone modifications dataset, showing a difference in performance well within the margin of error (Table 3). The synthetic dataset showed

Table 2. Completion times for the systemic lupus erythematosus dataset.*

Systemic lupus erythematosus

Threads	Time taken (seconds)	Comparative total time taken	Theoretical time per iteration	Comparative time taken
1	740	100.0%	74 (10)	100.0%
2	445	120.3%	89 (5)	60.1%
3	381	154.5%	95 (4)	51.5%
4	334	180.5%	111 (3)	45.1%
5	277	187.2%	138 (2)	37.4%
6	283	229.5%	141 (2)	38.2%
10	219	295.9%	219 (1)	29.6%

*Threads = the number of threads used for 10CV. Time Taken (seconds) = the system time needed to finish the computations, including the execution script. Comparative Total Time taken = the total time needed as the percentage of the execution time for one thread only. Theoretical Time per Iteration = the time needed per thread to finish one iteration. This accounts for the decreased efficiency of the computing system when using all resources simultaneously. Theoretical Time per Iteration = The amount of time spent on each iteration. This is an average value as iterations can vary in efficiency. Comparative Time Taken = the real time needed as a percentage of the time for using one thread. This is the practical outcome of using more threads for the computation.

stark differences, scaling better in the optimal configurations of two, five, and ten, threads, likely due to the minimal amount of data needed to represent the binary values of the dataset (Table 4). This indicates a strong correlation between memory throughput and thread performance.

The maximum memory usage differed based on features and objects (Table 5). The large number of objects in the Credit Card Payment Default dataset led to a high amount of initial memory needed as well as the steepest scaling with threads (Fig. 3). The more balanced dataset, Histone Modifications, showed a similar trend with smaller numbers. The dataset for SLE, with 33,007 features and 628 objects, did not significantly increase in memory usage with the number of threads, but had the highest starting point of memory use. The Synthetic dataset showed a linear increase in maximum memory use, likely due to the limited complexity of the data itself.

Compared to the theoretical gain, three datasets perform well (Fig. 2). The Credit Card Payment Default, Histone Modifications, and Synthetic datasets are close to the optimal performance, with a minor deviation around three and four threads.

The rules generated by the process do not differ based on multi-threading. A single-threaded execution generated the same rulesets in cross-validation as a ten-threaded execution. Accuracy, ROC curve, and deviations were consistent

Table 3. Completion times for the credit card payment default dataset.*

Credit card payment default				
Threads	Time taken (seconds)	Comparative total time taken	Theoretical time per iteration	Comparative time taken
1	174,133	100.0%	17,413 (10)	100.0%
2	93,902	107.9%	18,780 (5)	53.9%
3	67,337	116.0%	20,201 (4)	38.7%
4	56,399	129.6%	22,560 (3)	32.4%
5	41,796	120.0%	20,898 (2)	24.0%
6	41,592	143.3%	24,955 (2)	23.9%
10	28,613	164.3%	28,613 (1)	16.4%

*Threads = the number of threads used for 10CV. Time Taken (seconds) = the system time needed to finish the computations, including the execution script. Comparative Total Time taken = the total time needed as a percentage of the execution time for one thread only. Theoretical Time per Iteration = the time needed per thread to finish one iteration. This accounts for the decreased efficiency of the computing system when using all resources simultaneously. Theoretical Time per Iteration = The amount of time spent on each iteration. This is an average value as iterations can vary in efficiency. Comparative Time Taken = the real time needed as a percentage of the time for using one thread. This is the practical outcome of using more threads for the computation.

between executions, and all rulesets were exactly the same regardless of the number of threads used in the execution (Table 6, Table 7).

3.1 Comparisons

There are other programs for running multi-core Rough Set classifications.

WEKA [12] is a popular classification tool that can implement RSESLib [3], and the library itself also offers a scheduling tool called Simple Grid Manager (SGM). WEKA handles multi-thread execution differently from ||-ROSETTA, needing an external scheduler (WEKA Server) to coordinate concurrent executions over multiple processes as well as multiple compute nodes. SGM fills a similar role as WEKA Server, coordinating and collating independent algorithm executions over multiple compute nodes as long as all the nodes have the program client installed and running. ||-ROSETTA is so far the only program to offer a generalized Rough Set parallel execution feature that requires neither setup nor external programs or networks to collect and collate the classification results. This also comes with the restriction that ||-ROSETTA offers multi-thread computations, but not multi-node computations, something that both WEKA Server and SGM can facilitate with the proper configuration. This makes ||-ROSETTA better suited for workstation use. The single machine model

Table 4. Completion times for the synthetic dataset.*

Synthetic dataset

Threads	Time taken (seconds)	Comparative total time taken	Theoretical time per iteration	Comparative time taken
1	61,851	100.0%	6,185 (10)	100.0%
2	32,872	106.3%	6,574 (5)	53.1%
3	26,559	128.8%	6,640 (4)	42.9%
4	19,402	125.5%	6,467 (3)	31.4%
5	13,286	107.4%	6,643 (2)	21.5%
6	12,980	125.9%	6,490 (2)	21.0%
10	6,752	109.2%	6,752 (1)	10.9%

*Threads = the number of threads used for 10CV. Time Taken (seconds) = the system time needed to finish the computations, including the execution script. Comparative Total Time taken = the total time needed as a percentage of the execution time for one thread only. Theoretical Time per Iteration = the time needed per thread to finish one iteration. This accounts for the decreased efficiency of the computing system when using all resources simultaneously. Theoretical Time per Iteration = The amount of time spent on each iteration. This is an average value as iterations can vary in efficiency. Comparative Time Taken = the real time needed as a percentage of the time for using one thread. This is the practical outcome of using more threads for the computation.

Table 5. Memory usage by the different datasets.

Peak memory usage

Threads	Credit card payment default	Systemic lupus erythematosus	Histone modifications	Synthetic dataset
1	1,082	1,859	202	689
2	2,049	1,869	372	1,005
3	3,079	1,897	567	1,058
4	4,115	1,994	761	1,417
5	5,135	2,096	955	1,688
6	5,400	2,227	1,028	1,696
7	5,665	2,107	1,100	1,966
8	5,994	2,179	1,172	2,237
9	6,196	2,123	1,245	2,508
10	6,560	2,155	1,402	2,774

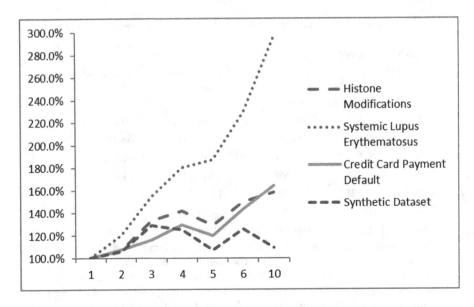

Fig. 1. Comparative total time taken. The comparative time needed of each added thread when compared to the single-threaded execution. The y-axis depicts the total computing time needed as compared to a single thread (100%) and the x-axis shows the number of threads used for attaining that efficacy. A low value on the y-axis indicates a low overhead cost of adding threads from memory throughput and time spent on loading data. The Histone Modifications and Credit Card datasets are similar in efficacy over all thread-teams even though the datasets dimensions differ notably ($115 \times 7{,}258$ and $24 \times 20{,}000$, respectively). The Systemic Lupus dataset scales worse with its drastic dimensions ($33{,}007 \times 628$), though a factor 2 difference between 24 features and 33,007 features shows a beneficial scaling coefficient. The fast computation of binary data in the synthetic dataset shows the speed attainable when memory throughput is not an issue.

makes ||-ROSETTA more likely to run at optimal speed given that all threads run on the same hardware, reducing the likelihood that one computation will be slower than the rest and delay the final results.

4   Conclusions

The addition of parallel execution to ROSETTA has drastically increased the speed at which computations are handled, and the program can now fully make use of modern hardware. The optimal configuration for efficiency uses a number of threads divisible by the number of iterations needed for the cross-validation, in this case two, five, and ten. This places increased reliance on memory and drive lanes for optimal performance (Fig. 1), and the use of threads should in some capacity be matched to the width of lanes. The number of threads will impact the frequency of thread locks and atomic operations, resulting in a limited amount

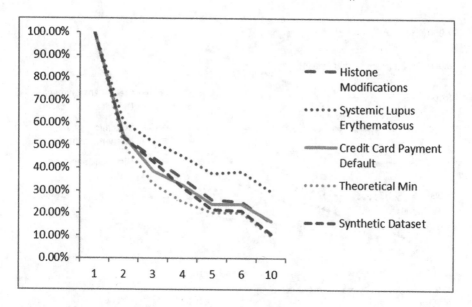

Fig. 2. Comparative time taken. The comparative time taken shows the time the pipeline needed to finish a ten-fold cross-validation (10CV) using the different thread numbers. The y-axis shows the time taken relative to using one thread (100%), and the x-axis shows the number of threads used. The most drastic gains were made from one to two threads, and gains were linear at two, three, four, five, and ten threads, corresponding to the decrease in iterations needed for the cross-validation. All datasets follow the same pattern. The lowest line is the theoretical minimum as computed from a single thread.

of overhead. There is no clear evidence of the effect of frequency boosting at lower thread counts as all four datasets show different increases in the effective time needed. The similarity in results between the Histone Modifications dataset and the Credit Card Payment Default dataset shows that the number of objects has little impact on the overhead generated by parallel execution (Fig. 2). Furthermore, it requires a significant number of features to negatively impact performance, as seen in the doubling of time taken per thread going from 115 to 33,007 features. The trivial increase per feature further suggests an efficient scaling in terms of dataset size. This scaling is dependent on the memory throughput as can be seen from the performance of the Synthetic dataset. When memory throughput is not an issue, threaded executions perform similarly in efficiency as single-threaded executions (Fig. 1). The size of the dataset does not matter in this regard. The independent nature of the implemented memory management has resulted in an expansion of memory use based on data size with respect to threading. While this approach is the most expensive for memory usage, it does allow for extending the framework without concern for the underlying memory architecture. A partially shared memory structure would need to be implemented on an algorithm basis as there is no consensus nor program standard for read and

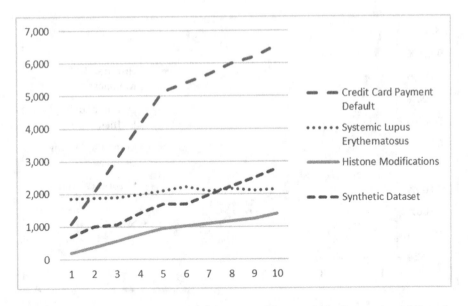

Fig. 3. Peak memory use of the three datasets with regards to the number of threads. The y-axis shows the peak memory in MB used, and the x-axis shows the number of threads used for the same computation. The Credit Card Payment Default dataset and the Histone Modifications dataset show a similar trend of scaling, peaking at five threads and then tapering off the increase in memory usage. The SLE dataset has few objects and uses a similar amount of memory regardless of threads but starts with the highest amount. The consistent level indicates that memory use by thread number scales mostly with the number of objects. The Synthetic dataset has a stable increase in memory consistent with threads unhindered by memory throughput.

write requirements on the computed results. The shared structures would also require shared hashes for indexing the thread-specific results so that they can be reused without the need for searching. While this would be possible, placing the memory management within the algorithm would add a layer of complexity to extending the ROSETTA framework. The rulesets generated from the parallel version and the original do not show any difference at all. This holds regardless of what criterion is used for the filtering. This is due to the use of ordered and static schedules for thread workloads and the use of a singular random number generator that is copied into the private memory of each thread prior to executing the parallel section of the batch execution pipeline. This consistency is of considerable importance for reliable and reproducible research and application.

5   Availability

||-ROSETTA is freely available on GitHub https://github.com/komorowskilab.

Table 6. Top five rules with coverage at least 0.05 from two different classifications of the Histone Modifications dataset using one and ten threads.*

	One thread	Decisions		Ten threads	Decisions	
		EXPR(1)	EXPR(0)		EXPR(1)	EXPR(0)
H3K9me2(1) AND H3K9ac(1)						
Supp. (LHS)	248			248		
Supp. (RHS)		181	67		181	67
Acc. (RHS)		0.73	0.27		0.73	0.27
Cov. (LHS)	0.04			0.04		
Cov. (RHS)		0.05	0.02		0.05	0.02
Odds Ratio		2.39 (1.80–3.18)	0.41 (0.31–0.55)		2.39 (1.80–3.18)	0.41 (0.31–0.55)
Risk Ratio		1.37 (1.27–1.49)	0.57 (0.46–0.70)		1.37 (1.27–1.49)	0.57 (0.46–0.70)
H3K9me3(1)						
Supp. (LHS)	286			286		
Supp. (RHS)		207	79		207	79
Acc. (RHS)		0.72	0.28		0.72	0.28
Cov. (LHS)	0.04			0.04		
Cov. (RHS)		0.06	0.03		0.06	0.03
Odds Ratio		2.33 (1.79–3.04)	0.42 (0.32–0.55)		2.33 (1.79–3.04)	0.42 (0.32–0.55)
Risk Ratio		1.36 (1.26–1.47)	0.58 (0.48–0.70)		1.36 (1.26–1.47)	0.58 (0.48–0.70)
H3K36me1(1) AND H3K23ac(1)						
Supp. (LHS)	330			330		
Supp. (RHS)		233	97		233	97
Acc. (RHS)		0.71	0.29		0.71	0.29
Cov. (LHS)	0.05			0.05		
Cov. (RHS)		0.07	0.03		0.07	0.03
Odds Ratio		2.14 (1.68–2.73)	0.46 (0.36–0.59)		2.14 (1.68–2.73)	0.46 (0.36–0.59)
Risk Ratio		1.33 (1.24–1.43)	0.62 (0.52–0.73)		1.33 (1.24–1.43)	0.62 (0.52–0.73)
H3K23ac(1) AND H3K23ac.succ(0)						
Supp. (LHS)	262			262		
Supp. (RHS)		181	81		181	81
Acc. (RHS)		0.69	0.31		0.69	0.31
Cov. (LHS)	0.04			0.04		
Cov. (RHS)		0.05	0.03		0.05	0.03
Odds Ratio		1.97 (1.51–2.57)	0.50 (0.38–0.66)		1.97 (1.51–2.57)	0.50 (0.38–0.66)
Risk Ratio		1.30 (1.19–1.41)	0.65 (0.54–0.79)		1.30 (1.19–1.41)	0.65 (0.54–0.79)
H3K9me2.succ(1) AND H3K9ac(1)						
Supp. (LHS)	237			237		
Supp. (RHS)		178	59		178	59
Acc. (RHS)		0.75	0.25		0.75	0.25
Cov. (LHS)	0.04			0.04		
Cov. (RHS)		0.05	0.02		0.05	0.02
Odds Ratio		2.65 (1.97–3.58)	0.37 (0.27–0.50)		2.65 (1.97–3.58)	0.37 (0.27–0.50)
Risk Ratio		1.41 (1.30–1.52)	0.53 (0.42–0.66)		1.41 (1.30–1.52)	0.53 (0.42–0.66)

*The two groups, one thread and ten threads, indicate how many threads were used for the classifications from which the rules were generated. Each segment of the table consists of one feature pattern for which both classifiers had rules, followed by the statistics Support left-hand size, Support right-hand side, Accuracy right-hand side, Coverage left-hand side, Coverage right-hand side, Odds Ratio, and Risk Ratio. Where values are not specific to a decision class (the two possible classes are EXPR(1) and EXPR(0)) they are given in the first column, otherwise under their respective decision class column. No value differs between one and ten threads, and this result was consistent across the entire rulesets generated from classification.

Table 7. Second best five rules with coverage at least 0.05 from two different classifications of the Histone Modifications dataset using one and ten threads.*

	One thread	Decisions		Ten threads	Decisions	
		EXPR(1)	EXPR(0)		EXPR(1)	EXPR(0)
H3K9me2(1) AND H3K9ac(1)						
Supp. (LHS)	248			248		
Supp. (RHS)		186	62		186	62
Acc. (RHS)		0.75	0.25		0.75	0.25
Cov. (LHS)	0.04			0.04		
Cov. (RHS)		0.05	0.02		0.05	0.02
Odds Ratio		2.64 (1.97–3.54)	0.37 (0.28–0.50)		2.64 (1.97–3.54)	0.37 (0.28–0.50)
Risk Ratio		1.41 (1.30–1.52)	0.53 (0.42–0.66)		1.41 (1.30–1.52)	0.53 (0.42–0.66)
H3K9me2(1) AND H3K27me2.succ(1)						
Supp. (LHS)	267			267		
Supp. (RHS)		192	75		192	75
Acc. (RHS)		0.72	0.28		0.72	0.28
Cov. (LHS)	0.04			0.04		
Cov. (RHS)		0.05	0.02		0.05	0.02
Odds Ratio		2.25 (1.71–2.95)	0.44 (0.33–0.581)		2.25 (1.71–2.95)	0.44 (0.33–0.581)
Risk Ratio		1.35 (1.25–1.46)	0.60 (0.49–0.72)		1.35 (1.25–1.46)	0.60 (0.49–0.72)
H3K9me3(1)						
Supp. (LHS)	286			286		
Supp. (RHS)		203	83		203	83
Acc. (RHS)		0.71	0.29		0.71	0.29
Cov. (LHS)	0.04			0.04		
Cov. (RHS)		0.06	0.03		0.06	0.03
Odds Ratio		2.15 (1.66–2.79)	0.46 (0.35–0.60)		2.15 (1.66–2.79)	0.46 (0.35–0.60)
Risk Ratio		1.33 (1.23–1.44)	0.61 (0.51–0.74)		1.33 (1.23–1.44)	0.61 (0.51–0.74)
H3K36me1(1) AND H3K23ac(1)						
Supp. (LHS)	331			331		
Supp. (RHS)		232	99		232	99
Acc. (RHS)		0.70	0.30		0.70	0.30
Cov. (LHS)	0.05			0.05		
Cov. (RHS)		0.07	0.03		0.07	0.03
Odds Ratio		2.07 (1.62–2.63)	0.48 (0.37–0.61)		2.07 (1.62–2.63)	0.48 (0.37–0.61)
Risk Ratio		1.32 (1.22–1.42)	0.63 (0.53–0.75)		1.32 (1.22–1.42)	0.63 (0.53–0.75)
H3K9me1.prec(0) AND H4K20me1.prec(1) AND H3K4ac.succ(1)						
Supp. (LHS)	265			265		
Supp. (RHS)		77	188		77	188
Acc. (RHS)		0.29	0.71		0.29	0.71
Cov. (LHS)	0.04			0.04		
Cov. (RHS)		0.02	0.06		0.02	0.06
Odds Ratio		0.33 (0.25–0.44)	2.94 (2.25–3.86)		0.33 (0.25–0.44)	2.94 (2.25–3.86)
Risk Ratio		0.53 (0.43–0.64)	1.56 (1.44–1.69)		0.53 (0.43–0.64)	1.56 (1.44–1.69)

*The two groups, one thread and ten threads, indicate how many threads were used for the classifications from which the rules were generated. Each segment of the table consists of one feature pattern for which both classifiers had rules, followed by the statistics Support left-hand size, Support right-hand side, Accuracy right-hand side, Coverage left-hand side, Coverage right-hand side, Odds Ratio, and Risk Ratio. Where values are not specific to a decision class (the two possible classes are EXPR(1) and EXPR(0)) they are given in the first column, otherwise under their respective decision class column. No value differs between one and ten threads, and this result was consistent across the entire rulesets generated from classification.

Acknowledgments. The authors would like to thank Aleksander Öhrn for his previous work in developing the ROSETTA framework, Andrzej Skowron for making their code available in ROSETTA as well as Carl Nettelblad and Justin Pearson for invaluable conversations. We also thank the anonymous reviewers who helped us improve this article.

References

1. Baltzer, N., Sundström, K., Nygård, J.F., Dillner, J., Komorowski, J.: Risk stratification in cervical cancer screening by complete screening history: applying bioinformatics to a general screening population. Int. J. Cancer **141**(1), 200–209 (2017). https://doi.org/10.1002/ijc.30725

2. Banchereau, R., et al.: personalized immunomonitoring uncovers molecular networks that stratify lupus patients. Cell **165**(3), 551–565 (2016). https://doi.org/10.1016/j.cell.2016.03.008

3. Bazan, J.G., Szczuka, M.: RSES and RSESlib - a collection of tools for rough set computations. In: Ziarko, W., Yao, Y. (eds.) RSCTC 2000. LNCS (LNAI), vol. 2005, pp. 106–113. Springer, Heidelberg (2001). https://doi.org/10.1007/3-540-45554-X_12

4. Bulkowski, P., Ejdys, P., Grzegorz, G.: Projekt i implementacja jezyka do obliczen rozproszonych (A project and implementation of a language supporting distributed computations). Ph.D. thesis, Poland (1999)

5. Dheeru, D., Karra Taniskidou, E.: UCI machine learning repository. University of California, Irvine, School of Information and Computer Sciences (2017). http://archive.ics.uci.edu/ml

6. Dougherty, J., Kohavi, R., Sahami, M.: Supervised and unsupervised discretization of continuous features. In: Prieditis, A., Russell, S. (eds.) Machine Learning Proceedings 1995, pp. 194–202. Morgan Kaufmann, San Francisco (CA) (1995). https://doi.org/10.1016/B978-1-55860-377-6.50032-3. http://www.sciencedirect.com/science/article/pii/B9781558603776500323

7. Enroth, S., Bornelöv, S., Wadelius, C., Komorowski, J.: Combinations of histone modifications mark exon inclusion levels. PLoS One **7**(1), e29911 (2012)

8. Garbulowski, M., et al.: R.rosetta: a package for analysis of rule-based classification models. bioRxiv p. 625905, January 2019. https://doi.org/10.1101/625905

9. Johnson, D.S.: Approximation algorithms for combinatorial problems. J. Comput. Syst. Sci. **9**(3), 256–278 (1974). https://doi.org/10.1016/S0022-0000(74)80044-9

10. Makosa, E.: Rule tuning, pp. 1–51. Uppsala University, Sweden (2005)

11. Press, W.H., Teukolsky, S.A., Vetterling, W.T., Flannery, B.P.: Numerical Recipes in C. The art of Scientific Computing, 2nd edn. Cambridge University Press, Cambridge (1992)

12. Witten, I.H., Frank, E., Hall, M.A., Pal, C.J.: Data Mining: Practical Machine Learning Tools and Techniques, 4th edn. Morgan Kaufmann Publishers Inc., Burlington (2016)

13. Yeh, I.C., Lien, C.H.: The comparisons of data mining techniques for the predictive accuracy of probability of default of credit card clients. Expert Syst. Appl. **36**, 2473–2480 (2009)

14. Øhrn, A., Komorowski, J.: Rosetta-a rough set toolkit for analysis of data. Citeseer (1997)

Sequences of Refinements of Rough Sets: Logical and Algebraic Aspects

Stefania Boffa[1]($\boxtimes$) and Brunella Gerla[2]

[1] University of Milano-Bicocca, Milan, Italy
stefania.boffa@unimib.it
[2] University of Insubria, Varese, Italy

Abstract. In this thesis, a generalization of the classical *Rough set theory* [83] is developed considering the so-called *sequences of orthopairs* that we define in [20] as special sequences of rough sets.

Mainly, our aim is to introduce some *operations between sequences of orthopairs*, and to discover how to generate them starting from the operations concerning standard rough sets (defined in [32]). Also, we prove several *representation theorems* representing the class of *finite centered Kleene algebras with the interpolation property* [31], and some classes of *finite residuated lattices* (more precisely, we consider *Nelson algebras* [87], *Nelson lattices* [23], *IUML-algebras* [73] and *Kleene lattice with implication* [27]) as sequences of orthopairs.

Moreover, as an application, we show that a sequence of orthopairs can be used to represent *an examiner's opinion on a number of candidates applying for a job*, and we show that opinions of two or more examiners can be combined using operations between sequences of orthopairs in order to get a final decision on each candidate.

Finally, we provide the original *modal logic SO_n* with semantics based on sequences of orthopairs, and we employ it to describe the knowledge of an agent that increases over time, as new information is provided. Modal logic SO_n is characterized by the sequences $(\Box_1, \ldots, \Box_n)$ and $(\bigcirc_1, \ldots, \bigcirc_n)$ of n modal operators corresponding to a sequence $(t_1, \ldots, t_n)$ of consecutive times. Furthermore, the operator $\Box_i$ of $(\Box_1, \ldots, \Box_n)$ represents the knowledge of an agent at time t_i, and it coincides with the *necessity modal operator* of S5 logic [29]. On the other hand, the main innovative aspect of modal logic SO_n is the presence of the sequence $(\bigcirc_1, \ldots, \bigcirc_n)$, since $\bigcirc_i$ establishes whether an agent is *interested in knowing* a given fact at time t_i.

Keywords: Rough sets · Orthopairs · Refinements · Many-valued logic · Modal logic

1 Introduction

> *We can only see a short distance ahead, but we can see plenty there that needs to be done.*
>
> Alan Turing

© Springer-Verlag GmbH Germany, part of Springer Nature 2020
J. F. Peters and A. Skowron (Eds.): TRS XXII, LNCS 12485, pp. 26–122, 2020.
https://doi.org/10.1007/978-3-662-62798-3_3

Rough sets and orthopairs are mathematical tools that are used to deal with vague, imprecise and uncertain information. Rough set theory was introduced by the Polish mathematician Zdzislaw Pawlak in 1980 [82–84], and successively numerous researchers of several fields have contributed to its development. The rough set approach appears of fundamental importance in many research domains, for example in artificial intelligence and cognitive sciences, especially in the areas of machine learning, knowledge acquisition, decision analysis, knowledge discovery from databases, expert systems, inductive reasoning and pattern recognition [54, 77, 85, 109]. Also, rough set theory has been applied to solve many real-life problems in medicine, pharmacology, engineering, banking, finance, market analysis, environment management, etc. (see [52, 91, 94] for some examples). On the other hand, rough sets are also explored in mathematical logic for their relationship with three-valued logics [34, 89, 100]. Rough set philosophy is founded on the assumption that each object of the universe of discourse is described by some information, some data, or knowledge. Objects characterized by the same data are *indiscernible* in view of the available information about them. In this way, an *indiscernibility relation* between objects is generated, and it is the mathematical basis of rough set theory. The set of all indiscernible objects is named *elementary set*, and we can say that it is the *basic granule of knowledge* about the universe. Indiscernibility relations are equivalence relations, and elementary sets are their equivalence classes. Then, given an equivalence relation R defined on U, the *rough set* of a subset X of the universe U is the pair $(\mathcal{L}_R(X), \mathcal{U}_R(X))$ consisting respectively of the union of all equivalence classes fully contained in X, named *lower approximation* of X with respect to R, and the union of all the equivalence classes that have at least one element in common with X, named *upper approximation* of X with respect to R. Therefore, the rough set $(\mathcal{L}_R(X), \mathcal{U}_R(X))$ is the approximation of X with respect to the relation R. The set $\mathcal{B}_R(X)$ is called the R-boundary region of X, and it is the set $\mathcal{U}_R(X) \setminus \mathcal{L}_R(X)$. The objects of $\mathcal{B}_R(X)$ cannot be classified as belonging to X with certainty.

In this dissertation, we focus on *orthopairs* generated by an equivalence relation. They are equivalent to rough sets and are defined as follows. Let R be an equivalence relation on U, and let X be a subset of U, the *orthopair* of X determined by R is the pair $(\mathcal{L}_R(X), \mathcal{E}_R(X))$, where $\mathcal{L}_R(X)$ is the lower approximation and $\mathcal{E}_R(X)$, called *impossibility domain* or *exterior region* of X with respect to R, is the union of equivalence classes of R with no elements in common with X [32]. Orthopairs and rough sets are obtained from one another; indeed, the impossibility domain coincides with the complement of the upper approximation with respect to the universe. A pair (A, B) of disjoint subsets of a universe U can be viewed as the orthopair of a subset of U generated by an equivalence relation on U; in this case, we can say that (A, B) is an orthopair on U. We can view any orthopair (A, B) on the universe U as a three-valued function $f : U \mapsto \{0, \frac{1}{2}, 1\}$ such that, let $x \in U$, $f(x) = 1$ if $x \in A$, $f(x) = 0$ if $x \in B$ and $f(x) = \frac{1}{2}$ otherwise. Conversely, the three-valued function $f : U \mapsto \{0, \frac{1}{2}, 1\}$ determines the orthopair (A, B) on U, where $A = \{x \in U | f(x) = 1\}$ and $B = \{x \in U | f(x) = 0\}$.

Several kinds of operations between rough sets have been considered [34]. They correspond to connectives in three-valued logics. Logical approaches to some of these connectives have been given, such as Lukasiewicz, Nilpotent Minimum, Nelson and Gödel connectives [4,9,13,81].

Several authors generalized the definitions of rough sets and orthopairs by considering binary relations that are not equivalence relations, since the latter are not usually suitable to describe the real-world relationships between elements [93,107]. We consider orthopairs generated by a tolerance relation, that is a reflexive and symmetric binary relation [92]. Given a tolerance relation R defined on U and an element x of U, by *tolerance class* of x with respect to R, we mean the set of elements of U indiscernible to x with respect to R. The set of all tolerance classes of R is a covering of U, that is a set of subsets of U whose union is U. Moreover, if R is an equivalence relation, then the set of all equivalence classes is a partition of U (a partition is a set of subsets of U that are pairwise disjoint and whose union is U). Therefore, we can define rough sets and orthopairs determined by a covering (or a partition) instead of a tolerance relation (or an equivalence relation).

In this thesis, we focus on *sequences of orthopairs* generated by refinement sequences of coverings [19,20]. A *refinement sequence* of a universe U is a finite sequence $(C_1, \ldots, C_n)$ of coverings of U such that C_i is finer than C_j (each block of C_i is included at least in a block of C_j) for each $j \leq i$. Clearly, for each subset X of U, the refinement sequence $(C_1, \ldots, C_n)$ generates the sequence

$$((\mathcal{L}_1(X), \mathcal{E}_1(X)), \ldots, (\mathcal{L}_n(X), \mathcal{E}_n(X))),$$

where $(\mathcal{L}_i(X), \mathcal{E}_i(X))$ is the orthopair of X determined by C_i. Furthermore, we deal with sequences of *partial coverings*. These are coverings that do not fully cover the universe, and they are suitable for describing situations in which some information is lost during the refinement process [39]. Refinement sequences of partial coverings are obtained starting from *incomplete information tables*, that are tables where a set of objects is described by a set of attributes, but some information is lost or not available [66]. It is interesting to notice that when $(C_1, \ldots, C_n)$ consists of all partitions of U, the pair $(U, (C_1, \ldots, C_n))$ is an *Aumann structure*, that is a mathematical structure used by economists and game theorists to represent the knowledge [6,7]. Refinement sequences can be represented as partially ordered sets. Hence, sequences of orthopairs generated by refinement sequences can be represented as pairs of upward closed subsets of such partially ordered sets. By using this correspondence, we give a concrete representation of some finite algebraic structures related with Kleene algebras. *Kleene algebras* form a subclass of *De Morgan algebras*. The latter were introduced by Moisil [74], and successively, they were explored by several authors, in particular, by Kalman [63] (under the name of *distributive i-lattices*), and by Bialynicki-Birula and Rasiowa, which called them *quasi-Boolean algebras* [12]. The notation that is still used was introduced by Monteiro [75]. We are interested in the family of *finite centered Kleene algebras with the interpolation property*, studied by the Argentinian mathematician Roberto Cignoli. In particular, in

[31], he proved that centered Kleene algebras with the interpolation property are represented by *bounded distributive lattices* [86]. By Birkhoff representation, each bounded distributive lattice is characterized as a set of upsets of a partially ordered set with set intersection and union [14]. In this thesis, we prove that each finite centered Kleene algebra with the interpolation property is isomorphic to the set of sequences of orthopairs generated by a refinement sequence with operations obtained extending the *Kleene operations* between orthopairs (see [34]) to the sequences of orthopairs. We obtain a similar result for some other finite structures that are *residuated lattices* [100], and having as reduct a centered Kleene algebras with the interpolation property. More exactly, we show that some subclasses of *Nelson algebras, Nelson lattices* and *IUML-algebras* are represented as sequences of orthopairs in which the residuated operations are respectively obtain by extending *Nelson implication, Łukasiewicz conjunction and implication*, and *Sobociński conjunction and implication* between orthopairs (listed in [34]) to sequences of orthopairs. In Table 1 each structure is associated with its orthopaired operations.

Table 1. Structures and operations between orthopairs

Structures	Operations between orthopairs
Nelson algebras	Kleene conjunction and Nelson implication
Nelson lattices	Łukasiewicz conjunction and implication
IUML-algebras	Sobociński conjunction and implication

Nelson algebras were introduced by Rasiowa [87], under the name of N-lattices, as the algebraic counterparts of the constructive logic with strong negation considered by Nelson and Markov [22,88]. The centered Nelson algebras with the interpolation property are represented by Heyting algebras [11]. Nelson lattices are involutive residuated lattices, and are equationally equivalent to centered Nelson algebras [23]. IUML-algebras are the algebraic models of the logic IUML, which is a substructural fuzzy logic that is an axiomatic extension of the multiplicative additive intuitionistic linear logic MAILL [73]. IUML-algebras can also be defined as *bounded odd Sugihara monoids*, where a Sugihara monoid is the equivalent algebraic semantics for the relevance logic RM^t of R-mingle as formulated with Ackermann constants. In [49], a dual categorical equivalence is shown between IUML-algebras and suitable topological spaces defined starting from Kleene spaces. In this dissertation we focus only on finite IUML-algebras, and we refer to [3] and [73].

Moreover, we investigate the relationship between sequences of orthopairs and some finite lattices with implication. The latter are more general than Nelson lattices and form a subclass of *algebras with implication*, (DLI-algebras for short) [28]. We find a pair of operations that allows us to consider sequences of orthopairs as Kleene lattices with implication, but they coincide with no pair of three-valued operations. Consequently, we can introduce new operations between orthopairs, and so between rough sets.

On the other hand, some three-valued algebraic structures have been represented as rough sets generated by one covering [4, 40, 60–62]. Our results are more general, since many-valued algebraic structures correspond to sequences of rough sets determined by a sequence of coverings.

An important application of rough set theory is to partition a given universe into three pairwise disjoint regions: the *acceptance region* (i.e. the lower approximation), the *rejection region* (i.e. the impossibility domain), and the *uncertain region* (i.e. the boundary region). This classification is at the basis of the *three-way decision theory* [103], which allows us to make a decision on each object by considering the region to which it belongs. In this framework, we use a sequence of orthopairs to represent an examiner's opinion on a number of candidates applying for a job. Moreover, we show that the opinions of two or more examiners can be combined using operations between sequences of orthopairs in order to get a final decision on each candidate. On the other hand, we also show that sequences of orthopairs are identified as *decision trees* with three outcomes. Decision trees are graphical models widely used in machine learning for describing sequential decision problems [48].

Rough sets can be interpreted as the *necessity* and *possibility* operators in modal logic $S5$ [8, 80]. Moreover, the relationships between modal logic and many generalizations of rough set theory have been examined by several authors [69, 106]. In Sect. 5, we present a new modal logic, named SO_n logic, with semantics based on sequences of orthopairs. Modal logic SO_n is characterized by two families of modal operators, $(\Box_1, \ldots, \Box_n)$ and $(\bigcirc_1, \ldots, \bigcirc_n)$, which are semantically interpreted through the Kripke frame $(U, (R_1, \ldots, R_n))$, where $(R_1, \ldots, R_n)$ is a sequence of equivalence relations defined on the domain U, such that $R_j(u) \subseteq R_i(u)$, for each $i \leq j$ and $u \in U$.

Modal logic SO_n can also be viewed as an epistemic logic. More precisely, SO_n can represent the knowledge of an agent that increases over time, as new information is provided. Epistemic logic is the logic of knowledge and belief [58]. Epistemic modal logic provides models to formalize and describe the process of accumulating knowledge by individual knowers and groups of knowers by using modal logic [16, 46, 59]. Its applications include addressing numerous complex problems in philosophy, artificial intelligence, economics, linguistics and in other fields [57, 95]. Therefore, the sequences $(\Box_1, \ldots, \Box_n)$ and $(\bigcirc_1, \ldots, \bigcirc_n)$ correspond to a sequence $(t_1, \ldots, t_n)$ of consecutive instants of time. The operator $\Box_i$ of $(\Box_1, \ldots, \Box_n)$ represents the knowledge of an agent at time t_i, and it coincides with the *necessity modal operator* of S5 logic [56]. The main innovative aspect of our logic is the presence of $(\bigcirc_1, \ldots, \bigcirc_n)$, since its element $\bigcirc_i$ establishes whether the agent is *interested in knowing* the truth or falsity of the sentences at time t_i.

Contents of the Thesis

We conclude this introductory chapter by briefly describing the contents of the following chapters.

Section 2 reviews the basic notions and the notation that we will use throughout the thesis along with some simple preliminary results. Specially, we will focus on rough set theory, partial order theory and lattice theory.

In *Sect.* 3, we introduce the definition of *refinement sequences of partial coverings* as special sequences of coverings representing situations where new information is gradually provided on ever smaller sets of objects. We provide examples of environments in which refinement sequences arise; in detail, we obtain refinement sequences starting from incomplete information tables and formal contexts. Some families of sequences are defined considering how much the blocks of their coverings overlap. We identify refinement sequences as partially ordered sets. Moreover, the notion of *sequences of orthopairs* is introduced in order to generalize the rough set theory. We represent each sequence of orthopairs as a pair of disjoint upsets of a partially ordered set, or equivalently, as a labelled poset. Finally, we view sequences of orthopairs as decision trees with only three outcomes.

Preliminary versions of this chapter appeared in [1, 2, 19, 20].

In *Sect.* 4, we equip sets of sequences of orthopairs with some operations in order to obtain finite many-valued algebraic structures. Furthermore, we prove theorems wherewith to represent such structures as sequences of orthopairs. We show that, when sequences of orthopairs are generated by one covering, our operations coincide with some operations between orthopairs listed in [34]. Also, we discover how to generate operations between sequences of orthopairs starting from those concerning individual orthopairs. Finally, we use a sequence of orthopairs to represent an examiner's opinion on a number of candidates applying for a job. Moreover, we show that opinions of two or more examiners can be combined using our operations in order to get a final decision on each candidate.

Some results shown in this chapter can be found in [1, 2, 19, 20].

In *Sect.* 5, we recall some basic notions of modal logic and the existing connections between modal logic and rough sets. Then, we develop the original modal logic SO_n, defining its language, introducing its Kripke models, and providing its axiomatization. Moreover, we investigate the properties of our logic system, such as the consistency, the soundness and the completeness with respect to Kripke's semantics. We explore the relationships between modal logic SO_n and sequences of orthopairs. We consider the operations between orthopairs and between sequences of orthopairs from the logical point of view. Eventually, we employ modal logic SO_n to represent the knowledge of an agent that increases over time, as new information is provided.

We conclude this dissertation with *Sect.* 6, in which we briefly summarize the results that we have obtained, and we discuss their potential further developments along with new research objectives.

2   Preliminaries

> *That language is an instrument of human reason, and*
> *not merely a medium for the expression of thought, is a*
> *truth generally admitted.*
>
> George Boole

In this chapter, we introduce the basic notions and the notation that we will use throughout the thesis along with some simple preliminary results. Briefly, in Sect. 2.1, we recall the main definitions of rough set theory. In Sect. 2.2, we list several operations between orthopairs that are found in [34]; moreover, we show the connection between these operations and three-valued connectives. Finally, Sect. 2.3 focuses on some important contents of partial order theory and lattice theory.

2.1   Rough Sets and Orthopairs

Rough set theory, developed by Pawlak [82,83], is a mathematical tool used to deal with imprecise and vague information of datasets, and it finds numerous applications in several areas of science, such as, for instance chemistry [65], medicine [98], marketing [52], social network [18,41], etc. Rough sets provide approximations of sets with respect to equivalence relations.

Definition 1 (Equivalence relation). *An equivalence relation R of U is a subset on $U \times U$ such that*

1. $(x, x) \in R$ *(reflexivity),*
2. *if $(x, y) \in R$, then $(y, x) \in R$ (symmetry),*
3. *if $(x, y) \in R$ and $(y, z) \in R$, then $(x, z) \in R$ (transitivity),*

for each $x, y, z \in U$.

Moreover, let $x \in U$, we set $R(x) = \{y \in U \mid (x, y) \in R\}$, and we call $R(x)$ equivalence class *of x with respect to R.*

Definition 2 (Rough set). *Let R be an equivalence relation on U, and let $X \subseteq U$. Then, the* rough set *of X determined by R is the pair $(\mathcal{L}_R(X), \mathcal{U}_R(X))$, where*

$$\mathcal{L}_R(X) = \{x \in U \mid R(x) \subseteq X\} \text{ and}$$
$$\mathcal{U}_R(X) = \{x \in U \mid R(x) \cap X \neq \emptyset\}.$$

$\mathcal{L}_R(X)$ *and $\mathcal{U}_R(X)$ are respectively called* lower approximation *and* upper approximation *of X with respect to R.*

We write $(\mathcal{L}(X), \mathcal{U}(X))$ *instead of $(\mathcal{L}_R(X), \mathcal{U}_R(X))$, when R is clear from the context.*

Also, we call the *R-boundary region* of X the set $\mathcal{B}_R(X) = \mathcal{U}_R(X) \setminus \mathcal{L}_R(X)$.

Remark 1. Let R be an equivalence relation on U, and let $X \subseteq U$. Then,

$$\mathcal{L}_R(X) \subseteq X \subseteq \mathcal{U}_R(X) \text{ and } \mathcal{U}_R(X) = \mathcal{L}_R(X) \cup \mathcal{B}_R(X).$$

Definition 3 (Orthopair). *Let R be an equivalence relation on U, and let $X \subseteq U$. Then, the* orthopair *of X determined by R is the pair $(\mathcal{L}_R(X), \mathcal{E}_R(X))$, where*

$\mathcal{L}_R(X)$ is the lower approximation given in Definition 2, and
$\mathcal{E}_R(X) = \{x \in U \mid R(x) \cap X = \emptyset\}$.

$\mathcal{E}_R(X)$ *is called* impossibility domain *or* exterior domain *of X. We write $(\mathcal{L}(X), \mathcal{E}(X))$ instead of $(\mathcal{L}_R(X), \mathcal{E}_R(X))$, when R is clear from the context.*

Remark 2. Let R be an equivalence relation on U, and let $X \subseteq U$. Then,

$$\mathcal{L}_R(X) \cap \mathcal{E}_R(X) = \emptyset \text{ and } \mathcal{E}_R(X) = U \setminus \mathcal{U}_R(X).$$

The lower and upper approximations, the R-boundary region and the impossibility domain are depicted in Fig. 1. The blocks, that cover the universe U (the largest rectangle), represent the equivalence classes with respect to an equivalence relation R on U. Moreover, if X is represented by the oval shape, then $\mathcal{L}(X)$ is the union of green blocks, $\mathcal{U}(X)$ is the union of green and white blocks, $\mathcal{B}(X)$ is the union of white blocks, and $\mathcal{E}(X)$ is the union of red blocks.

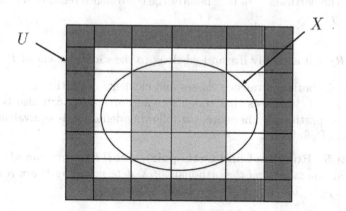

Fig. 1. Graphic representation of $\mathcal{L}(X)$, $\mathcal{U}(X)$, $\mathcal{B}(X)$ and $\mathcal{E}(X)$ (Color figure online)

In Rough set theory, given an equivalence relation R on the universe U, the pair (U, R) is called *Pawlak space*.

Remark 3. Let U be a universe, we denote the power set of U (i.e. the set of all subsets of U) with 2^U. Then, the structure $(2^U, \cap, \cup, \neg, \emptyset, U)$ is a *Boolean algebra* [101], where $\cap$, $\cup$ and $\neg$ are the usual set-theoretic operators. On the other hand, lower and upper approximations can be defined as unary operators on 2^U satisfying some properties [71], and so they are also named *approximation operators*. Thus, given an equivalence relation R on U, the system $(2^U, \cap, \cup, \neg, \mathcal{L}_R, \mathcal{U}_R, \emptyset, U)$, called *Pawlak rough set algebra*, is a topological algebra [88], which is an extension of the Boolean algebra $(2^U, \cap, \cup, \neg, \emptyset, U)$. This means that we can regard Rough set theory as an extension of Set theory with the additional approximation operators [105].

We can observe that equivalence relations are equivalent to partitions that are defined as follows.

Definition 4 (Partition). *By* partition P *of the universe* U, *we mean a set* $\{b_1, \ldots, b_n\}$ *such that*

1. $b_1, \ldots, b_n \subseteq U$,
2. $b_i \cap b_j = \emptyset$, *for each* $i \neq j$,
3. $b_1 \cup \ldots \cup b_n = U$.

Therefore, a partition of U is a set of subsets of U that are pairwise disjoint and whose union is U.

Remark 4. The equivalence relation R of U determines the partition P_R of U made of all equivalence classes of R, namely

$$P_R = \{R(x) \mid x \in U\};$$

vice-versa, the partition P of U generates the equivalence relation R_P on U such that, let $x, y \in U$,

$$x \ R_P \ y \text{ if and only if } x \text{ and } y \text{ belong to the same element of } P.$$

We call blocks both equivalence classes and elements of partitions.

By Remark 4, it follows that rough sets and orthopairs can also be defined starting from partitions. Therefore, the following definition is equivalent to Definition 2 and Definition 3.

Definition 5 (Rough set and Orthopair). *Let* P *be a partition of* U, *and let* $X \subseteq U$. *The* rough set *and the* orthopair *of* X *determined by* P *are respectively the pairs* $(\mathcal{L}_P(X), \mathcal{U}_P(X))$ *and* $(\mathcal{L}_P(X), \mathcal{E}_P(X))$, *where*

$\mathcal{L}_P(X) = \cup\{b \in P \mid b \subseteq X\}$,
$\mathcal{U}_P(X) = \cup\{b \in P \mid b \cap X \neq \emptyset\}$, *and*
$\mathcal{E}_P(X) = \cup\{b \in P \mid b \cap X = \emptyset\}$.

Several authors generalize the classical definitions of rough sets and orthopairs, by considering binary relations that are not equivalence relations, since the latter are not usually suitable to describe the real-world relationships between elements (e.g. [93, 107]).

In this thesis, we consider orthopairs generated by tolerance relations [70, 92], or equivalently by coverings [33, 36].

Definition 6 (Tolerance relation). *A tolerance relation R on U is a subset of $U \times U$ such that*

1. *$(x, x) \in R$ (reflexivity),*
2. *if $(x, y) \in R$, then $(y, x) \in R$ (symmetry),*

for each $x, y, z \in U$.

  Moreover, let $x \in U$, we set $R(x) = \{y \in U \mid (x, y) \in R\}$ and we call $R(x)$ the tolerance class *of x with respect to R.*

Trivially, an equivalence relation is also a tolerance relation. Moreover, tolerance relations generate coverings that are defined as follows.

Definition 7 (Covering). *By* covering C *of the universe U, we mean a set $\{b_1, \ldots, b_n\}$ such that*

1. *$b_1, \ldots, b_n \subseteq U$,*
2. *$b_1 \cup \ldots \cup b_n = U$.*

We can say that a partition is a covering that satisfies the additional property to have blocks pairwise disjoint.

2.2 Operations Between Orthopairs

In this section, we focus on some operations between orthopairs corresponding to three-valued connectives; moreover, here, by orthopair on U, we mean any pair of disjoint subsets of U, which may not even be the approximation of a subset of U with respect to a relation on U (see Definition 3).

  The relationship between orthopairs and three-valued logics is based on the idea expressed in the following observation.

Remark 5. The orthopair (A, B) on the universe U generates the three-valued function $f_{(A,B)} : U \mapsto \{0, \frac{1}{2}, 1\}$ such that, let $x \in U$,

$$f_{(A,B)}(x) = \begin{cases} 1 & if\, x \in A, \\ 0 & if\ x \in B, \\ \frac{1}{2} & if\ x \in U \setminus (A \cup B). \end{cases}$$

Conversely, the three-valued function $f : U \mapsto \{0, \frac{1}{2}, 1\}$ determines the orthopair (A_f, B_f) on U, where

$$A_f = \{x \in U \mid f(x) = 1\} \text{ and } B_f = \{x \in U \mid f(x) = 0\}.$$

The most simple operations between orthopairs are defined as follows.

Definition 8. *Let (A, B) and (C, D) be two orthopairs on the universe U, we set*

$(A, B) \wedge_{\mathcal{K}} (C, D) = (A \cap C, B \cup D)$ *and*
$(A, B) \vee_{\mathcal{K}} (C, D) = (A \cup C, B \cap D)$.

Table 2. Kleene conjunction

$\wedge$	0	$\frac{1}{2}$	1
0	0	0	0
$\frac{1}{2}$	0	$\frac{1}{2}$	$\frac{1}{2}$
1	0	$\frac{1}{2}$	1

Table 3. Kleene disjunction

$\vee$	0	$\frac{1}{2}$	1
0	0	$\frac{1}{2}$	1
$\frac{1}{2}$	$\frac{1}{2}$	$\frac{1}{2}$	1
1	1	1	1

Theorem 1 states that $\wedge_{\mathcal{K}}$ and $\vee_{\mathcal{K}}$ are respectively obtained from the *Kleene conjunction* and the *Kleene disjunction* on $\{0, \frac{1}{2}, 1\}$. The latter are defined by Table 2 and Table 3, respectively.

Also, we notice that $\wedge$ and $\vee$ are the minimum and the maximum on $\{0, \frac{1}{2}, 1\}$, respectively.

Theorem 1. *Let (A, B) and (C, D) be orthopairs on U. Then,*

$$(A, B) \wedge_{\mathcal{K}} (C, D) = (E, F) \quad and \quad (A, B) \vee_{\mathcal{K}} (C, D) = (G, H),$$

where

$$E = \{x \in U \mid f_{(A,B)}(x) \wedge f_{(C,D)}(x) = 1\},$$
$$F = \{x \in U \mid f_{(A,B)}(x) \wedge f_{(C,D)}(x) = 0\},$$
$$G = \{x \in U \mid f_{(A,B)}(x) \vee f_{(C,D)}(x) = 1\} \ and$$
$$H = \{x \in U \mid f_{(A,B)}(x) \vee f_{(C,D)}(x) = 0\}.$$

Proof. Let $x \in U$. By Remark 5, $x \in A \cap C$ if and only if $f_{(A,B)}(x) = 1$ and $f_{(C,D)}(x) = 1$, namely $f_{(A,B)}(x) \wedge f_{(C,D)}(x) = 1$ (see Table 2). Similarly, we can prove that $x \in B \cup D$ if and only if $f_{(A,B)}(x) \wedge f_{(C,D)}(x) = 0$. By Remark 5 and starting from Table 3, we can prove that

$$x \in A \cup C \quad \text{if and only if} \quad f_{(A,B)}(x) \vee f_{(C,D)}(x) = 1, \text{ and}$$
$$x \in B \cap D \quad \text{if and only if} \quad f_{(A,B)}(x) \vee f_{(C,D)}(x) = 0.$$

The next operations between orthopairs are equivalent to some three-valued connectives belonging to the families of conjunctions and implications on $\{0, \frac{1}{2}, 1\}$. Now, we recall the definitions of conjunction and implication that are based on some intuitive properties in scope of modelling incomplete information.

Definition 9 (Conjunction). *A conjunction on $\{0, \frac{1}{2}, 1\}$ is a map*

$$* : \left\{0, \frac{1}{2}, 1\right\} \times \left\{0, \frac{1}{2}, 1\right\} \mapsto \left\{0, \frac{1}{2}, 1\right\}$$

satisfying the following properties: let $x, y, z \in \{0, \frac{1}{2}, 1\}$,

1. *if $x \leq y$, then $x * z \leq y * z$,*
2. *if $x \leq y$, then $z * x \leq z * y$,*
3. *$0 * 0 = 0 * 1 = 1 * 0 = 0$ and $1 * 1 = 1$.*

Table 4. Łukasiewicz conjunction

$\circledast_{\mathcal{L}}$	0	$\frac{1}{2}$	1
0	0	0	0
$\frac{1}{2}$	0	0	$\frac{1}{2}$
1	0	$\frac{1}{2}$	1

Table 5. Sobociński conjunction

$\circledast_{\mathcal{S}}$	0	$\frac{1}{2}$	1
0	0	0	0
$\frac{1}{2}$	0	$\frac{1}{2}$	1
1	0	1	1

Example 1. Among the conjunctions listed in [34], we only consider the Kleene conjunction, the *Łukasiewicz conjunction* and the *Sobociński conjunction* [96]. The latter two are defined by Table 4 and Table 5.

Definition 10 (Implication). *An* implication *on* $\{0, \frac{1}{2}, 1\}$ *is a map*

$$\rightarrow : \left\{0, \frac{1}{2}, 1\right\} \times \left\{0, \frac{1}{2}, 1\right\} \mapsto \left\{0, \frac{1}{2}, 1\right\}$$

satisfying the following properties: let $x, y \in \{0, \frac{1}{2}, 1\}$,

1. *if* $x \le y$, *then* $y \rightarrow z \le x \rightarrow z$,
2. *if* $x \le y$, *then* $z \rightarrow x \le z \rightarrow y$,
3. $0 \rightarrow 0 = 1 \rightarrow 1 = 1$ *and* $1 \rightarrow 0 = 0$.

Example 2. Among the implications listed in [34], we consider the *Nelson implication*, the *Łukasiewicz implication* and the *Sobociński implication*. They are defined by the following tables, respectively (Tables 6, 7 and 8).

Table 6. Nelson implication

$\Rightarrow_{\mathcal{N}}$	0	$\frac{1}{2}$	1
0	1	1	1
$\frac{1}{2}$	1	1	1
1	0	$\frac{1}{2}$	1

Table 7. Łukasiewicz implication

$\Rightarrow_{\mathcal{L}}$	0	$\frac{1}{2}$	1
0	1	1	1
$\frac{1}{2}$	$\frac{1}{2}$	1	1
1	0	$\frac{1}{2}$	1

Table 8. Sobociński implication

$\Rightarrow_{\mathcal{S}}$	0	$\frac{1}{2}$	1
0	1	1	1
$\frac{1}{2}$	0	$\frac{1}{2}$	1
1	0	0	1

Now, we regard two multiplications between orthopairs defined as follows.

Definition 11. *Let* (A, B) *and* (C, D) *be orthopairs on* U*, we set*

1. $(A, B) *_{\mathcal{L}} (C, D) = (A \cap C, (U \setminus (A \cup C)) \cup B \cup D)$,
2. $(A, B) *_{\mathcal{S}} (C, D) = ((A \setminus D) \cup (C \setminus B), B \cup D)$.

We can prove that $*_{\mathcal{L}}$ and $*_{\mathcal{S}}$ are respectively equivalent to the three-valued conjunctions $\circledast_{\mathcal{L}}$ and $\circledast_{\mathcal{S}}$. More precisely, the following theorem holds.

Theorem 2. *Let* (A, B) *and* (C, D) *be orthopairs on* U*. Then,*

$$(A, B) *_{\mathcal{L}} (C, D) = (E, F) \text{ and } (A, B) *_{\mathcal{S}} (C, D) = (G, H),$$

where

$$E = \{x \in U \mid f_{(A,B)}(x) \circledast_{\mathcal{L}} f_{(C,D)}(x) = 1\},$$
$$F = \{x \in U \mid f_{(A,B)}(x) \circledast_{\mathcal{L}} f_{(C,D)}(x) = 0\},$$
$$G = \{x \in U \mid f_{(A,B)}(x) \circledast_{\mathcal{S}} f_{(C,D)}(x) = 1\} \text{ and}$$
$$H = \{x \in U \mid f_{(A,B)}(x) \circledast_{\mathcal{S}} f_{(C,D)}(x) = 0\}.$$

Proof. The proof is similar to that of Theorem 1.

Finally, we consider the following implications between orthopairs.

Definition 12. *Let* (A, B) *and* (C, D) *be orthopairs on* U*, then*

1. $(A, B) \to_{\mathcal{N}} (C, D) = ((U \setminus A) \cup C, A \cap D)$,
2. $(A, B) \to_{\mathcal{L}} (C, D) = (((U \setminus A) \cup C) \cap (B \cup (U \setminus D)), A \cap D)$,
3. $(A, B) \to_{\mathcal{S}} (C, D) = (B \cup C, U \setminus [(((U \setminus A) \cup C) \cap (A \cup (U \setminus D)))])$.

The previous implications are respectively obtained from the three-valued implications $\Rightarrow_{\mathcal{N}}$, $\Rightarrow_{\mathcal{L}}$ and $\Rightarrow_{\mathcal{S}}$. More precisely, the following theorem holds.

Theorem 3. *Let* (A, B)*,* (C, D) *and* (E, F) *be orthopairs on* U*. Then,*

$(A, B) \to_{\mathcal{N}} (C, D) = (E, F)$, *where*
$E = \{x \in U \mid f_{(A,B)}(x) \Rightarrow_{\mathcal{N}} f_{(C,D)}(x) = 1\}$ *and*
$F = \{x \in U \mid f_{(A,B)}(x) \Rightarrow_{\mathcal{N}} f_{(C,D)}(x) = 0\}.$

$(A, B) \to_{\mathcal{L}} (C, D) = (G, H)$, *where*
$G = \{x \in U \mid f_{(A,B)}(x) \Rightarrow_{\mathcal{L}} f_{(C,D)}(x) = 1\}$ *and*
$H = \{x \in U \mid f_{(A,B)}(x) \Rightarrow_{\mathcal{L}} f_{(C,D)}(x) = 0\},$

$(A, B) \to_{\mathcal{S}} (C, D) = (I, J)$,
$I = \{x \in U \mid f_{(A,B)}(x) \Rightarrow_{\mathcal{S}} f_{(C,D)}(x) = 1\}$ *and*
$J = \{x \in U \mid f_{(A,B)}(x) \Rightarrow_{\mathcal{S}} f_{(C,D)}(x) = 0\}.$

Proof. The proof is similar to that of Theorem 1.

On the other hand, there is an equivalent way to describe the relationship between the three-valued connectives $\wedge$, $\vee$, $\circledast_\mathcal{L}$, $\circledast_\mathcal{S}$, $\Rightarrow_\mathcal{N}$, $\Rightarrow_\mathcal{L}$ and $\Rightarrow_\mathcal{S}$, and the operations defined in Definitions 8, 11 and 12. It is provided by using the next definition and the next theorem.

Definition 13. *Let C be a covering of the universe U, and let $X \subseteq U$, we can define the function $F_X^C : C \mapsto \{0, \frac{1}{2}, 1\}$, where*

$$F_X^C(N) = \begin{cases} 1 & \text{if } N \subseteq X, \\ 0 & \text{if } N \cap X = \emptyset, \\ \frac{1}{2} & \text{otherwise.} \end{cases} \tag{1}$$

for each $N \in C$. We denote F_X^C with F_X, when C is clear from the context.

The following theorem states that each operation between orthopairs is obtained from the respective three-valued connective, by using function 1.

Theorem 4. *Let C be a covering of U, and let $X, Y \subseteq U$. Suppose that the operation $\circ$ belongs to $\{\wedge_K, \vee_K, *_\mathcal{L}, *_\mathcal{S}, \rightarrow_\mathcal{N}, \rightarrow_\mathcal{L}, \rightarrow_\mathcal{S}\}$, then*

$$(\mathcal{L}(X), \mathcal{E}(X)) \circ (\mathcal{L}(Y), \mathcal{E}(Y))$$

is the orthopair (A, B) such that

$$A = \bigcup \{N \in C \mid F_X(N) \odot F_Y(N) = 1\}$$

and

$$B = \bigcup \{N \in C \mid F_X(N) \odot F_Y(N) = 0\},$$

where $\odot$ respectively belongs to $\{\wedge, \vee, \circledast_\mathcal{L}, \circledast_\mathcal{S}, \Rightarrow_\mathcal{N}, \Rightarrow_\mathcal{L}, \Rightarrow_\mathcal{S}\}$.

Proof. We provide the proof only for the operation $*_\mathcal{S}$, since the remaining cases can be similarly demonstrated.

Let $x \in U$ and suppose that $(\mathcal{L}(X), \mathcal{E}(X)) *_\mathcal{S} (\mathcal{L}(Y), \mathcal{E}(Y)) = (A, B)$. By Definition 11, $x \in A$ if and only if $x \in (\mathcal{L}(X) \setminus \mathcal{E}(Y)) \cup (\mathcal{L}(Y) \setminus \mathcal{E}(X))$, namely $x \in \mathcal{L}(X) \setminus \mathcal{E}(Y)$ or $x \in \mathcal{L}(Y) \setminus \mathcal{E}(X)$. This is equivalent to affirm that x belongs to a node N of C such that

- $N \subseteq X$ and $N \cap Y = \emptyset$, or
- $N \subseteq X$ and $N \cap Y = \emptyset$.

Then, $F_X(N) = 1$ and $F_Y(N) \neq 0$, or $F_Y(N) = 1$ and $F_X(N) \neq 0$. We conclude that $F_X(N) \circledast_\mathcal{S} F_Y(N) = 1$, since $\circledast_\mathcal{S}$ is the Sobociński conjunction.

Similarly, $x \in B$ if and only if $x \in \mathcal{E}(X) \cup \mathcal{E}(Y)$, by 11; namely, x belongs to a node N of C such that $N \cap X = \emptyset$ or $N \cap Y = \emptyset$. Then, $F_X(N) = 0$ or $F_Y(N) = 0$. Hence, $F_X(N) \circledast_\mathcal{S} F_Y(N) = 0$.

Remark 6. However, the previous operations can be also defined by considering orthopairs that correspond to rough sets (see Definition 3). In this case, it is necessary to introduce some closure properties in order to ensure that operations between rough sets always generate a rough set. But, it will be done in the next sections.

Moreover, in Sect. 4.5, we extend the operations defined in Definitions 8, 11 and 12 to sequences of orthopairs in order to obtain many-valued algebraic structures.

2.3   Ordered Structures

Partial Orders and Lattices. This section contains some important contents of partial order theory and lattice theory. Partial order and lattice theory play an important role in many disciplines of computer science and engineering [14,53].

Definition 14 (Partially ordered set). *A partially ordered set, more briefly a poset, is a pair* $(P, \leq)$*, where P is a non empty set and* $\leq$ *is a binary relation on P satisfying the following properties.*

1. $x \leq x$ *(reflexivity),*
2. *if* $x \leq y$ *and* $y \leq x$*, then* $x = y$ *(antisymmetry),*
3. *if* $x \leq y$ *and* $y \leq z$*, then* $x \leq z$ *(transitivity),*

for each $x, y, z \in L$*.*
    Moreover, if $(P, \leq)$ *is a poset, then* $(S, \leq)$ *is also a poset, for each* $S \subseteq P$*.*

An example of partially ordered set is the set 2^U of all subsets of U with the set inclusion $\subseteq$.
    Let $(P, \leq)$ be a poset, and $x, y \in P$, we say that y is the *successor* of x in P, if $x < y$ and there is no $z \in P$ such that $x < z < y$. Furthermore, P has a *maximum* (or *greatest*) element if there exists $x \in P$ such that $y \leq x$ for all $y \in P$. An element $x \in P$ is *maximal* if there is no element $y \in P$ with $y > x$. Minimum and minimal elements are dually defined. P has a *minimum* (or *least*) element if there exists $x \in P$ such that $x \leq y$ for all $y \in P$. An element $x \in P$ is *minimal* if there is no element $y \in P$ with $y < x$.
    We can draw the *Hasse diagram* of each finite poset $(P, \leq)$: the elements of P are represented by points in the plane, and a line is drawn from x up to y, when y is a successor of x. Smaller elements are drawn under their successors.

Definition 15 (Chain). *A partially ordered set* $(P, \leq)$ *is a chain if and only if* $x \leq y$ *or* $y \leq x$*, for each* $x, y \in P$*.*

Definition 16 (Downset and Upset). *Let* $(P, \leq)$ *be a partially ordered set, and let* $S \subseteq P$*. Then, S is a downset of P if and only if satisfies the following property:*

$$for\ any\ y \in P,\ if\ y \leq x\ and\ x \in S,\ then\ y \in S.$$

Dually, S is an upset *of P if and only if satisfies the following property:*

$$for \ any \ y \in P, \ if \ x \leq y \ and \ x \in S, \ then \ y \ \in \ S.$$

Moreover, we set

$\downarrow S = \{y \in P \mid y \leq x \ for \ some \ x \in \ S\}$ *and*
$\uparrow S = \{y \in P \mid x \leq y \ for \ some \ x \in \ S\}.$

Definition 17 (Forest). *A partially ordered set $(P, \leq)$ is a forest if and only if the downset of each element of P is a chain.*

Definition 18 (Tree). *A tree $(P, \leq)$ is a forest that has minimum.*

Example 3. Consider the following binary relation on the set $\mathbb{N}$ of positive integers defined as follows: let $x, y \in \mathbb{N}$,

$$x \preccurlyeq y \ \text{if and only if} \ x \ \text{divides} \ y. \tag{2}$$

Then, the Hasse diagrams of the partially ordered sets

$$(\{1, 2, 3\}, \preccurlyeq), \ (\{1, 2, 5, 10\}, \preccurlyeq) \ \text{and} \ (\{2, 7, 14\}, \preccurlyeq)$$

are respectively represented as follows.

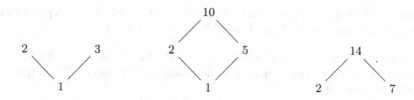

Fig. 2. Partially ordered sets

The poset $(\uparrow \{7\}, \preccurlyeq)$ is a chain. The poset $(\{1, 2, 3\}, \preccurlyeq)$ is a forest.

Minimal elements of a forest are called *roots*, while maximal elements are called *leaves*. A map $f : F \mapsto G$ between forests is *open* if, for $a \in G$ and $b \in F$, whenever $a \leq f(b)$ there exists $c \in F$ with $c \leq b$ such that $f(c) = a$. Equivalently, open maps carry upsets to upsets.

Let P be a poset, and let S be a subset of P. We say that an element $x \in P$ is an *upper bound* for S if $x \geq s$ for each $s \in S$. We can say that x is the *least upper bound* for S if x is an upper bound for S and $x \leq y$, for every upper bound y of S. Dually, x is a *lower bound* for S if $s \leq x$ for each $s \in S$; x is the *greatest lower bound* for S if x is a lower bound for S and $y \leq x$, for every lower bound y of S. If the least upper bound and the greatest lower bound of S exist, then they are unique.

Definition 19 (Lattice). *A lattice is a partially ordered set in which every pair of elements x and y has a least upper bound and a greatest lower bound, denoted with $x \wedge y$ and $x \vee y$, respectively.*

Lattices can also be defined as algebraic structures.

Definition 20 (Lattice). *[78] A lattice is an algebra $(L, \wedge, \vee)$ that satisfies the following proprieties.*

1. *$x \wedge x = x$ and $x \vee x = x$ (idempotent laws),*
2. *$x \wedge y = y \wedge x$ and $x \vee y = y \vee x$ (commutative laws),*
3. *$x \wedge (y \wedge z) = (x \wedge y) \wedge z$ and $x \vee (y \vee z) = (x \vee y) \vee z$ (associative laws),*
4. *$x \wedge (x \vee y) = x$ and $x \vee (x \wedge y) = x$ (absorption law),*

for each $x, y, z \in L$.

Remark 7. The latter two definitions are equivalent. Indeed, suppose that $(L, \leq)$ is a lattice, and $x \wedge y$ and $x \vee y$ denote the least upper bound and a greatest lower bound of x and y, respectively. Then, $(L, \wedge, \vee)$ satisfies the all proprieties of Definition 20.

Moreover, given a lattice $(L, \wedge, \vee)$, we can consider the following binary relation $\leq$ on L: let $x, y \in L$

$$x \leq y \text{ if and only if } x \wedge y = x \text{ (or } x \vee y = y).$$

We can prove that $(L, \leq)$ is a partially ordered set, in which every pair of elements has a greatest lower bound and a least upper bound.

An example of lattice is the structure $(2^U, \cap, \cup)$ of all subsets of a set U, with the usual set operations of intersection and union, or equivalently $(2^U, \subseteq)$, where $\subseteq$ is the set inclusion.

We are interested in *bounded distributive lattices* having the following definition.

Definition 21 (Bounded lattice). *A bounded lattice is a structure*

$$(L, \wedge, \vee, 0, 1)$$

such that $(L, \wedge, \vee)$ is a lattice, 0 is the identity element for $\vee$ ($x \vee 0 = 0$) and 1 is the identity element for $\wedge$ ($x \wedge 1 = x$). Moreover, 0 and 1 are called bottom *and* top *of L, respectively.*

Definition 22 (Distributive lattice). *[45] A lattice $(L, \wedge, \vee)$ is distributive if and only if the operations $\wedge$ and $\vee$ distribute over each other, namely*

1. *$x \wedge (y \vee z) = (x \wedge y) \vee (x \wedge z)$ and*
2. *$x \vee (y \wedge z) = (x \vee y) \wedge (x \vee z)$*

for each $x, y, z \in L$.

In 1937, the mathematician Garrett Birkhoff proved that there exists a one-to-one correspondence between distributive lattices and partial orders [15]. Namely, elements of a distributive lattice can be viewed as upsets, and the lattices operations correspond to intersection and union between sets.

Theorem 5 (Birkhoff's representation theorem). *Let $(P, \leq)$ be a partially ordered set, then the structure $(Up(P), \cap, \cup, \emptyset, P)$, where $Up(P)$ is the set of all upsets of P, and the operations $\cap$ and $\cup$ are respectively the intersection and the union between sets, is a bounded distributive lattice; furthermore, if $(L, \wedge, \vee, 0, 1)$ is a bounded distributive lattice, then there exists a partially ordered set $(P, \leq)$ such that $(Up(P), \cap, \cup, \emptyset, P)$ is isomorphic to $(L, \wedge, \vee, 0, 1)$.*

Definition 23 (Residuated lattice). *A residuated lattice is a structure*

$$(L, \wedge, \vee, *, \rightarrow, e, 0, 1)$$

such that

1. $(L, \wedge, \vee, 0, 1)$ *is a bounded lattice,*
2. $(L, *, e)$ *is a monoid,*
3. $x * y \leq z$ *if and only if $x \leq z \rightarrow y$, for each $x, y, z \in L$ ($*$ and $\rightarrow$ satisfy the adjointness property).*

Kleene Algebras. Kleene algebras are a subclass of *De Morgan algebras.* The latter were introduced by Moisil [74] without the restriction including 0 and 1. Successively, they were studied by several authors, in particular, by Kalman [63] (under the name of *distributive i-lattices*), and by Bialynicki-Birula and Rasiowa, which called them *quasi-Boolean algebras* [12]. The notation that is still used was introduced by Monteiro [75].

Definition 24 (De Morgan algebra). *A De Morgan algebra is a structure $(A, \wedge, \vee, \neg, 0, 1)$, where*

1. $(A, \wedge, \vee, 0, 1)$ *is a bounded distributive lattice,*
2. $\neg(x \vee y) = \neg x \wedge \neg y$ *(the Morgan's law),*
3. $\neg\neg x = x$ *($\neg$ is an involution),*

for each $x, y \in A$.

Definition 25 (Kleene algebra). *[30] A Kleene algebra $(A, \wedge, \vee, \neg, 0, 1)$ is a De Morgan algebra such that the following property, called Kleene property, holds:*

$$x \wedge \neg x \leq y \vee \neg y \tag{3}$$

for each $x, y \in A$.

Kleene algebras are also called *normal i-lattices* by Kalman.

Example 4. The structure $(\{0, \frac{1}{2}, 1\}, \wedge, \vee, \neg, 0, 1)$ is a three-element Kleene algebra, where $\wedge$ and $\vee$ are respectively the Kleene conjunction and implication defined in Sect. 2.2, and $\neg x = 1 - x$ for each $x \in \{0, \frac{1}{2}, 1\}$.

Example 5. Let C be a partition of the finite universe U, and let O_C be the set of all orthopairs generated by C. Then, the structure

$$(O_C, \wedge_K, \vee_K, \neg, (\emptyset, U), (U, \emptyset))$$

is a Kleene algebra, where $\wedge_K$ and $\vee_K$ are given in Definition 8, and $\neg(A, B) = (B, A)$ for each $(A, B) \in O_C$.

We are interested in the family of *finite centered Kleene algebras with the interpolation property*, that are explored in [31].

From now on, we denote an algebraic structure having support A with $\mathbb{A}$.

Definition 26 (Centered Kleene algebra). *A Kleene algebra $\mathbb{A}$ is a centered Kleene algebra if there exists $c \in A$ such that $c = \neg c$. The element c is called* center *of A.*

By using the Kleene property (see Definition 25), it is easy to prove that if c is a center of A, then it is unique.

The following notion was introduced for the first time by Monteiro [76].

Definition 27. *Let $(A, \wedge, \vee, \neg, 0, 1)$ be a centered Kleene algebra. Let c be the center of A. We say that A has the* interpolation property *if and only if for every $x, y \geq c$ such that $x \wedge y \leq c$ there exists z such that $z \vee c = x$ and $\neg z \vee c = y$.*

In [27] the above definition is called (CK) property, but it is also noticed that it coincides with the interpolation property described in [31], so we will use this last name. Not every centered Kleene algebra has the interpolation property, see Example 5 in [27].

Definition 28. *As in [31], let $(A, \wedge, \vee, \neg, 0, 1)$ be a Kleene algebra, we set*

$$A^+ = \{x \in A \mid \neg x \leq x\} \quad and \quad A^- = \{x \in A \mid x \leq \neg x\}.$$

We call A^+ and A^- positive *and* negative cone, *respectively.*

We can observe that the structure $(A^+, \wedge, \vee)$ is a sublattice of $(A, \wedge, \vee)$ containing 1, and dually, $(A^-, \wedge, \vee)$ is a sublattice of $(A, \wedge, \vee)$ containing 0.

Kalman Construction. The following construction is due to Kalman [63]. Let $(L, \wedge, \vee, 0, 1)$ be a bounded distributive lattice, we consider

$$K(L) = \{(x, y) \in L \times L \mid x \wedge y = 0\} \tag{4}$$

and the operations $\sqcap, \sqcup$ and $\neg$ defined on $K(L)$ as follows:

$$(x, y) \sqcap (u, v) = (x \wedge u, y \vee v) \tag{5}$$

$$(x, y) \sqcup (u, v) = (x \vee u, y \wedge v) \tag{6}$$

$$\neg(x, y) = (y, x) \tag{7}$$

for each $(x, y), (u, v) \in \mathsf{K}(L)$. Then,

$$\mathbb{K}(L) = (\mathsf{K}(L), \sqcap, \sqcup, \neg, (0, 1), (1, 0)) \tag{8}$$

is a centered Kleene algebra, with center $(0, 0)$. Moreover,

$$\mathsf{K}(L)^+ = \{(x, 0) \mid x \in L\} \quad \text{and} \quad \mathsf{K}(L)^- = \{(0, x) \mid x \in L\}.$$

The following theorem, proved by Cignoli [31] states that centered Kleene algebras with the interpolation property are represented by bounded distributive lattices.

Theorem 6. *A Kleene algebra $\mathbb{A}$ is isomorphic to $\mathbb{K}(L)$ for some bounded distributive lattice $\mathbb{L}$ if and only if $\mathbb{A}$ is centered and satisfies the interpolation property. In this case $\mathbb{L}$ is isomorphic to the lattice $\mathbb{A}^+$.*

By Birkhoff representation theorem and by Theorem 6, the following result holds.

Theorem 7. *A Kleene algebra $\mathbb{A}$ is isomorphic to $\mathbb{K}(Up(P))$, for some partially ordered set $(P, \leq)$, if and only if $\mathbb{A}$ is centered and satisfies the interpolation property. In this case $(Up(P), \cap, \cup, \emptyset, P)$ is isomorphic to the lattice $\mathbb{A}^+$.*

Remark 8. Trivially, $\mathsf{K}(Up(P))$ is the set of all pairs of disjoint upsets of P, and the operations 5 and 6 are the following: let $(X^1, X^2), (Y^1, Y^2) \in \mathsf{K}(Up(P))$, then

$$(X^1, X^2) \sqcap (Y^1, Y^2) = (X^1 \cap Y^1, X^2 \cup Y^2), \tag{9}$$

$$(X^1, X^2) \sqcup (Y^1, Y^2) = (X^1 \cup Y^1, X^2 \cap Y^2). \tag{10}$$

In this thesis, we focus on some structures having Kleene algebras as reduct. Namely, they are Nelson algebras, Nelson lattices, Kleene lattices with implication and IUML-algebras. Moreover, we will require that they are centered and satisfy the interpolation property.

Nelson Algebras. Nelson algebras were introduced by Rasiowa [87], under the name of N-lattices, as the algebraic counterparts of the constructive logic with strong negation considered by Nelson and Markov [88]. The centered Nelson algebras with the interpolation property are represented by Heyting algebras, that are defined as follows.

Definition 29 (Pseudo-complement). [31] *Let $(L, \wedge, \vee, 0, 1)$ be a bounded distributive lattice, and let $x, y \in L$. Then, the pseudo-complement of x with respect to y, denoted with $x \rightarrow y$, is an element of L satisfying the following proprieties:*

1. $x \wedge x \rightarrow y \leq y$ and
2. if $x \wedge z \leq y$, then $z \leq x \rightarrow y$, for each $z \in L$.

Notice that, given a bounded distributive lattice $(L, \wedge, \vee, 0, 1)$, the pseudo-complement of x with respect to y does not always exist.

Definition 30 (Heyting algebra). *An* Heyting algebra *is a structure*

$$(H, \wedge, \vee, \rightarrow, 0, 1),$$

where the reduct $(H, \wedge, \vee, 0, 1)$ *is a bounded residuated lattice, and* $x \rightarrow y$ *is the pseudo-complement of* x *with respect to* y *given in Definition 29.*

The next theorem affirms that there exists a one-to-one correspondence between finite Heyting algebras and finite partially ordered sets.

Theorem 8. [15] *For each finite Heyting algebra* $\mathbb{H}$*, there exists a finite poset* $(P, \leq)$ *such that* $\mathbb{H}$ *is isomorphic to* $(Up(P), \cap, \cup, \rightarrow_P, \emptyset, P)$*, where*

$$X \rightarrow_P Y = P \setminus \downarrow (X \setminus Y), \tag{11}$$

for each $X, Y \in Up(P)$*.*

Definition 31 (Quasi-Nelson algebra). *A* quasi-Nelson algebra *is a structure*

$$(A, \wedge, \vee, \neg, \Rightarrow, 0, 1)$$

such that

1. $(A, \wedge, \vee, \neg, 0, 1)$ *is a Kleene algebra, and*
2. *for each* $x, y \in A$*, the pseudo-complement of* x *with respect to* $\neg x \vee y$*, denoted with* $x \Rightarrow y$*, exists.*

Definition 32 (Nelson algebra). *A* Nelson algebra *is a quasi Nelson algebra* $(A, \wedge, \vee, \neg, \Rightarrow, 0, 1)$*, that satisfies the following property: let* $x, y, z \in A$

$$(x \wedge y) \Rightarrow z = x \Rightarrow (y \Rightarrow z).$$

Example 6. The structure $(\{0, \frac{1}{2}, 1\}, \wedge, \vee, \neg, \Rightarrow_\mathcal{N}, 0, 1)$, where $\neg x = 1 - x$ for each $x \in \{0, \frac{1}{2}, 1\}$, and $\Rightarrow_\mathcal{N}$ is the Nelson implication on $\{0, \frac{1}{2}, 1\}$ defined in Sect. 2.2, is a three-element Nelson algebra.

Example 7. Let C be a partition of the finite universe U, and let O_C be the set of all orthopairs generated by C. Then, the structure

$$(O_C, \wedge_K, \vee_K, \neg, \rightarrow_\mathcal{N}, (\emptyset, U), (U, \emptyset))$$

is a finite Nelson algebra, where $\rightarrow_\mathcal{N}$ is given in Definition 12.

Manuel M. Fidel [47] and Dimiter Vakarelov [99] have shown independently that if $(H, \wedge, \vee, \rightarrow, 0, 1)$ is an Heyting algebra, then $(\mathbb{K}(H), \Rightarrow)$, that is the structure $(K(H), \sqcap, \sqcup, \neg, \Rightarrow, (\emptyset, H), (H, \emptyset))$, is a Nelson algebra, where

$$(x, y) \Rightarrow (u, v) = (x \rightarrow u, x \wedge v) \tag{12}$$

for each $(x, y), (u, v) \in K(H)$.

Moreover, Cignoli [31] proved the following result.

Theorem 9. *A finite Nelson algebra $\mathbb{A}$ is isomorphic to $(\mathbb{K}(H), \Rightarrow)$ for some finite Heyting algebra $\mathbb{H}$ if and only if $\mathbb{A}$ is centered and satisfies the interpolation property.*

By Theorem 8, Eq. 12 and Theorem 9, the following result holds.

Theorem 10. *Let $\mathbb{A}$ be a Nelson algebra. Then, $\mathbb{A}$ is a finite centered Nelson algebra with the interpolation property if and only if there exists a finite poset $(P, \leq)$ such that $A \cong (\mathbb{K}(Up(P)), \rightarrow_1)$, where*

$$(X^1, X^2) \rightarrow_1 (Y^1, Y^2) = (P\backslash \downarrow (X^1 \setminus Y^1), X^1 \cap Y^2), \tag{13}$$

for each $(X^1, X^2), (Y^1, Y^2) \in K(Up(P))$.

Nelson Lattices. Nelson lattices are algebraic models of constructive logic with strong negation [97]. They are particular involutive residuated lattices. Moreover, finite centered Nelson lattices are represented by Heyting algebras.

Definition 33 (Involutive residuated lattice). *An* involutive residuated lattice *is a bounded, integral and commutative residuated lattice*

$$(A, \wedge, \vee, *, \rightarrow, e, 0, 1)$$

such that the operation $\neg$, defined by $\neg x = x \rightarrow 0$ for each $x \in A$, is an involution.

The operations $*$ and $\rightarrow$ of an involutive residuated lattice with support A can be obtained one from each other as follows: let $x, y \in A$, then

$$x * y = \neg(x \rightarrow \neg y) \tag{14}$$

and

$$x \rightarrow y = \neg(x * \neg y). \tag{15}$$

Definition 34 (Nelson lattice). *A* Nelson lattice *is an involutive residuated lattice*

$$(A, \wedge, \vee, *, \rightarrow, e, 0, 1),$$

where the following inequality holds: let $x^2 = x * x$,

$$(x^2 \rightarrow y) \wedge ((\neg y^2) \rightarrow \neg x) \leq x \rightarrow y,$$

for each $x, y \in A$.

Example 8. The structure $(\{0, \frac{1}{2}, 1\}, \wedge, \vee, \circledast_{\mathcal{L}}, \Rightarrow_{\mathcal{L}}, \frac{1}{2}, 0, 1)$ is a three-element Nelson lattice, where $\circledast_{\mathcal{L}}$ and $\Rightarrow_{\mathcal{L}}$ are respectively the Łukasiewicz conjunction and implication on $\{0, \frac{1}{2}, 1\}$ defined in Sect. 2.2.

Example 9. Let C be a partition of the finite universe U, and let O_C be the set of all orthopairs generated by C. Then, the structure

$$(O_C, \wedge_K, \vee_K, *_L, \to_L, (\emptyset, \emptyset), (\emptyset, U), (U, \emptyset)),$$

where $*_L$ and $\to_L$ are defined in Sect. 2.2, is a finite Nelson lattice.

Remark 9. Centered Nelson algebras and Nelson lattices are equationally equivalent, namely they are obtained one from the other as follows [23].
If $(A, \wedge, \vee, \neg, \Rightarrow, 0, 1)$ is a centered Nelson algebra, then $(A, \wedge, \vee, *, \to, 0, 1)$ is a Nelson lattice, where

$$x * y = \neg(x \Rightarrow \neg y) \vee \neg(y \Rightarrow \neg x) \text{ and } x \to y = (x \Rightarrow y) \wedge (\neg y \Rightarrow \neg x),$$

for each $x, y, z \in A$. Vice-versa, if $(A, \wedge, \vee, *, \to, 0, 1)$ is a Nelson lattice, then $(A, \wedge, \vee, \neg, \Rightarrow, 0, 1)$ is a centered Nelson algebra, where

$$\neg x = x \to 0 \text{ and } x \Rightarrow y = x^2 \to y,$$

for each $x, y \in A$.

We can notice that if $(H, \wedge, \vee, \to, 0, 1)$ is an Heyting algebra, then

$$(\mathbb{K}(H), *, \Rightarrow),$$

where $(\mathbb{K}(H), *, \Rightarrow)$ denotes $(\mathsf{K}(H), \sqcap, \sqcup, *, \Rightarrow, (\emptyset, \emptyset), (\emptyset, H), (H, \emptyset))$, is a Nelson lattice, such that

$$(x, y) * (u, v) = (x \wedge u, (x \to v) \wedge (u \to y)) \tag{16}$$

and

$$(x, y) \Rightarrow (u, v) = ((x \to u) \wedge (v \to y), x \wedge v), \tag{17}$$

for each $x, y, u, v \in H$.

Finite centered Nelson lattices with the interpolation property are represented by finite Heyting algebras [27].

Theorem 11. *A finite Nelson lattice $\mathbb{A}$ is isomorphic to $(\mathbb{K}(H), *, \Rightarrow)$ for some finite Heyting algebra $\mathbb{H}$ if and only if $\mathbb{A}$ is centered and satisfies the interpolation property.*

By Theorem 8, Eq. 16, Eq. 17 and Theorem 11, the following result holds.

Theorem 12. *Let $\mathbb{A}$ be a Nelson lattice. Then, $\mathbb{A}$ is a finite centered Nelson lattice with the interpolation property if and only if there exists a finite poset $(P, \leq)$ such that $A \cong (\mathbb{K}(Up(P)), \star_2 \to_2)$, where*

$$(X^1, X^2) \star_2 (Y^1, Y^2) = (X^1 \cap Y^1, P \setminus (\downarrow (X^1 \setminus Y^2) \cup \downarrow (Y^1 \setminus X^2))), \tag{18}$$

$$(X^1, X^2) \to_2 (Y^1, Y^2) = (P \setminus (\downarrow (X^1 \setminus Y^1) \cup \downarrow (Y^2 \setminus X^2)), X^1 \cap Y^2), \tag{19}$$

for each $(X^1, X^2), (Y^1, Y^2) \in \mathsf{K}(Up(P))$.

IUML-algebras. IUML-algebras are the algebraic counterpart of the logic IUML, which is a substructural fuzzy logic that is an axiomatic extension of the multiplicative additive intuitionistic linear logic MAILL [73]. IUML-algebras can also be defined as *bounded odd Sugihara monoids*, where a Sugihara monoid is the equivalent algebraic semantics for the relevance logic RM^t of R-mingle as formulated with Ackermann constants. In [49] a dual categorical equivalence is shown between IUML-algebras and suitable topological spaces defined starting from Kleene spaces. In this dissertation, we focus only on finite IUML-algebras refers to [3] and [73].

Definition 35 (IUML-algebra). *An* idempotent uninorm mingle logic algebra (IUML-algebra) [73] *is an idempotent commutative bounded residuated lattice*

$$(A, \wedge, \vee, *, \rightarrow, e, \bot, \top),$$

satisfying the following properties:

1. $(x \rightarrow y) \vee (y \rightarrow x) \geq e$, *and*
2. $(x \rightarrow e) \rightarrow e = x$,

for every $x, y \in A$.

In any IUML-algebra, if we define the unary operation $\neg$ as $\neg x = x \rightarrow e$, then $\neg \neg x = x$ ($\neg$ is involutive) and $x \rightarrow y = \neg(x * \neg y)$.

Example 10. The structure $(\{0, \frac{1}{2}, 1\}, \wedge, \vee, \circledast_S, \Rightarrow_S, \frac{1}{2}, 0, 1)$ is a three-element IUML-algebra, , where $\circledast_S$ and $\Rightarrow_S$ are respectively the Sobociński conjunction and implication on $\{0, \frac{1}{2}, 1\}$ defined in Sect. 2.2.

Example 11. Let C be a partition of the finite universe U, and let O_C be the set of all orthopairs generated by C. Then, the structure

$$(O_C, \wedge_K, \vee_K, *_S, \rightarrow_S, (\emptyset, \emptyset), (\emptyset, U), (U, \emptyset)),$$

where $*_S$ and $\rightarrow_S$ are defined in Sect. 2.2, is a finite IUML-algebra.

Moreover, in [3] a dual categorical equivalence is described between finite forests F with order preserving open maps and finite IUML-algebras with homomorphisms.

Definition 36. *For any finite forest F, we consider $K(Up(F))$, that is the set of pairs of disjoint upsets of F (it is the set defined by Eq. 4 starting from the lattice $(Up(F), \cap, \cup, \emptyset, F)$, and we define the following operations: if (X^1, X^2) and (Y^1, Y^2) belong to $K(Up(F))$, we set:*

$$(X^1, X^2) \star_3 (Y^1, Y^2) = ((X^1 \cap Y^1) \cup (X \diamond Y), (X^2 \cup Y^2) \setminus (X \diamond Y)) \quad (20)$$

where, for each $U = (U^1, U^2), V = (V^1, V^2) \in K(Up(F))$, letting $U^0 = F \setminus (U^1 \cup U^2)$, we set

$$U \diamond V = \uparrow ((U^0 \cap V^1) \cup (V^0 \cap U^1)).$$

$$(X^1, X^2) \rightarrow_3 (Y^1, Y^2) = \neg((X^1, X^2) \star_3 (Y^2, Y^1)). \quad (21)$$

Theorem 13. [3] *For every finite forest F, the structure*

$$(\mathbb{K}(Up(F)), \star_3, \rightarrow_3) = (K(Up(F)), \sqcap, \sqcup, \star_3, \rightarrow_3, (\emptyset, \emptyset), (\emptyset, F), (F, \emptyset))$$

is an IUML-algebra. Vice-versa, for each finite IUML-algebra $\mathbb{A}$ there is a finite forest F_A such that $\mathbb{A}$ is isomorphic with $(\mathbb{K}(Up(F_A)), \star_3, \rightarrow_3)$.

Kleene Lattices with Implication. Kleene lattices with implication are a class of Kleene algebras where an additional operation of implication can be defined in such a way to make them *DLI*-algebras, (i.e. *algebras with implication*). The latter generalize the Heyting algebras and are defined in [28].

Definition 37 (DLI-algebra). *A DLI-algebra is a structure*

$$(H, \vee, \wedge, \rightarrow, 0, 1),$$

where $(H, \wedge, \vee, 0, 1)$ is a bounded distributive lattice and the following properties hold: let $x, y, z \in A$

1. $(x \rightarrow y) \wedge (x \rightarrow z) = x \rightarrow (y \wedge z)$,
2. $(x \rightarrow z) \wedge (y \rightarrow z) = (x \vee y) \rightarrow z$,
3. $0 \rightarrow x = 1$,
4. $x \rightarrow 1 = 1$.

Furthermore, a DLI^+-algebra is a DLI-algebra $(H, \vee, \wedge, \rightarrow, 0, 1)$ where the following inequality holds: $a \wedge (a \rightarrow b) \leq b$, for each $a, b \in H$.

It is easy to prove that each Heyting algebra is also a DLI^+-algebra.

Definition 38 (DLI*-algebra). *A DLI^*-algebra is a structure*

$$(H, \wedge, \vee, \rightarrow, 0, 1),$$

where $(H, \wedge, \vee, 0, 1)$ is a bounded distributive lattice and $\rightarrow$ is defined as follows: let $x, y \in H$,

$$x \rightarrow y = \begin{cases} 1 & \text{if } x = 0, \\ y & \text{if } x \neq 0. \end{cases} \tag{22}$$

Proposition 1. *A DLI^*-algebra is a DLI^+-algebra.*

By Theorem 5, the following result holds.

Theorem 14. *The structure $(H, \wedge, \vee, \rightarrow, 0, 1)$ is a DLI^*-algebra if and only if $\mathbb{H} \cong (Up(P), \cap, \cup, \rightarrow_P^*, \emptyset, P)$, where*

$$X \rightarrow_P^* Y = \begin{cases} P & \text{if } X = \emptyset, \\ Y & \text{if } X \neq \emptyset, \end{cases} \tag{23}$$

for each $X, Y \in P$.

Definition 39 (Kleene lattice with implication). *A Kleene lattice with implication is a structure*

$$(A, \wedge, \vee, \neg, *, \rightarrow, 0, 1)$$

such that $(A, \wedge, \vee, \neg, 0, 1)$ *is a centered Kleene algebra and the following conditions hold: let c be the center of A and let* $x, y \in A$

1. $(A, \wedge, \vee, \rightarrow, 0, 1)$ *is a DLI-algebra,*
2. $(x \wedge (x \rightarrow y)) \vee c \leq y \vee c$,
3. $c \rightarrow c = 1$,
4. $(x \rightarrow y) \wedge c = (\neg x \vee y) \wedge c$,
5. $(x \rightarrow \neg y) \vee c = ((x \rightarrow (\neg x \vee c)))$.

By Eq. 14, we can define the operation $*$ *from* $\rightarrow$. *Vice-versa, by Eq. 15,* $\rightarrow$ *is obtained from* $*$.

It is easy to prove that each Nelson algebra is also a Kleene lattice with implication.

Let $(H, \wedge, \vee, \rightarrow, 0, 1)$ be a DLI^+-algebra, then $(\mathbb{K}(H), \star, \Rightarrow)$ is a Kleene lattice with implication, where $\Rightarrow$ is defined by 17 and $x \star y = \neg(x \Rightarrow \neg y)$. Moreover, the following theorem holds.

Theorem 15. *A Kleene lattice with implication A is isomorphic to the structure* $(\mathbb{K}(H), \star, \Rightarrow)$ *for some* DLI^+-*algebra* $\mathbb{H}$ *if and only if it has the interpolation property.*

Definition 40 (KLI*-algebra). *A KLI*-algebra is a structure*

$$(A, \wedge, \vee, \neg, *, \rightarrow, 0, 1),$$

where $(A, \wedge, \vee, \neg, 0, 1)$ *is a centered Kleene algebra and the operations* $*$ *and* $\rightarrow$ *are defined as follows: let c be the center of A, and let* $x, y \in A$

$$x \rightarrow y = \begin{cases} 1, & \text{if } x \leq c \text{ and } y \geq c; \\ \neg x, & \text{if } x \leq c \text{ and } y \ngeq c; \\ y, & \text{if } x \nleq c \text{ and } y \geq c; \\ ((y \vee c) \wedge \neg x) \vee ((\neg x \vee c) \wedge y), & \text{if } x \nleq c \text{ and } y \ngeq c; \end{cases} \qquad (24)$$

and $x * y = \neg(x \rightarrow y)$.

Proposition 2. [27] *A KLI*-algebra is a Kleene lattice with implication.*

The next result follows by Theorem 14 and Theorem 15.

Theorem 16. *The structure* $(A, \wedge, \vee, \neg, *, \rightarrow, 0, 1)$ *is a KLI*-algebra with the interpolation property if and only if* $\mathbb{A} \cong (\mathbb{K}(Up(P)), \star_4, \rightarrow_4)$, *where* $\star_4$ *and* $\rightarrow_4$ *are defined as follows.*

$$(X^1, X^2) \star_4 (Y^1, Y^2) = \begin{cases} (\emptyset, P), & \text{if } X^1 = \emptyset \text{ and } Y^1 = \emptyset; \\ (X^1, X^2), & \text{if } X^1 = \emptyset \text{ and } Y^1 \neq \emptyset; \\ (Y^1, Y^2), & \text{if } X^1 \neq \emptyset \text{ and } Y^1 = \emptyset; \\ (X^1 \cap Y^1, X^2 \cap Y^2), & \text{if } X^1 \neq \emptyset \text{ and } Y^1 \neq \emptyset; \end{cases} \qquad (25)$$

and

$$(X^1, X^2) \rightarrow_4 (Y^1, Y^2) = \begin{cases} (P, \emptyset), & \text{if } X^1 = \emptyset \text{ and } Y^2 = \emptyset; \\ (X^2, X^1), & \text{if } X^1 = \emptyset \text{ and } Y^2 \neq \emptyset; \\ (Y^1, Y^2), & \text{if } X^1 \neq \emptyset \text{ and } Y^2 = \emptyset; \\ (Y^1 \cap X^2, X^1 \cap Y^2), & \text{if } X^1 \neq \emptyset \text{ and } Y^2 \neq \emptyset; \end{cases}$$

$$(26)$$

for each $(X^1, X^2), (Y^1, Y^2) \in \mathsf{K}(Up(P))$.

3 Sequences of Refinements of Orthopairs

> *Mathematical objects are not so directly given as physical objects. They are something between the ideal world and the empirical world.*
>
> Kurt Gödel

In this chapter, we introduce the definition of *refinement sequences of partial coverings* as special sequences of coverings representing situations where new information is gradually provided on ever smaller sets of objects. We provide examples of environments in which refinement sequences arise; in detail, we obtain refinement sequences starting from incomplete information tables and formal contexts. We identify some families of sequences considering how much the blocks of their coverings overlap. We identify refinement sequences as partially ordered sets. Moreover, we introduce the notion of *sequences of orthopairs*, in order to generalize the rough set theory. We represent each sequence of orthopairs as a pair of disjoint upsets of a partially ordered set, or equivalently, as a labelled poset. Finally, we provide a theorem that is fundamental to prove the results of Sect. 4. Preliminary versions of this chapter appeared in [1,2,19,20].

3.1 Refinement Sequences

In this section, we introduce the notion of *refinement sequence* of a universe.

Refinement sequences are special sequences of partial coverings of a given universe (a partial covering of U is a subset of 2^U, i.e. any set of subsets of U [35]). More precisely, the refinements sequences are defined as follows.

Definition 41. *A sequence* $\mathcal{C} = (C_1, \ldots, C_n)$ *of partial coverings of U is a refinement sequence of U if each element of C_i is contained in an element of C_{i-1}, for $i = 2, \ldots, n$.*

For simplicity, we omit to specify on which universe the refinement sequence is defined, when it is clear.

Example 12. Suppose that $U = \{a, b, c, d, e, f, g\}$ and that C_1 and C_2 are partial coverings of U respectively defined as follows:

- $C_1 = \{\{a, b, c, d\}, \{d, e, f, g\}\}$;
- $C_2 = \{\{a, b, c\}, \{c, d\}, \{d, e\}, \{f, g\}\}$.

Then, (C_1, C_2) is a refinement sequence of U.

Remark 10. We notice that a partial covering of U naturally defines a tolerance relation on a subset of U and the vice-versa also holds. Moreover, we call blocks both the elements of a partial covering and the tolerance classes. Therefore, a refinement sequence $(C_1, \ldots, C_n)$ of partial coverings of U corresponds to a sequence $(R_1, \ldots, R_n)$ of tolerance relations respectively defined on the subsets $U_1, \ldots, U_n$ of U, where

- U_i is the union of the blocks of C_i, for each $i \in \{1, \ldots, n\}$;
- $U_i \subseteq U_j$, for each $j \leq i$;
- $R_i(u) \subseteq R_j(u)$, for each $j \leq i$ and $u \in U_i$.

In this thesis, we also consider refinement sequences of *partial partitions* of a universe, where a partition corresponds to an equivalence relation, and it is a covering such that its blocks are disjoint with each others.

As shown in the following example, the refinement sequences can be used for ontology construction.

Example 13. Suppose to start from a set of rocks (first covering) and then to specify our interest in *magmatic rocks* and *sedimentary rocks* that form a partial covering of the initial set of rocks (the latter also contains several elements that are metamorphic rocks, then the covering made of magmatic and sedimentary rocks is partial). Then, we intend to refine such classification by considering two groups of *magmatic rock* (*intrusive rocks* and *extrusive rocks*) and two groups of *sedimentary rocks* (*Chemical rocks* and *Clastic rocks*). The refinement sequence of partial coverings can be represented as follows.

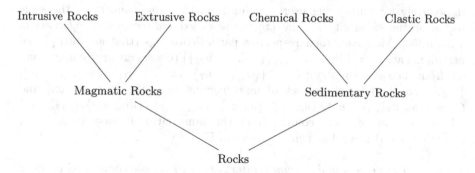

Fig. 3. Refinement sequence for rocks classification

The next example shows that a refinement sequence corresponds to an *incomplete information table*. The latter is a table where a set of objects is described by several attributes, but some data may be missing.

Example 14. Suppose that we have information about 22 users of Facebook, labelled with $u_1, \ldots, u_{22}$. In particular, we focus on information related to the place where each user declares to come from on its personal profile.

The available data are organized in the *information table* as in Table 9, (see [66]) where $U = \{u_1, \ldots, u_{22}\}$ is the universe and $\{Country, Region, City\}$ is the set of attributes.

Table 9. Information table of the users

	Country	Region	City		Country	Region	City
u_1	Italy	×	×	u_{12}	France	Brittany	Rennes
u_2	Italy	Lombardy	Varese	u_{13}	France	Brittany	Rennes
u_3	Italy	Lombardy	Varese	u_{14}	France	Brittany	×
u_4	Italy	Lombardy	Milan	u_{15}	France	Brittany	×
u_5	Italy	Lombardy	Milan	u_{16}	France	Grand Est	Strasbourg
u_6	Italy	Lombardy	Pavia	u_{17}	France	Grand Est	Strasbourg
u_7	Italy	Lombardy	Pavia	u_{18}	France	Grand Est	Mets
u_8	Italy	Campania	Naples	u_{19}	France	Grand Est	Mets
u_9	Italy	Campania	Naples	u_{20}	France	Grand Est	×
u_{10}	Italy	Campania	×	u_{21}	France	Grand Est	×
u_{11}	Italy	Campania	×	u_{22}	France	×	×

Observe that there are three equivalence relations between users determined respectively by considering users coming from the same country or the same region or the same city[1]. They are the so-called *indiscernibility relations* of Table 9 [66]. Moreover, their respective partial coverings (that are also partial partitions) are $C_1 = \{\{u_1, \ldots, u_{11}\}, \{u_{12}, \ldots, u_{22}\}\}$ (classes are sets of users coming from the same country); $C_2 = \{\{u_2, \ldots, u_7\}, \{u_8, \ldots, u_{11}\}, \{u_{12}, \ldots, u_{15}\}, \{u_{16}, \ldots, u_{21}\}\}$ (classes are set of users coming from the same region) and $C_3 = \{\{u_2, u_3\}, \{u_4, u_5\}, \{u_6, u_7\}, \{u_8, u_9\}, \{u_{12}, u_{13}\}, \{u_{16}, u_{17}\}, \{u_{18}, u_{19}\}\}$ (classes are set of users coming from the same city). It easy to see that $\mathcal{C} = (C_1, C_2, C_3)$ is a refinement sequence of U.

Refinement Sequences and Formal Context. There is a close connection between refinement sequences and *formal contexts*, which are mathematical structures used in *Formal Concept Analysis* and *Fuzzy Formal Concept Analysis* [26,50]. A formal context is a triple (X, Y, I), where X is a set of objects, Y is a set of

---

[1] The equivalence relations *coming from the same region* and *coming from the same city* are defined on proper subsets of U, for there are missing data for some users.

attributes, and I is a binary relation between X and Y. If I is a fuzzy relation, then (X, Y, I) is called *fuzzy formal context*, and $I(x, y)$ expresses the degree wherewith the object x has the attribute y. A formal context can be represented by a table with rows corresponding to objects, columns corresponding to attributes, and table entries containing each degree $I(x, y)$, with $x \in X$ and $y \in Y$. In particular, it is clear that if I is an ordinary relation, the table entries only contain the degrees 0 and 1. By using several techniques [10, 17], *formal concepts* are extracted from every formal context. Formal concepts are particular clusters which represent natural human-like concepts such as "organism living in water", "car with all wheel drive system", etc.

Given a refinement sequence $\mathcal{C} = (C_1, \ldots, C_n)$, we can view a block b of C_i as the set of all elements of U that have a specific attribute y_b. Thus, $\mathcal{C}$ corresponds to a formal context $(U, Y_{\mathcal{C}}, I)$, where $Y_{\mathcal{C}} = \cup \{y_b \mid b \in C_i \text{ and } i \in \{1, \ldots, n\}\}$ and "$(u, y_b) \in I$ if and only if $u \in b$". For example, let $\mathcal{C} = (C_1 = \{b_1, b_2\}, C_2 = \{b_3, b_4, b_5\})$ be the refinement sequence of $\{a, b, c, d, e, f, g\}$ such that $b_1 = \{a, b, c\}, b_2 = \{d, e, f, g\}, b_3 = \{a, b, c\}, b_4 = \{d, e\}$ and $b_5 = \{f, g\}$. Then, the formal context associated to $\mathcal{C}$ is represented by Table 10.

Table 10. Formal context of $\mathcal{C}$

I	y_{b_1}	y_{b_2}	y_{b_3}	y_{b_4}	y_{b_5}
a	1	0	1	0	0
b	1	0	1	0	0
c	1	0	1	0	0
d	0	1	0	1	0
e	0	1	0	1	0
f	0	1	0	0	1
g	0	1	0	0	1

Vice-versa, starting from a formal context, we can build a refinement sequence as follows. For each $y \in Y$, we set $b_y = \{x \in X \mid (x, y) \in I\}$. Let $s = |Y|$, if $s = 1$, then the refinement sequence assigned to (X, Y, I) is trivially made of only one covering. Suppose that $s > 1$, then we set $C_s = \{b_y \mid b_{y'} \not\subset b_y, \text{ for each } y' \in Y\}$ and, let $i < s$, $C_i = \{b_y \mid \text{ there exists } b_{y'} \in C_{i+1} \text{ such that } b_{y'} \subset b_y \text{ and } b_{y'} \subset b_{y''} \subset b_y \text{ does not hold for each } y'' \in Y\}$. Therefore, $\mathcal{C} = (C_k, C_{k+1}, \ldots, C_s)$ is the refinement sequence assigned to (X, Y, I), where $k = max\{i \in \{1, \ldots, s - 1\} \mid C_i \neq C_{i+1}\}$. For example, we consider the formal context

$$K = (\{a_1, a_2, a_3, a_4, a_5\}, \{\text{feline, cat, tiger}\}, I),$$

where $\{a_1, a_2, a_3, a_4, a_5\}$ represents a set of 5 animals and I is defined by Table 11.

Table 11. Formal context K

I	feline	cat	tiger
a_1	1	1	0
a_2	1	1	0
a_3	0	0	0
a_4	1	0	1
a_5	1	0	1

Then, the refinement sequence assigned to K is made of coverings C_1 and C_2 such that $C_1 = \{\{a_1, a_2, a_4, a_5\}\} = \{$animals that are felines$\}$ and $C_2 = \{\{a_1, a_2\}, \{a_4, a_5\}\} = \{\{$animals that are cats$\}, \{$animals that are tigers$\}\}$.

3.2 Refinement Sequences as Posets

In this section, we show that each refinement sequence is represented as a partially ordered set.

Definition 42. *Let $\mathcal{C} = (C_1, \ldots, C_n)$ be a refinement sequence of U. We assign the partially ordered set $(P_\mathcal{C}, \leq_\mathcal{C})$ to $\mathcal{C}$, where:*

- *$P_\mathcal{C} = \bigcup_{i=1}^{n} C_i$ (the set of nodes is the set of all subsets of U belonging to the coverings $C_1, \ldots, C_n$), and*
- *$N \leq_\mathcal{C} M$ if and only if $M \subseteq N$, for $N, M \in P_\mathcal{C}$ (the partial ordered relation is the reverse inclusion between sets).*

Example 15. Let (C_1, C_2, C_3) be a refinement sequence of $\{a, b, c, d, e, f, g, h\}$, where

- $C_1 = \{\{a, b, c, d, e, f, g\}\}$,
- $C_2 = \{\{a, b, c, d\}, \{c, d, e, f\}\}$ and
- $C_3 = \{\{c, d\}, \{d, e, f\}\}$.

The poset assigned to (C_1, C_2, C_3) is shown in the following figure.

Proposition 3. *If $\mathcal{C}$ is a refinement sequence of partial partitions of U, then $(P_\mathcal{C}, \leq_\mathcal{C})$ is a forest.*

Proof. Let $N, M \in \downarrow X$, with $X \in P_\mathcal{C}$. Then, $N, M \leq_\mathcal{C} X$. By Definition 42, $X \subseteq N \cap M$. Suppose that $N \in C_i$ and $M \in C_j$, with $i \leq j$. By Definition 41, there exists $\tilde{N} \in C_j$ such that $\tilde{N} \subseteq N$. Since C_j is a partial partition of U, we have that $\tilde{N} = M$ or $\tilde{N} \cap M = \emptyset$. On the other hand, both M and $\tilde{N}$ contain X. Consequently, $\tilde{N} = M$ and so $N \leq_\mathcal{C} M$.

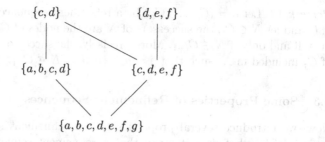

Fig. 4. Poset assigned to (C_1, C_2, C_3)

Example 16. If $\mathcal{C}$ is the refinement sequence of Example 14, then $(P_{\mathcal{C}}, \leq_{\mathcal{C}})$ is the following forest.

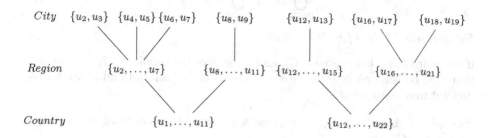

Fig. 5. Forest of the users

Remark 11. The maximal and minimal elements of $(P_{\mathcal{C}}, \leq_{\mathcal{C}})$ are all blocks of C_n and C_1, respectively.

Remark 12. The main difference between $\mathcal{C} = (C_1, \ldots, C_n)$ and the partially ordered set $P_{\mathcal{C}}$ is that the coverings $C_1, \ldots, C_n$ can also contain the same blocks, while each block appears only once in $P_{\mathcal{C}}$. For example, consider the refinement sequence $\mathcal{C} = (C_1, C_2)$ such that $C_1 = \{\{a, b\}, \{b, c, d, e\}\}$ and $C_2 = \{\{a, b\}, \{c, d\}\}$, then $P_{\mathcal{C}}$, that is represented by the following figure, has only one block $\{a, b\}$.

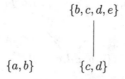

Fig. 6. Poset assigned to (C_1, C_2)

Remark 13. Let $C = (C_1, \ldots, C_n)$ be a refinement sequence of partial partitions of U and let $N \in C_i$, the successors of N are the nodes of C_{i+1} that are included in N if and only if $N \notin C_{i+1}$. More precisely, the successors of N are the blocks of C_j included in N, such that $j = min\{k > i \mid N \notin C_k\}$.

3.3   Some Properties of Refinement Sequences

Now, we introduce several properties that a refinement sequence could have; so, we define what does it mean that a refinement sequence is *complete, safe* and *pairwise overlapping*. Given a refinement sequence C, we denote by $\mathsf{K}(C)$ the set made of the pairs of disjoint upsets of P_C. We notice that $\mathsf{K}(C)$ coincides with the set $\mathsf{K}(Up(P_C))$ given by 4 (see Sect. 2.3) starting from the lattice $(Up(P_C), \cap, \cup, \emptyset, P)$.

Definition 43. *A refinement sequence C of a universe U is* complete *if and only if*

$$\bigcup_{N \in A} N \cap \bigcup_{N \in B} N = \emptyset \tag{27}$$

for each pair (A, B) of $\mathsf{K}(C)$.

If the pair (A, B) belongs to $\mathsf{K}(C)$, and it satisfies the condition 27, then we say that "(A, B) is a pair of *totally disjoint* upsets of P_C" and "A and B are *totally disjoint* from each other".

Example 17. Let $C = (C_1, C_2, C_3)$ be a refinement sequence of the universe $\{a, b, c, d, e, f\}$, where

- $C_1 = \{\{a, b, c, d, e, f\}\}$,
- $C_2 = \{\{a, b, c, d\}, \{d, e, f\}\}$ and
- $C_3 = \{\{a, b\}\}$.

Also, we consider the sets $A^1 = \{\{a, b, c, d\}, \{a, b\}\}$ and $A^2 = \{\{d, e, f\}\}$, which are upsets of P_C, and they are pairwise disjoint. We have that $\{d\}$ is the intersection between $\{a, b, c, d\} \cup \{a, b\}$ (the blocks of A^1) and $\{d, e, f\}$ (the only block of A^2). Indeed, the refinement sequence C is not complete.

Example 18. The refinement sequence of $\{a, b, c, d, e, f, g\}$ represented by the following forest is complete.

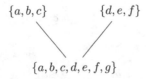

Fig. 7. Complete refinement sequence

Proposition 4. *Let $C = (C_1, \ldots, C_n)$ be a refinement sequence of U. If $C_1, \ldots,$ C_n are partial partitions of U, then C is complete.*

Proof. Let A^1 and A^2 be upsets of P_C such that $A^1 \cap A^2 = \emptyset$. Suppose that $b_1 \in A^1 \cap C_i$ and $b_2 \in A^2 \cap C_j$ with $i \leq j$. By Definition 41, there exists $\tilde{b}_2 \in C_i$ with $b_2 \subseteq \tilde{b}_2$. Since C_i is a partial partition, $b_1 \cap \tilde{b}_2 = \emptyset$ or $b_1 = \tilde{b}_2$. The equality $b_1 = \tilde{b}_2$ implies $b_2 \in A_1 \cap A_2$ which can not occur (A_2 is an upset). Consequently, $b_1 \cap \tilde{b}_2 = \emptyset$ and so $b_1 \cap b_2 = \emptyset$.

On the other hand, there exist complete refinement sequences made of coverings that are not partitions (see the following example).

Example 19. Let $C = (C_1, C_2, C_3)$ be the refinement sequence of the universe $\{a, b, c, d, e, f, g\}$ such that

- $C_1 = \{\{a, b, c, d, e\}, \{f, g\}\}$,
- $C_2 = \{\{a, b, c\}, \{a, b, d\}, \{f, g\}\}$ and
- $C_3 = \{\{a, b\}, \{f, g\}\}$.

Then, C is complete.

Definition 44. *A refinement sequence C is safe if for each $N \in P_C$ such that $N \subseteq N_1 \cup \ldots \cup N_r$ with $N_1, \ldots, N_r \in P_C$, there exists $j \in \{1, \ldots r\}$ such that $N \subseteq N_j$.*

Therefore, given a safe refinement sequence C, each node N of P_C is not included in the union of some other nodes of P_C that are all greater than N or disjoint with N.

The followings are two examples of refinement sequence: the first one is safe and the second one is not safe.

Example 20. Suppose that

$$C_1 = \{\{a, b, c, d, e\}, \{a, f, g, h\}\} \text{ and } C_2 = \{\{a, b, c\}, \{c, d\}, \{f, g\}\},$$

then the refinement sequence $C = (C_1, C_2)$ is safe.

Example 21. The refinement sequence $(\tilde{C}_1, \tilde{C}_2)$ with

$$\tilde{C}_1 = \{\{a, b, c, d, e\}, \{c, d, e, f, g, h\}\} \text{ and } \tilde{C}_2 = \{\{a, b, c\}, \{c, d\}, \{e, f, g\}\},$$

is not safe, since $\{a, b, c, d, e\} \subseteq \{a, b, c\} \cup \{c, d\} \cup \{e, f, g\}$.

The next remark provides a condition that all nodes of P_C must satisfy so that the complete refinement sequence C is also safe.

Remark 14. By Definition 44, if C is safe and $N \in P_C$, then there exists $x \in N$ such that $x \notin M$, for each $M \in P_C \backslash \downarrow \{N\}$.

The following proposition yields a condition on nodes of P_C, so that a complete refinement sequence C is also safe.

Proposition 5. *Let C be a complete refinement sequence of U. C is safe if and only if each node of P_C is not included in the union of its successors.*

Proof. ($\Rightarrow$). This implication is trivial, and it holds true even without the assumption that C is complete.

($\Leftarrow$). Suppose that $N \in P_C$ and $N \subseteq N_1 \cup \ldots \cup N_r$, with $N_1, \ldots, N_r \in P_C$ and $N_i \cap N \neq \emptyset$ for each $i \in \{1, \ldots, r\}$. Since C is complete, $N_i \subseteq N$ or $N \subseteq N_i$, for each $i \in \{1, \ldots, n\}$. By hypothesis, there exists $\tilde{N} \in \{N_1, \ldots, N_r\}$ such that $N_i \not\subseteq N$. Then, $N \subseteq N_i$.

By Proposition 4, we can say that a refinement sequence of partial partitions is safe if and only if each node of the respective forest is not equal the union of its successors.

Definition 45. *A refinement sequence $C = C_1, \ldots, C_n$ is pairwise overlapping if there are not disjoint blocks in C_i, for each $i \in \{1, \ldots, n\}$.*

Example 22. The refinement sequence of Examples 15 is pairwise overlapping, since the element d belongs to each block of C_1, C_2 and C_3.

A pairwise overlapping refinement sequence differs more from the sequences of partial partitions than the other refinement sequences. Furthermore, refinement sequences of partial partitions are pairwise overlapping if and only if the forests assigned with them are chains.

We also notice that refinement sequences that are associated to forests are not complete, when are pairwise overlapping. As a consequence, a complete refinement sequence cannot also be pairwise overlapping.

3.4   Sequences of Refinements of Orthopairs

The main aim of this section is to define sequences of refinements of orthopairs.

Definition 46. *Let $C = (C_1, \ldots, C_n)$ be a refinement sequence of U and $X \subseteq U$. The sequence of refinements of orthopairs of X determined by C is the sequence*

$$\mathcal{O}_C(X) = ((\mathcal{L}_1(X), \mathcal{E}_1(X)), \ldots, (\mathcal{L}_n(X), \mathcal{E}_n(X))),$$

where $(\mathcal{L}_i(X), \mathcal{E}_i(X))$ is the orthopair of X determined by C_i.

For short, $\mathcal{O}_C(X)$ is also called *sequence of orthopairs* of X determined by C.

Example 23. Let $U = \{a, b, c, d, e, f, g, h, i, j\}$ and $X = \{a, b, c, d, e\}$. If C is the refinement sequence of U made of $C_1 = \{\{a, b, c, d, e, f, g, h, i, j\}\}$, $C_2 = \{\{a, b, c, d, e\}, \{e, f, g, h, i\}\}$, $C_3 = \{\{a, b, c\}, \{c, d\}, \{e, f, g\}, \{g, h\}\}$, then

$$\mathcal{O}_C(X) = ((\emptyset, \emptyset), (\{\{a, b, c, d, e\}\}, \emptyset), (\{\{a, b, c\}, \{c, d\}\}, \{\{g, h\}\})).$$

Example 24. Suppose that we are interested to describe the set $X = \{u_1, u_8, u_9, u_{10}, u_{11}, u_{12}, u_{13}, u_{14}, u_{15}, u_{16}, u_{17}\}$ with respect to the refinement sequence $\mathcal{C}$ of Example 14. We know that X contains all users that have the attributes Campania (hence Naples), Brittany (hence Rennes) and Strasbourg; while users that come from Lombardy (hence Varese, Milan and Pavia) and Mets do not belong to X. This means that the sequence of orthopairs of X is $(\mathcal{O}_{C_1}(X), \mathcal{O}_{C_2}(X), \mathcal{O}_{C_3}(X))$ where $\mathcal{O}_{C_1}(X) = (\emptyset, \emptyset)$, $\mathcal{O}_{C_2}(X) = (\{u_8, \ldots, u_{15}\}, \{u_2, \ldots, u_7\})$ and $\mathcal{O}_{C_3}(X) = (\{u_8, u_9, u_{12}, u_{13}, u_{16}, u_{17}\}, \{u_2, \ldots, u_7, u_{18}, u_{19}\}))$.

We indicate the set of all sequences of orthopairs generated by $\mathcal{C}$ with $\mathsf{SO}(\mathcal{C})$; namely, we set

$$\mathsf{SO}(\mathcal{C}) = \{\mathcal{O}_{\mathcal{C}}(X) \mid X \subseteq U\}.$$

Given a refinement sequence $\mathcal{C} = (C_1, \ldots, C_n)$ of U, by Definition 46, the orthopair $(\mathcal{L}_i(X), \mathcal{E}_i(X))$ of $\mathcal{O}_{\mathcal{C}}(X)$ is generated by the covering C_i that is finer than C_{i-1}. Clearly, this does not imply that $(\mathcal{L}_i(X), \mathcal{E}_i(X))$ approximates better than $(\mathcal{L}_{i-i}(X), \mathcal{E}_{i-1}(X))$ the set X (we say that the orthopair $O(X) = (\mathcal{L}(X), \mathcal{E}(X))$ approximates better than the orthopair $\tilde{O}(X) = (\tilde{\mathcal{L}}(X), \tilde{\mathcal{E}}(X))$ the set X if and only if $\tilde{\mathcal{L}}(X) \subseteq \mathcal{L}(X)$ and $\tilde{\mathcal{E}}(X) \subseteq \mathcal{E}(X))$, since $X \cap U_i$ may be strictly included in $X \cap U_{i-1}$ (the sets $U_1, \ldots, U_n$ are defined in Remark 10).

Example 25. We consider the sequence of Example 24. We observe that $\mathcal{O}_{C_3}(X)$ is not a better approximation of X than $\mathcal{O}_{C_2}(X)$, despite C_3 is finer than C_2, since $u_{10}, u_{11}, u_{14}, u_{15}$ appear in $\mathcal{O}_{C_2}(X)$, but they do not appear in $\mathcal{O}_{C_3}(X)$. Trivially, this is the consequence of the fact that the sequence of partial coverings loses objects during the refinement process.

More precisely, the following proposition holds.

Proposition 6. *Let $\mathcal{C} = (C_1, \ldots, C_n)$ be a refinement sequence of U and $X \subseteq U$. Suppose that $a \in \mathcal{L}_{i-1}(X)$ (or $a \in \mathcal{E}_{i-1}(X)$), with $i \in \{2, \ldots, n\}$. Then, $a \in \mathcal{L}_i(X)$ if and only if $a \in U_i$; (or $a \in \mathcal{E}_i(X)$ if and only if $a \in U_i$).*

Moreover, it is clear that two different subsets of the given universe can have the same sequences of orthopairs.

Example 26. Let $\mathcal{C} = (C_1, C_2)$ be the refinement sequence of Example 18. Suppose that $X = \{a, b, c, d\}$ and $Y = \{a, b, c, e\}$, then

$$\mathcal{O}_{\mathcal{C}}(X) = \mathcal{O}_{\mathcal{C}}(Y) = ((\emptyset, \emptyset), (\{a, b, c\}, \emptyset)).$$

At this is point, in order to show that each sequence of orthopairs is represented by a pair of disjoint upsets of the poset assigned to the given refinement sequence, we give the following definition.

Definition 47. *Let* $\mathcal{C} = (C_1, \ldots, C_2)$ *be a refinement sequence of* U *and* $X \subseteq U$. *We set*

$$(X_\mathcal{C}^1, X_\mathcal{C}^2) = (\{N \in P_\mathcal{C} \mid N \subseteq X\}, \{N \in P_\mathcal{C} \mid N \cap X = \emptyset\}).$$

Moreover, we set $\mathsf{K}_O(\mathcal{C}) = \{(X_\mathcal{C}^1, X_\mathcal{C}^2) \mid X \subseteq U\}$.

From now, we write (X^1, X^2) instead of $(X_\mathcal{C}^1, X_\mathcal{C}^2)$, when $\mathcal{C}$ is clear from the context.

The following theorem shows that there is a one-to-one correspondence between the elements of $\mathsf{SO}(\mathcal{C})$ and $\mathsf{K}_O(\mathcal{C})$.

Theorem 17. *Given a refinement sequence* $\mathcal{C} = (C_1, \ldots, C_n)$ *of a universe* U, *the map*

$$\alpha : \mathcal{O}_\mathcal{C}(X) \in \mathsf{SO}(\mathcal{C}) \mapsto (X^1, X^2) \in \mathsf{K}_O(\mathcal{C})$$

is a bijection.

Proof. First of all, we prove that α is well defined and injective, namely $\mathcal{O}_\mathcal{C}(X) = \mathcal{O}_\mathcal{C}(Y)$ if and only if $(X^1, X^2) = (Y^1, Y^2)$.

($\Rightarrow$). We observe that $N \in X^1$ if and only if $N \in C_i$ and $N \subseteq X$ for some $i \in \{1, \ldots, n\}$, namely $N \in C_i$ and $N \subseteq \mathcal{L}_i(X)$. Consequently $N \in Y^1$, since $\mathcal{L}_i(X) = \mathcal{L}_i(Y)$. Dually, $N \in X^2$ if and only if $N \in Y^2$, since $\mathcal{E}_i(X) = \mathcal{E}_i(Y)$ for each $i \in \{1, \ldots, n\}$.

($\Leftarrow$). Let $i \in \{1, \ldots, n\}$. $x \in \mathcal{L}_i(X)$ if and only if there is $N \in P_\mathcal{C}$ such that $x \in N$ and $N \subseteq X$. By hypothesis, $N \subseteq Y$. Then, $x \in \mathcal{L}_i(Y)$. Dually, we can prove that $\mathcal{E}_i(X) = \mathcal{E}_i(Y)$ for each $i \in \{1, \ldots, n\}$, since $X^2 = Y^2$.

Surjectivity follows by the definition of $\mathsf{K}_O(\mathcal{C})$. Hence, α is a bijection.

Remark 15. Definition 42 and Theorem 17 allow us to view a sequence of orthopairs as a labelled poset. Indeed, we can graphically represent sequences of orthopairs. More precisely, given a refinement sequence $\mathcal{C}$, the sequence $\mathcal{O}_\mathcal{C}(X)$ corresponds to the poset $P_\mathcal{C}$ that has labels associated with its nodes through the function $l_X : P_\mathcal{C} \mapsto \{\bullet, \circ, ?\}$ such that

$$l_X(N) = \begin{cases} \bullet & \text{if } N \in X^1; \\ \circ & \text{if } N \in X^2; \\ ? & \text{if } N \in P_\mathcal{C} \setminus \{X^1 \cup X^2\}. \end{cases} \tag{28}$$

For example, consider the refinement sequence of Example 18. Assume that $X = \{d, e, f, g\}$, $Y = \{a, b, c, d, e, f\}$ and $Z = \{a\}$, then the sequences $\mathcal{O}_\mathcal{C}(X) = ((\emptyset, \emptyset), (\{d, e, f\}, \{a, b, c\}))$, $\mathcal{O}_\mathcal{C}(Y) = ((\emptyset, \emptyset), (\{a, b, c, d, e, f\}, \emptyset))$ and $\mathcal{O}_\mathcal{C}(Z) = ((\emptyset, \emptyset), (\emptyset, \{d, e, f\}))$ have the following labelled posets, respectively.

Trivially, by Eq. 28, if $l_X(N) = \bullet$ and $N \leq_\mathcal{C} M$, then $l_X(M) = \bullet$. Similarly, if $l_X(N) = \circ$ and $N \leq_\mathcal{C} M$, then $l_X(M) = \circ$. On the other hand, $l_X(M)$ can be anyone between $\bullet$, $\circ$ and $?$, when $l_X(N) = ?$ and $N \leq_\mathcal{C} M$.

Fig. 8. Labelled posets

Sequences of Orthopairs and Decision Trees. Sequences of orthopairs correspond to *decision trees*. These are graphical models widely used in machine learning for describing sequential decision problems. A decision tree generates a classification procedure that recursively partitions a universe into smaller subdivisions on the basis of a set of tests defined at each branch (or node) in the tree [48]. The tree is made of a *root node* (the universe), a set of *internal nodes* (splits), and a set of *terminal nodes* (leaves). A test is applied for the universe and for each internal node in order to split the set of objects into successively smaller groups. The terminal nodes are labelled with values corresponding to the final decisions. An example of decision tree can be viewed in Fig. 9, where the labels A, B, C and D represent the final outcomes of the decision-making process.

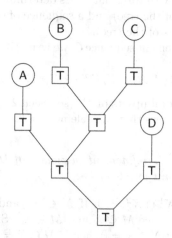

Fig. 9. Decision tree

Let C be a refinement sequence of *partial partitions* of U, and let $X \subseteq U$. The sequence of orthopairs $\mathcal{O}_C(X)$ determines three pairwise disjoint subsets of U: $\cup\{N \in P_C \mid l_X(N) = \bullet\}$, $\cup\{N \in P_C \mid l_X(N) = \circ\}$ and $\cup\{N \in P_C \mid l_X(N) = ?\}$. This also corresponds to result produced by the decision tree $(\mathcal{T}_C(X), \leq_C)$ such that

– $\mathcal{T}_\mathcal{C}(X) = (P_\mathcal{C} \cup \{U\}) \setminus H$, where

$H = \{N \in P_\mathcal{C} \mid$ if $M \in P_\mathcal{C}$ and $M \leq_\mathcal{C} N$ then $l_X(M) \in \{\bullet, \circ\}\}$, and

– let N be a leaf of $\mathcal{T}_\mathcal{C}(X)$, then the label of N is $l_X(N)$.

Trivially, $\mathcal{T}_\mathcal{C}(X)$ can have three outcomes at most, which are $\bullet$, $\circ$ and ?. Hence, if $\mathcal{O}_\mathcal{C}(X)$ is the sequence of orthopairs having labelled poset as in Fig. 10. Then, the tree decision $\mathcal{T}_\mathcal{C}(X)$ is shown in Fig. 11.

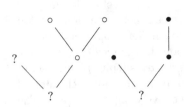
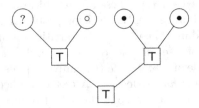

Fig. 10. Labelled poset of $\mathcal{O}_\mathcal{C}(X)$ **Fig. 11.** Decision tree $\mathcal{T}_\mathcal{C}(X)$

Clearly, a decision tree with three outcomes determines a refinement sequence (by considering all nodes of the tree) and a sequence of orthopairs (by considerings all nodes and all labels of the tree).

From now, given a refinement sequence $\mathcal{C}$, we write $\mathsf{K}(\mathcal{C})$ to denote $\mathsf{K}(Up(P_\mathcal{C}))$, that is

$$\mathsf{K}(Up(P_\mathcal{C})) = \{(A, B) \in Up(P_\mathcal{C}) \times Up(P_\mathcal{C}) \mid A \cap B = \emptyset\},$$

where $Up(P_\mathcal{C})$ is the set of all upsets of $P_\mathcal{C}$ (see Sect. 2.3).

The next proposition shows that each element of $\mathsf{K}_O(\mathcal{C})$ also belongs to $\mathsf{K}(\mathcal{C})$.

Proposition 7. *Let $\mathcal{C}$ be a refinement sequence of U and $X \subseteq U$. Then, (X^1, X^2) is a pair of disjoint upsets of $P_\mathcal{C}$.*

Proof. By Definition 47, $X^1 \cap X^2 = \emptyset$. If $N \in X^1$ and $N \leq_\mathcal{C} M$, then $M \subseteq N \subseteq X$ (by Definition 47) hence $M \subseteq X$ and $M \in X^1$. Similarly, if $N \in X^2$ and $N \leq_\mathcal{C} M$ then $M \subseteq N$ and $N \cap X = \emptyset$, hence $M \cap X = \emptyset$ and $M \in X^2$.

By Proposition 7, $\mathsf{K}_O(\mathcal{C}) \subseteq \mathsf{K}(\mathcal{C})$. However, the opposite does not always hold.

Example 27. Consider the refinement sequence $\mathcal{C}$, where $P_\mathcal{C}$ is represented in Fig. 12.

We have that $(\{\{a, b, c\}\}, \{\{c, d\}\}) \in \mathsf{K}_O(\mathcal{C})$, but $(\{\{a, b, c\}\}, \{\{c, d\}\}) \notin \mathsf{K}(\mathcal{C})$.

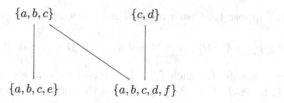

Fig. 12. Poset of $\mathcal{C}$

The next theorem (Theorem 18) provides the condition that a pair of disjoint upsets of $P_{\mathcal{C}}$ must have in order to belong to $\mathsf{K}_O(\mathcal{C})$, when $\mathcal{C}$ is safe. To prove Theorem 18, we need the following proposition.

Proposition 8. *Let $\mathcal{C}$ be a safe refinement sequence of U and let A be an upset of $P_{\mathcal{C}}$. Suppose that $N \in P_{\mathcal{C}}$ and*

$$N \subseteq \bigcup_{M \in A} M.$$

Then, $N \in A$.

Proof. Since $\mathcal{C}$ is safe (see Definition 44), there exists $M \in A$ such that $N \subseteq A$. However, A is an upset of $P_{\mathcal{C}}$, then $N \in A$.

From now on, we only consider coverings that do not contain singletons, which are blocks with only one element. We stress that the imposition of this constraint concerns the very relations between coverings and orthopairs as approximation of sets, as shown in the following example.

Example 28. Let $U = \{a, b, c, d, e\}$ and consider the covering of U given by $C = \{\{a, b\}, \{c\}, \{d, e\}\}$. Then, $(X^1, X^2) = (\{a, b\}, \{d, e\})$ is an orthopair made of blocks of C, but (X^1, X^2) does not approximate any subset X of U, since either $c \in X$, and then $c \in X^1$ or $c \in X$, and then $c \in X^2$. More generally, each orthopair such that $\{c\}$ is not contained in one of the components of the pair does not approximate any subset of U.

In order to state the next theorem, we recall that two upsets A and B of a given poset are *totally disjoint* if and only if all blocks of A are disjoint from all blocks of B.

Theorem 18. *Let $\mathcal{C}$ be a safe refinement sequence of U and let $(A, B) \in \mathsf{K}(\mathcal{C})$. Then, $(A, B) \in \mathsf{K}_O(\mathcal{C})$ if and only if A and B are totally pairwise disjoint.*

Proof. ($\Rightarrow$). By Definition 47, if $(A, B) \in \mathsf{K}_O(\mathcal{C})$, then there exists $X \subseteq U$ such that $N \subseteq X$ for each $N \in A$ and $N \cap M = \emptyset$ for each $M \in B$. Trivially, each node of A is disjoint with each node of B, since there is not $x \in U$ such that $x \in X$ and $x \notin X$.

($\Leftarrow$). Suppose that each node of A is disjoint with each node of B. We set

$$D = \{N \in P_{\mathcal{C}} \setminus (A \cup B) \mid N \cap M = \emptyset \text{ for each } M \in A \text{ and if } M >_{\mathcal{C}} N \text{ then } M \in B\}.$$

Since $\mathcal{C}$ is safe, for each $N \in D$, we can pick an element $x_N \in N$ such that $x_N \notin M$, for each $M \in P_{\mathcal{C}} \setminus \{\downarrow N\}$ (see Remark 14). Then, we set

$$X = \bigcup_{N \in A} N \cup \{x_N \mid N \in D\}.$$

We prove that $(A, B) = (X^1, X^2)$. It is trivial that $A \subseteq X^1$ and $B \subseteq X^2$. Now, we suppose that $N \in X^1$, and we intend to prove that $N \in A$. Let $x \in N$. Then, $x = x_M$ with $M \in D$ or x belongs to some node of A. If $x = x_M$ with $M \in D$, then $N \in \downarrow M$ (see Remark 14), and so $M \subseteq N$. Now, two cases can happen. If M is not a maximal element of $P_{\mathcal{C}}$, then M contains some elements of the nodes of B. However, by the hypothesis that A and B are totally pairwise disjoint, this is an absurd. In the other case, namely, if M is a maximal element of $P_{\mathcal{C}}$, then it contains at least another element that is not equal to x_M (we assumed that the blocks of refinement sequences are not singletons). By definition of D, such element is not in A and it is different from other elements x_N. It is clear that it is an absurd, since N is included in X, by hypothesis. We can conclude N is included in the union of blocks of A. Therefore, by Proposition 8, since $\mathcal{C}$ is safe, we have that $N \in A$. Now, we suppose that $N \in X^2$, and we intend to prove that $N \in B$. if $N \in X^2$, then $N \cap M = \emptyset$, for each $M \in A \cup D$. Consequently, $N \notin (\downarrow A) \cup (\downarrow D)$. Moreover, we can notice that $B = P_{\mathcal{C}} \setminus \{(\downarrow A) \cup (\downarrow D)\}$. Then, we can state that $N \in B$.

Theorem 18 permits us to prove the following result, which is relevant to regard sequences of orthopairs as Kleene algebras.

Theorem 19. *Let $\mathcal{C}$ be a complete and safe refinement sequence of U. Then,* $K_O(\mathcal{C}) = K(\mathcal{C})$.

Proof. We have that $K_O(\mathcal{C}) \subseteq K(\mathcal{C})$, by Proposition 7. Moreover, Let $(A, B) \in K(\mathcal{C})$, then A and B are totally pairwise disjoint, since $\mathcal{C}$ is complete. By hypothesis that $\mathcal{C}$ is safe and by Theorem 18, $(A, B) \in K_O(\mathcal{C})$.

As a consequence of the previous theorem, we can define several operations on sequences of orthopairs, using the operations already defined on sets of pairs of disjoint upsets of posets (see Sect. 2.3). However, we will explore this topic in the next chapter.

4   Sequences of Orthopairs as Kleene Algebras

Mathematics is the art of giving the same name to different things.

Henrie Poincaré

In this chapter, we equip sets of sequences of orthopairs with some operations in order to obtain finite many-valued algebraic structures (those are defined in Sect. 2.3). Furthermore, we prove theorems providing to represent such structures as sequences of orthopairs. We show that, when sequences of orthopairs are generated by one covering, our operations coincide with operations between orthopairs listed in Sect. 2.2. Also, we discover how to generate operations between sequences of orthopairs starting from those concerning individual orthopairs. Finally, we use a sequence of orthopairs to represent an examiner's opinion on a number of candidates applying for a job. Moreover, we show that opinions of two or more examiners can be combined using our operations in order to get a final decision on each candidate.

4.1   From a Safe Refinement Sequence to a Kleene Algebra

In the previous chapter, given a refinement sequence $\mathcal{C}$, we proved that each element of $\mathsf{K}_O(\mathcal{C})$ is a pair of disjoint upsets of $P_\mathcal{C}$ (see Proposition 7), and that $\mathsf{K}_O(\mathcal{C})$ coincides with $\mathsf{K}(\mathcal{C})$ if and only if $\mathcal{C}$ is safe and complete (see Example 27 and Theorem 19). As a consequence, we can equip $\mathsf{K}_O(\mathcal{C})$ with the operations $\sqcap$, $\sqcup$ and $\neg$ defined by Definition 9, 10 and 7 (see Sect. 2.3), respectively, and so we can consider the following structure

$$\mathbb{K}_O(\mathcal{C}) = (\mathsf{K}_O(\mathcal{C}), \sqcap, \sqcup, \neg, (P_\mathcal{C}, \emptyset), (\emptyset, P_\mathcal{C})).$$

Unfortunately, $\mathbb{K}_O(\mathcal{C})$ is not always a lattice, since $\mathsf{K}_O(\mathcal{C})$ could not be closed under $\sqcap$ and $\sqcup$, when $\mathsf{K}_O(\mathcal{C}) \subset \mathsf{K}(\mathcal{C})$.

Example 29. Let $U = \{a, b, c, d\}$ and $\mathcal{C} = (C_1, C_2)$, where

- $C_1 = \{\{a, b, c, d\}\}$ and
- $C_2 = \{\{a, b\}, \{c, d\}\})$.

Then, it occurs that

- $(\emptyset, \{\{a, b\}\}) \sqcap (\emptyset, \{\{c, d\}\}) = (\emptyset, \{\{a, b\}, \{c, d\}\})$ and
- $(\{\{a, b\}\}, \emptyset) \sqcup (\{\{c, d\}\}, \emptyset) = (\{\{a, b\}, \{c, d\}\}, \emptyset)$.

However, $(\emptyset, \{\{a, b\}, \{c, d\}\}), (\{\{a, b\}, \{c, d\}\}, \emptyset) \notin \mathsf{K}_O(\mathcal{C})$.

On the other hand, the following theorem states that requiring that refinement sequences be safe is sufficient to obtain finite centered Kleene algebras.

Theorem 20. *Let $\mathcal{C}$ be a safe refinement sequence of U. Then,*

1. *$\mathsf{K}_O(\mathcal{C}) \supseteq \mathsf{K}^+(\mathcal{C})$ and*
2. *$\mathbb{K}_O(\mathcal{C})$ is a centered Kleene subalgebra of $\mathbb{K}(\mathcal{C})$ (see Definition 26), where*

$$\mathbb{K}(\mathcal{C}) = (\mathsf{K}(\mathcal{C}), \sqcap, \sqcup, \neg, (\emptyset, P_\mathcal{C}), (P_\mathcal{C}, \emptyset)),$$

and the center is $(\emptyset, \emptyset)$.

Proof. 1. Let $(A, B) \in \mathsf{K}^+(\mathcal{C})$, then $B = \emptyset$. Consequently, A and B are totally disjoint, namely satisfy Condition 27. Certainly, $(A, B) \in \mathsf{K}_O(\mathcal{C})$, by Theorem 18.

2. Since $\mathsf{K}^+(\mathcal{C}) \subseteq \mathsf{K}_O(\mathcal{C})$, we have that $(\emptyset, \emptyset) \in \mathsf{K}_O(\mathcal{C})$. Moreover, $\mathsf{K}_O(\mathcal{C})$ is closed under all operations of $\mathbb{K}(\mathcal{C})$, since both $(X^1 \cap Y^1, X^2 \cup Y^2)$ and $(X^1 \cup Y^1, X^2 \cap Y^2)$ are pairs of totally disjoint upsets of $P_{\mathcal{C}}$. Then, by Theorem 18, both belong to $\mathsf{K}_O(\mathcal{C})$.

Remark 16. Clearly, when $\mathcal{C}$ is a safe refinement sequence of U, then $\mathsf{K}^-(\mathcal{C})$ is also included in $\mathsf{K}_O(\mathcal{C})$.

When a safe refinement sequence $\mathcal{C}$ is also complete or pairwise overlapping, $\mathbb{K}_O(\mathcal{C})$ satisfies properties that are additional to those of Theorem 20. More precisely, the following theorem holds.

Theorem 21. *Let $\mathcal{C}$ be a safe refinement sequence of U,*

1. *if $\mathcal{C}$ is complete, then $\mathbb{K}_O(\mathcal{C})$ is a finite centered Kleene algebra with the interpolation property,*
2. *if $\mathcal{C}$ is pairwise overlapping, then $\mathsf{K}_O(\mathcal{C}) = \mathsf{K}^+(\mathcal{C}) \cup \mathsf{K}^-(\mathcal{C})$.*

Proof. 1. By Theorem 19, $\mathsf{K}_O(\mathcal{C}) = \mathsf{K}(\mathcal{C})$. Moreover, the structure $\mathbb{K}(\mathcal{C})$ is a centered Kleene algebra with the interpolation property (see Theorem 7).

2. By Definition 47, if $(A, B) \in \mathsf{K}_O(\mathcal{C})$, then A and B are totally disjoint. However, since $\mathcal{C}$ is pairwise overlapping, Vice-versa, by Theorem 20, if (A, B) is in $\mathsf{K}^+(\mathcal{C})$ or $\mathsf{K}^-(\mathcal{C})$, then belongs to $\mathsf{K}_O(\mathcal{C})$, also.

In the next example, we take three different refinement sequences such that their posets are isomorphic, and we show that the Hasse diagrams of their respective Kleene algebras are not isomorphic.

Example 30. We consider the refinement sequences $\mathcal{C} = (C_1, C_2)$ and $\mathcal{C}' = (C'_1, C'_2)$ of $\{a, b, c, d, e, f\}$, where

- $C_1 = \{\{a, b, c, d, e\}, \{c, d, f\}\}$,
- $C_2 = \{\{a, b\}, \{c, d\}\}$,
- $C'_1 = \{\{a, b, d, e, f\}, \{c, d, e\}\}$ and
- $C'_2 = \{\{b, d\}, \{d, e\}\}$.

As shown in Fig. 13 and Fig. 14, $P_{\mathcal{C}}$ and $P_{\mathcal{C}'}$ have the same Hasse diagram. Then, $\mathbb{K}(\mathcal{C}) \cong \mathbb{K}(\mathcal{C}')$.

We set $b_1 = \{a, b, c, d, e\}$, $b_2 = \{c, d, f\}$, $b_3 = \{a, b\}$, $b_4 = \{c, d\}$, $b'_1 = \{a, b, d, e, f\}$, $b'_2 = \{c, d, e\}$, $b'_3 = \{b, d\}$ and $b'_4 = \{d, e\}$. Then, the Hasse diagrams of $\mathbb{K}_O(\mathcal{C})$ and $\mathbb{K}_O(\mathcal{C}')$ are represented in Fig. 15 and Fig. 16, respectively.

Notice that $\mathsf{K}_O(\mathcal{C}) = \mathsf{K}(\mathcal{C})$, since $\mathcal{C}$ is safe and complete. Instead, since $\mathcal{C}'$ is safe but not complete, $\mathsf{K}_O(\mathcal{C}') \subset \mathsf{K}(\mathcal{C}')$ and $(\{b'_3\}, \{b'_4\}), (\{b'_4\}, \{b'_3\}),$ $(\{b'_3\}, \{b'_2, b'_4\}), (\{b'_2, b'_4\}, \{b'_3\}) \notin \mathsf{K}_O(\mathcal{C}')$. We stress that $\mathbb{K}_O(\mathcal{C}) \not\cong \mathbb{K}_O(\mathcal{C}')$, despite $P_{\mathcal{C}} \cong P_{\mathcal{C}'}$.

Fig. 13. Hasse diagram of P_C **Fig. 14.** Hasse diagram of P'_C

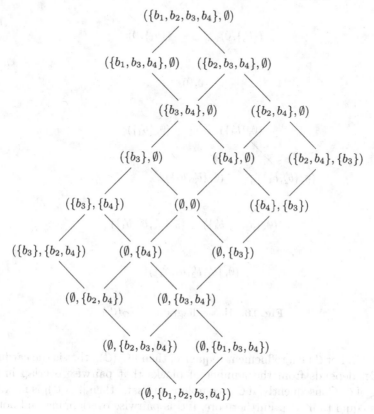

Fig. 15. Hasse diagram of $\mathbb{K}_O(\mathcal{C})$

Now, we consider the refinement sequence $\tilde{\mathcal{C}} = (\tilde{C}_1, \tilde{C}_2)$, where

- $\tilde{C}_1 = \{\{a, b, c, d, e\}, \{c, d, f\}\}$ and
- $\tilde{C}_2 = \{\{a, b, c\}, \{c, d\}\}$.

Clearly, $\tilde{\mathcal{C}}$ is a safe and pairwise overlapping refinement sequence. If we set $\tilde{b}_1 = \{a, b, c, d, e\}$, $\tilde{b}_2 = \{c, d, f\}$, $\tilde{b}_3 = \{a, b, c\}$ and $\tilde{b}_4 = \{c, d\}$, then the Hasse diagrams of $P_{\tilde{\mathcal{C}}}$ and $\mathbb{K}_O(\tilde{\mathcal{C}})$ are respectively represented in Fig. 17 and Fig. 18.

We can observe that $\mathbb{K}_O(\tilde{\mathcal{C}}) = \mathsf{K}(\tilde{\mathcal{C}})^+ \cup \mathsf{K}(\tilde{\mathcal{C}})^-$. Moreover, $\mathbb{K}_O(\tilde{\mathcal{C}}) \ncong \mathbb{K}_O(\mathcal{C})$ and $\mathbb{K}_O(\tilde{\mathcal{C}}) \ncong \mathbb{K}_O(\mathcal{C}')$, despite $P_{\tilde{\mathcal{C}}} \cong P_C$ and $P_{\tilde{\mathcal{C}}} \cong P_{C'}$.

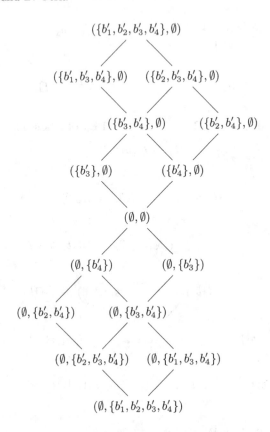

Fig. 16. Hasse diagram of $\mathbb{K}_O(\mathcal{C}')$

Remark 17. Let $\mathcal{C}$ be a refinement sequence, then $|\mathsf{K}_O(\mathcal{C})|$, that is the cardinality of $\mathsf{K}_O(\mathcal{C})$, depends from the number of blocks that pairwise overlap in every covering of $\mathcal{C}$. Consequently, if $\mathcal{C}$ is complete and safe, then $|\mathsf{K}_O(\mathcal{C})|$ is maximum, and it is equal to $|\mathsf{K}(\mathcal{C})|$. Furthermore, if $\mathcal{C}$ is pairwise overlapping and not safe, then $|\mathsf{K}_O(\mathcal{C})| \geq |\mathsf{K}(\mathcal{C})^+ \cup \mathsf{K}(\mathcal{C})^-|$.

We can extend the results shown in Theorem 21, by considering the operation $\rightarrow_1$ and the pairs of operations $(\star_2, \rightarrow_2)$, $(\star_3, \rightarrow_3)$ and $(\star_4, \rightarrow_4)$, defined in Sect. 2.3 (more exactly, see the Eqs. 13, 18, 19, 20, 21, 25 and 26), on the set $\mathsf{K}_O(\mathcal{C})$. Then, let $i \in \{1, \ldots, 4\}$, we can use the notation $\mathbb{K}_O^i(\mathcal{C})$ to denote the structure $\mathbb{K}_O(\mathcal{C})$ with the additional operations $\star_i$ and $\rightarrow_i$.

Corollary 1. *If $\mathcal{C}$ is a safe and complete refinement sequence, then*

- $\mathbb{K}_O^1(\mathcal{C})$ *is a finite Nelson algebra,*
- $\mathbb{K}_O^2(\mathcal{C})$ *is a finite Nelson lattice and*
- $\mathbb{K}_O^4(\mathcal{C})$ *is a finite KLI* algebra.*

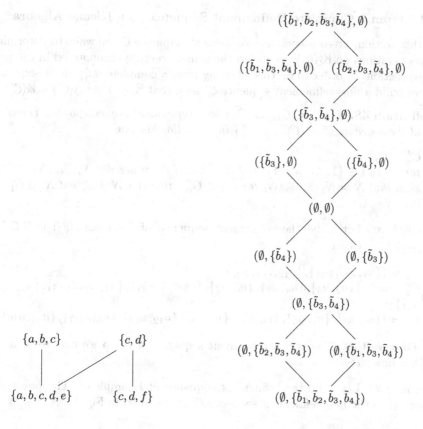

Fig. 17. Hasse diagram of $P_{\tilde{C}}$ **Fig. 18.** Hasse diagram of $\mathbb{K}_O(\tilde{C})$

Regarding $\mathbb{K}_O^3(\mathcal{C})$, we need to add the extra condition that $\mathcal{C}$ must be composed by partial partitions.

Corollary 2. *If $\mathcal{C}$ is a safe refinement sequence of partial partitions, then $\mathbb{K}_O^3(\mathcal{C})$ is a finite IUML-algebra.*

If some coverings of $\mathcal{C}$ are not partitions, then the operations $\star_i$ and $\rightarrow_i$ cannot be defined on $\mathsf{K}_O(\mathcal{C})$. Clearly, this is a consequence that such operations are defined between pairs of disjoint upsets of a forest (see Eqs. 20 and 21), and they can not be extended between pairs of disjoint upsets of a poset.

Example 31. Let $\mathcal{C}$ be the refinement sequence defined in Example 30. $\mathcal{C}$ is safe and complete, but

$$(\{b_3\}, \{b_2, b_4\}) \star_3 (\{b_1, b_3, b_4\}, \emptyset) = (\{b_1, b_3, b_4\}, \{b_2\})$$

and

$$(\{b_3\}, \{b_2, b_4\}) \rightarrow_3 (\emptyset, \{b_1, b_3, b_4\}) = (\{b_2\}, \{b_1, b_3, b_4\})$$

that do not belong to $\mathsf{K}(\mathcal{C})$.

4.2 From a Complete Refinement Sequence to a Kleene Algebra

In this section, given a complete refinement sequence $\mathcal{C}$, we want to determine new operations on $\mathsf{K}_O(\mathcal{C})$, to obtain the same structure encountered in the previous section. In order to do this, starting from a complete refinement sequence $\mathcal{C}$, we build a new refinement sequence $\mathcal{C}'$ such that $\mathsf{K}_O(\mathcal{C}) = \mathsf{K}_O(\mathcal{C}') = \mathsf{K}(\mathcal{C}')$.

Definition 48. *Let $\mathcal{C} = (C_1, \ldots, C_n)$ be a refinement sequence of U. Then, we build the sequence $\mathcal{C}' = (C_1', \ldots, C_n')$ in the following way.*

- $C_n' = C_n$,
- *for every $i \in \{1, \ldots, n-1\}$ and $N \in C_i$, if there are not $N_1, \ldots, N_l \in C_{i+1}'$ such that $N = N_1 \cup \ldots \cup N_l$ then $N \in C_i'$, otherwise $N \notin C_i'$ but $N_j \in C_i'$ for each $j = 1, \ldots, l$.*

Example 32. Let $\mathcal{C}$ be the refinement sequence of Example 14. Then, $\mathcal{C}' = (C_1', C_2', C_3')$, where

$C_1' = \{\{u_1, \ldots, u_{11}\}, \{u_{12}, \ldots, u_{22}\}\}$;
$C_2' = \{\{u_2, u_3\}, \{u_4, u_5\}, \{u_6, u_7\}, \{u_8, \ldots, u_{11}\}, \{u_{12}, \ldots, u_{15}\}, \{u_{16}, \ldots,$
$u_{21}\}\}$;
$C_3' = \{\{u_2, u_3\}, \{u_4, u_5\}, \{u_6, u_7\}, \{u_7, u_8\}, \{u_{12}, u_{13}\}, \{u_{16}, u_{17}\}, \{u_{18}, u_{19}\}.$

Observe that $\mathcal{C}'$ is still a refinement sequence of U, so we can associate it with a poset $P_{\mathcal{C}'}$.

Example 33. Let $\mathcal{C}$ be the refinement sequence of Example 14. The poset $P_{\mathcal{C}'}$ assigned to the new refinement sequence $\mathcal{C}'$ represented in Fig. 19.

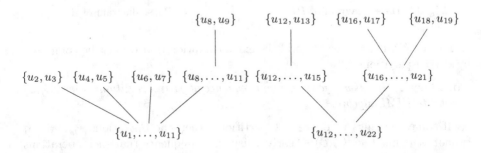

Fig. 19. Forest of the users

We notice that the node $\{u_2, \ldots, u_7\}$ of $P_{\mathcal{C}}$ (see Example 16) does not belong to $P_{\mathcal{C}'}$, and it is equal to the union of its successors $\{u_2, u_3\}, \{u_4, u_5\}$ and $\{u_6, u_7\}$.

Remark 18. In general, $P_{\mathcal{C}'}$ is obtained by removing from $P_{\mathcal{C}}$ all the nodes equal to the union of their successors (cfr. the operation of elimination in [24]). That is, we delete *reducible* elements, according to the terminology given in [108], in the covering generated by all sets in the forest $P_{\mathcal{C}}$.

By the previous remark follows this proposition.

Proposition 9. *Let $\mathcal{C}$ be a refinement sequence of U and let $N \in P_{\mathcal{C}}$. Then, $N \in P_{\mathcal{C}'}$ if and only if $N \neq N_1 \cup \ldots \cup N_r$, where $N_1, \ldots, N_r$ are the successors of N in $P_{\mathcal{C}}$.*

Clearly, $\mathsf{K}_O(\mathcal{C}') \subseteq \mathsf{K}_O(\mathcal{C})$. Moreover, it is clear that the following proposition holds.

Proposition 10. *Let $\mathcal{C}$ be a complete refinement sequence. Then, $\mathcal{C}'$ is also complete.*

The following proposition shows that there exists an order isomorphism between $\mathsf{K}_O(\mathcal{C})$ and $\mathsf{K}_O(\mathcal{C}')$, when $\mathcal{C}$ is complete.

Theorem 22. *Let $\mathcal{C} = (C_1, \ldots, C_n)$ be a complete refinement sequence of U. If $\mathcal{C}'$ is the refinement sequence of U built in Definition 48, then the function*

$$\beta : \mathsf{K}_O(\mathcal{C}) \mapsto \mathsf{K}_O(\mathcal{C}'),$$

where $\beta((X_{\mathcal{C}}^1, X_{\mathcal{C}}^2)) = (X_{\mathcal{C}'}^1, X_{\mathcal{C}'}^2)$ for each $X \subseteq U$, is an order isomorphism.

Proof. – The function β is injective. Let $X, Y \subseteq U$, we suppose that

$$\beta((X_{\mathcal{C}}^1, X_{\mathcal{C}}^2)) = \beta((Y_{\mathcal{C}}^1, Y_{\mathcal{C}}^2)).$$

Then,

$$(X_{\mathcal{C}'}^1, X_{\mathcal{C}'}^2) = (Y_{\mathcal{C}'}^1, Y_{\mathcal{C}'}^2). \tag{29}$$

Firstly, we intend to prove that $X_{\mathcal{C}}^1 = Y_{\mathcal{C}}^1$. By Definition 48, each node N of $P_{\mathcal{C}}$ is equal to $N_1 \cup \ldots \cup N_r$, where $N_1 \cup \ldots \cup N_r \in P_{\mathcal{C}'}$. Let $N \in X_{\mathcal{C}}^1$, then $N = N_1 \cup N_r \subseteq X$ and so $N_i \subseteq X$ for each $i \in \{1, \ldots, r\}$. Therefore, $N_1, \ldots, N_r \in X_{\mathcal{C}'}^1 = Y_{\mathcal{C}'}^1$. Consequently, N is included in Y and so belongs to $Y_{\mathcal{C}}^1$. The proof that $X_{\mathcal{C}}^2 = Y_{\mathcal{C}}^2$ is analogous.
 – The function β is surjective. Let $X \subseteq U$ and $(X_{\mathcal{C}'}^1, X_{\mathcal{C}'}^2) \in \mathsf{K}_O(\mathcal{C}')$. We consider the set

$$H = \{N \in P_{\mathcal{C}} : N = N_1 \cup \ldots \cup N_r, \text{ where } N_i \in X_{\mathcal{C}'}^1 \text{ for each } i \in \{1, \ldots, r\}\}$$

and

$$K = \{N \in P_{\mathcal{C}} : N = N_1 \cup \ldots \cup N_r, \text{ where } N_i \in X_{\mathcal{C}'}^2 \text{ for each } i \in \{1, \ldots, r\}\}.$$

Since $\mathcal{C}$ is complete, we have that $(X_{\mathcal{C}'}^1 \cup H, X_{\mathcal{C}'}^2 \cup K)$ belongs to $\mathsf{K}_O(\mathcal{C})$. Moreover, it is clear that $\beta((X_{\mathcal{C}'}^1 \cup H, X_{\mathcal{C}'}^2 \cup K)) = (X_{\mathcal{C}'}^1, X_{\mathcal{C}'}^2)$.
 – It is trivial that $(X_{\mathcal{C}}^1, X_{\mathcal{C}}^2) \leq (Y_{\mathcal{C}}^1, Y_{\mathcal{C}}^2)$ if and only if $(X_{\mathcal{C}'}^1, X_{\mathcal{C}'}^2) \leq (Y_{\mathcal{C}'}^1, Y_{\mathcal{C}'}^2)$ (we remember that, let (X^1, X^2) and (Y^1, Y^2) be two pairs of disjoint upsets, then $(X^1, X^2) \leq (Y^1, Y^2)$ if and only if $X^1 \subseteq Y^1$ and $Y^2 \subseteq Y^1$).

By Proposition 5 and Proposition 9, the next result follows.

Proposition 11. *Let $\mathcal{C}$ be a complete refinement sequence, then $\mathcal{C}'$ is safe.*

Consequently, by Theorem 19, $\mathsf{K}_O(\mathcal{C}')$ coincides with $\mathsf{K}(\mathcal{C}')$. Therefore, we can consider $\mathsf{K}_O(\mathcal{C}')$ equipped with the operations defined in the previous section. By using this result and Theorem 22, we can introduce the following new operations on $\mathsf{K}_O(\mathcal{C})$.

Definition 49. *Let $\mathcal{C}$ be a complete refinement sequence of U and let β be the function defined in Theorem 22. Then, we set*

- $(X_\mathcal{C}^1, X_\mathcal{C}^2) \cap_{\mathsf{K}_O} (Y_\mathcal{C}^1, Y_\mathcal{C}^2) := \beta^{-1}((X_{\mathcal{C}'}^1, X_{\mathcal{C}'}^2) \sqcap (Y_{\mathcal{C}'}^1, Y_{\mathcal{C}'}^2))$,
- $(X_\mathcal{C}^1, X_\mathcal{C}^2) \cup_{\mathsf{K}_O} (Y_\mathcal{C}^1, Y_\mathcal{C}^2) := \beta^{-1}((X_{\mathcal{C}'}^1, X_{\mathcal{C}'}^2) \sqcup (Y_{\mathcal{C}'}^1, Y_{\mathcal{C}'}^2))$,
- $\neg_{\mathsf{K}_O}(X_\mathcal{C}^1, X_\mathcal{C}^2) := \beta^{-1}(\neg(X_{\mathcal{C}'}^1, X_{\mathcal{C}'}^2))$,
- $(X_\mathcal{C}^1, X_\mathcal{C}^2) \star_{\mathsf{K}_O}^i (Y_\mathcal{C}^1, Y_\mathcal{C}^2) := \beta^{-1}((X_{\mathcal{C}'}^1, X_{\mathcal{C}'}^2) \star_i (Y_{\mathcal{C}'}^1, Y_{\mathcal{C}'}^2))$, *for each $i \in \{2, 3, 4\}$,*
- $(X_\mathcal{C}^1, X_\mathcal{C}^2) \to_{\mathsf{K}_O}^i (Y_\mathcal{C}^1, Y_\mathcal{C}^2) := \beta^{-1}((X_{\mathcal{C}'}^1, X_{\mathcal{C}'}^2) \to_i (Y_{\mathcal{C}'}^1, Y_{\mathcal{C}'}^2))$, *for each $i \in \{1, 2, 3, 4\}$.*

As a consequence of the previous definition and the results of Sect. 4.1, we obtain the following theorem.

Theorem 23. *Let $\mathcal{C}$ be a complete refinement sequence of U, then*

$$\mathbb{K}_O'(\mathcal{C}) = (\mathsf{K}_O(\mathcal{C}), \cap_{\mathsf{K}_O}, \sqcup_{\mathsf{K}_O}, \neg_{\mathsf{K}_O}, (\emptyset, P_{\mathcal{C}'}), (P_{\mathcal{C}'}, \emptyset))$$

is a centered Kleene algebra with the interpolation property and if $\mathcal{C}$ is pairwise overlapping, then $\mathsf{K}_O(\mathcal{C}) \cong \mathsf{K}(\mathcal{C}')^+ \cup \mathsf{K}(\mathcal{C}')^-$. Moreover,

- $(\mathbb{K}_O'(\mathcal{C}), \to_{\mathsf{K}_O}^1)$ *is a finite Nelson algebra;*
- $(\mathbb{K}_O'(\mathcal{C}), \star_{\mathsf{K}_O}^2, \to_{\mathsf{K}_O}^2)$ *is a finite Nelson lattice;*
- $(\mathbb{K}_O'(\mathcal{C}), \star_{\mathsf{K}_O}^4, \to_{\mathsf{K}_O}^4)$ *is a finite KLI*-algebra.*

If $\mathcal{C}$ is a refinement sequence of partial partitions, then

- $(\mathbb{K}_O'(\mathcal{C}), \star_{\mathsf{K}_O}^3, \to_{\mathsf{K}_O}^3)$ *is a finite IUML-algebra.*

Remark 19. Trivially, if $\mathcal{C}$ is also safe, then $\mathcal{C} = \mathcal{C}'$ and so $\mathsf{K}_O(\mathcal{C}) = \mathbb{K}_O'(\mathcal{C})$.

Example 34. Let $\mathcal{C}$ be the refinement sequence defined in Example 29. Trivially, $\mathcal{C}' = \{\{a, b\}, \{c, d\}\}$. The Hasse diagram of $\mathbb{K}(\mathcal{C})$, $\mathbb{K}_O(\mathcal{C})$ and $\mathbb{K}_O(\mathcal{C}')$ (which is the same as that of $\mathbb{K}(\mathcal{C}')$) are respectively represented in Fig. 20, Fig. 21 and Fig. 22.

Now, we consider $(\{\{a, b\}\}, \emptyset)$ and $(\{\{c, d\}\}, \emptyset)$ in $\mathsf{K}_O(\mathcal{C})$. Then $(\{\{a, b\}\}, \emptyset) \sqcup (\{\{c, d\}\}, \emptyset)$ is equal to $(\{\{a, b\}, \{c, d\}\}, \emptyset)$, which does not belong to $\mathsf{K}_O(\mathcal{C})$. However, $(\{\{a, b\}\}, \emptyset) \cup_{\mathsf{K}_O} (\{\{c, d\}\}, \emptyset) = \beta^{-1}((\{\{a, b\}, \{c, d\}\}, \emptyset)) = (\{\{a, b, c, d\}, \{a, b\}, \{c, d\}\}, \emptyset) \in \mathsf{K}_O(\mathcal{C})$.

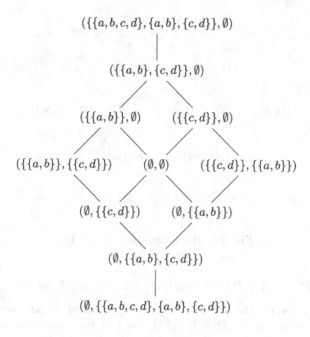

Fig. 20. Hasse diagram of $\mathbb{K}(\mathcal{C})$

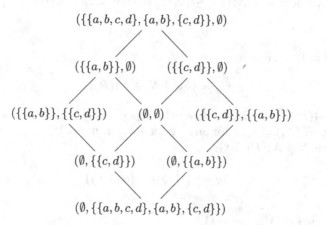

Fig. 21. Hasse diagram of $\mathbb{K}_O(\mathcal{C})$

4.3   From a Kleene Algebra to a Refinement Sequence

In this section, we associate a finite Kleene algebra with a refinement sequence and the respective sequences of orthopairs.

Let $(P, \leq)$ be a finite partially ordered set, and let n be the maximum number of elements of a chain in P. For each $i \in \{1, \ldots, n\}$ we define the i-th level of P as

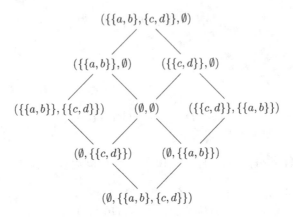

Fig. 22. Hasse diagram of $\mathbb{K}_O(\mathcal{C}')$

$$P^i = \{ N \in P \mid i = max\{|h| \mid h \text{ is a chain of } \downarrow N\} \}. \tag{30}$$

We denote by $\mathcal{M}(P)$ the set of maximal elements of P and we set $U_P = \{x_1, \ldots, x_m\}$, where $m = |P| + |\mathcal{M}(P)|$. We call *maximal sequence* of P the sequence $\mathcal{C} = (C_1, \ldots, C_n)$ built as follows. Suppose $\mathcal{M}(P)$ consists of nodes $N_1, \ldots, N_u$, where $u = |\mathcal{M}(P)| \leq \lfloor m/2 \rfloor$ since $u < 2u \leq |\mathcal{M}(P)| + |P| = m$. We set

$$b_{N_i} = \{x_{2i-1}, x_{2i}\} \tag{31}$$

for every $i = 1, \ldots, u$ and

$$C_n = \{b_{N_i} \mid N_i \in \mathcal{M}(P)\}. \tag{32}$$

Since $|P \backslash \mathcal{M}(P)| = m - 2u$, we denote by $N_{u+1}, \ldots, N_{m-u}$ the nodes of $P \backslash \mathcal{M}(P)$ and we set $\alpha_P(N_i) = x_{i+u}$ for any $i \in \{u+1, \ldots, m-u\}$.

For each $N \notin \mathcal{M}(P)$, let

$$b_N = \bigcup_{M > N} b_M \cup \{\alpha_P(N)\} \tag{33}$$

and, for each $j \in \{1, \ldots, n-1\}$,

$$C_j = \{b_N \mid N \in P^j\} \cup \{b_M \mid M \in \mathcal{M}(P) \text{ and } \downarrow M \cap P^j = \emptyset\}. \tag{34}$$

It is trivial to see that for each $N, M \in P$

$$b_N \cap b_M = \cup\{b_L \mid L \in \uparrow N \cap \uparrow M\}. \tag{35}$$

Example 35. Let P be the partially ordered set with the following Hasse diagram.

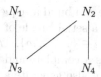

Fig. 23. Hasse diagram of P

$U_P = \{x_1, \dots, x_6\}$, since $6 = 4 + 2$, where $|P| = 4$ and $|\mathcal{M}(P)| = 2$. We have $\alpha_P(N_3) = x_5$ and $\alpha_P(N_4) = x_6$. Then, we have $b_{N_1} = \{x_1, x_2\}$, $b_{N_2} = \{x_3, x_4\}$, $b_{N_3} = \{x_1, x_2\} \cup \{x_3, x_4\} \cup \{\alpha_P(N_3)\} = \{x_1, x_2, x_3, x_4, x_5\}$ and $b_{N_4} = \{x_3, x_4\} \cup \{\alpha_P(N_4)\} = \{x_3, x_4, x_6\}$. Moreover, $n = 2$, then the maximal sequence is made of two partial coverings of $\{x_1, \dots, x_6\}$ that are $C_1 = \{\{x_1, x_2, x_3, x_4, x_5\}, \{x_3, x_4, x_6\}\}$ and $C_2 = \{\{x_1, x_2\}, \{x_3, x_4\}\}$.

Proposition 12. *Let P be a finite partially ordered set. Then, the maximal sequence C of P is a complete and safe refinement sequence of U_P and $SO(C) \cong K(U_P(P))$.*

Proof. Firstly, we prove that C is a refinement sequence of U_P. Then, suppose that $b \in C^i$ with $i > 1$, we have $b = b_N$ where $N \in P$. Since $b_N \in C^i$, two cases are possible: if $N \in P^i$, then there exists at least a node M of P^{i-1} such that $M < N$ (see Eq. 30), hence $b_M \in C^{i-1}$ (see Eq. 34) and $b_N \subset b_M$ (see Eq. 33); if $N \notin P^i$, then $N \in \mathcal{M}(P)$ and $\downarrow N \cap P^i = \emptyset$. In this latter case, we have two subcases to consider: $\downarrow N \cap P^{i-1} = \emptyset$ which implies $b_N \in C^{i-1}$ and $\downarrow N \cap P^{i-1} \neq \emptyset$ which implies that there exists $M \in P^{i-1}$ with $M \leq N$, hence $b_N \subseteq b_M$ where $b_M \in C^{i-1}$.

C is complete, since if $b_N \cap b_M \neq \emptyset$ with $b_N, b_M \in P_C$, then $b_N \cap b_M \supseteq b_L$ with $L \in \uparrow N \cap \uparrow M$ (see Eq. 35), hence b_N and b_M can not belong to two upsets that are disjoint. To prove that C is safe, we consider the blocks $b_N, b_{N_1}, \dots, b_{N_k}$ of coverings of C with $b_N \subseteq b_{N_1} \cup \dots \cup b_{N_k}$. Then, we pick a subset $\{b_{N'_1}, \dots, b_{N'_h}\}$ of $\{b_{N_1}, \dots, b_{N_k}\}$ such that $b_N \subseteq b_{N'_1} \cup \dots \cup b_{N'_h}$ and $b_N \cap b_{N'_i} \neq \emptyset$ for each $i \in \{1, \dots, h\}$. Trivially, $b_N \cap b \neq \emptyset$ if and only if $b_N \subseteq b$, when $N \in \mathcal{M}(P)$. Otherwise, if $N \notin \mathcal{M}(P)$, by Eq. 33 we have that $\alpha_P(N) \in b_N$, hence $\alpha_P(N)$ belongs to b'_{N_i} for some $i \in \{1, \dots, h\}$, then $b_N \subseteq b_{N'_i}$ since $N'_i \leq N$ (see Eq. 33).

By Proposition 7, $K_O(C) \subseteq K(C)$. Vice-versa, let $(A, B) \in K(C)$, then $A^* \cap B^* = \emptyset$, since otherwise, by Eq. 35, there exist $N, M, L \in P$ such that $b_L \subseteq b_N \cap b_M$, then $b_L \in A \cap B$ that is an absurd. By Theorem 19, $(A, B) \in K_O(C)$. Therefore, $K(C) \subseteq K_O(C)$.

Furthermore, observe that if $C = (C_1, \dots, C_n)$ is the maximal sequence of the poset P, then C_n is a partial partition of the respective universe U_P.

We remark that the maximal sequence $C = (C_1, \dots, C_n)$ of a given partially ordered set P is not the only complete and safe refinement sequence having the assigned poset isomorphic to P. We can generate such sequences in addressing numerous ways. For example, we can build a sequence C^* by adopting the previous procedure, but by assigning a set A_i made of at least three elements to the maximal node N_i of P, for each $i \in \{1, \ldots, m\}$. Trivially, if the sets $A_1, \ldots, A_m$ are pairwise disjoints, then C^* is a complete and safe refinement sequence satisfying $P_{C^*} \cong P_C$. Clearly, we can also generate a safe and complete refinement with its poset isomorphic to P by starting from the maximal sequence C. For example, we can add a finite set disjoint with U_P to each block of an upsets of C. On the other hand, we observe that the universe covered by any safe and complete refinement sequence with its poset isomorphic to P has cardinality grater that $|U_P|$.

By Theorem 9 and Proposition 12, the following theorem holds.

Theorem 24. *Let P be a partially ordered set and C its maximal sequence. Then, $\mathbb{K}_O(C)$ is a centered Kleene algebra that satisfies the interpolation property.*

4.4   Representation Theorems

Considering that $\mathsf{K}_O(C)$ coincides with the set of sequences of orthopairs generated by C (see Theorem 17), we can define on $\mathsf{SO}(C)$ the following operations.

Definition 50. *Let C be a refinement sequence of U and let α be the function defined in Theorem 17. Then, let $X, Y \subseteq U$, we set*

- $\mathcal{O}(X) \curlywedge \mathcal{O}(Y) := \alpha^{-1}((X^1, X^2) \cap_{K_O} (Y^1, Y^2));$
- $\mathcal{O}(X) \curlyvee \mathcal{O}(Y) := \alpha^{-1}((X^1, X^2) \cup_{K_O} (Y^1, Y^2));$
- $\sim \mathcal{O}(X) := \alpha^{-1}(\neg_{K_O}(X^1, X^2));$
- $\mathcal{O}(X) \odot_i \mathcal{O}(Y) := \alpha^{-1}((X^1, X^2) \star^i_{K_O} (Y^1, Y^2)),$ *for* $i \in \{2, 3, 4\};$
- $\mathcal{O}(X) \hookrightarrow_i \mathcal{O}(Y) := \alpha^{-1}((X^1, X^2) \to^i_{K_O} (Y^1, Y^2)),$ *for* $i \in \{1, 2, 3, 4\}.$

Moreover, given a refinement sequence $C = (C_1, \ldots, C_n)$, we set

$$\bot_C = (\bot_1, \ldots, \bot_n) \text{ and } \top_C = \sim \bot_C,$$

where $\bot_i = (\emptyset, \{x \in b \mid b \in C_i\})$, for each i from 1 to n. Then, it is clear that $\bot_C$ and $\top_C$ are respectively the minimum and the maximum of $\mathsf{SO}(C)$. Moreover, we set $e_C = ((\emptyset, \emptyset), \ldots, (\emptyset, \emptyset))$, that is $\alpha^{-1}((\emptyset, \emptyset))$.

Theorem 25. *Let $\mathbb{S}$ be a Kleene algebra. $\mathbb{S}$ is a finite centered Kleene algebra with interpolation property if and only if*

$$\mathbb{S} \cong (\mathsf{SO}(C), \curlywedge, \curlyvee, \sim, \bot_C, \top_C),$$

where C is a complete refinement sequence of a finite universe U.

Proof. ($\Rightarrow$). If $\mathbb{S}$ is a centered Kleene algebra with interpolation property, then there exists a bounded distributive lattice $L_{\mathbb{S}}$ such that $\mathbb{S} \cong K(L_{\mathbb{S}})$, by Theorem 9. By Birkhoff representation theorem, there exists a poset $P_{L_{\mathbb{S}}}$ such that $L_{\mathbb{S}} \cong U(P_{L_{\mathbb{S}}})$. Consequently, $\mathbb{S} \cong \mathsf{K}(U(P_{L_{\mathbb{S}}}))$. By Proposition 12, C is the maximal sequence of $P_{L_{\mathbb{S}}}$, that is a complete and safe refinement sequence of $U_{P_{L_{\mathbb{S}}}}$.

($\Leftarrow$). By the Theorems 17 and 23, if C is complete, then $(\mathsf{SO}(C), \curlywedge, \curlyvee, \sim, \bot_C, \top_C)$ is a centered Kleene algebra with the interpolation property.

Similarly, by using the theorems of Sect. 2.3, we can present some classes of finite many-valued structures such that their reduct is a centered Kleene algebra with the interpolation property as sequences of orthopairs. More precisely, the following theorems hold.

Theorem 26. *Let $\mathbb{S}$ be a Nelson algebra. $\mathbb{S}$ is a finite centered Nelson algebra with interpolation property if and only if*

$$\mathbb{S} \cong (\mathsf{SO}(\mathcal{C}), \curlywedge, \curlyvee, \sim, \odot_1, \hookrightarrow_1, \bot_\mathcal{C}, \top_\mathcal{C}),$$

where $\mathcal{C}$ is a complete refinement sequence of a finite universe U.

Theorem 27. *Let $\mathbb{S}$ be a Nelson lattice. $\mathbb{S}$ is a finite centered Nelson lattice with interpolation property if and only if*

$$\mathbb{S} \cong (\mathsf{SO}(\mathcal{C}), \curlywedge, \curlyvee, \sim, \odot_2, \hookrightarrow_2, e_\mathcal{C}, \bot_\mathcal{C}, \top_\mathcal{C}),$$

where $\mathcal{C}$ is a complete refinement sequence of a finite universe U.

Theorem 28. *Let $\mathbb{S}$ be a IUML-algebra. $\mathbb{S}$ is a finite IUML-algebra if and only if*

$$\mathbb{S} \cong (\mathsf{SO}(\mathcal{C}), \curlywedge, \curlyvee, \sim, \odot_3, \hookrightarrow_3, e_\mathcal{C}, \bot_\mathcal{C}, \top_\mathcal{C}),$$

where $\mathcal{C}$ is a refinement sequence of partial partitions of a finite universe U.

Theorem 29. *Let $\mathbb{S}$ be a KLI^*-algebra. $\mathbb{S}$ is finite and satisfies the interpolation property if and only if*

$$\mathbb{S} \cong (\mathsf{SO}(\mathcal{C}), \curlywedge, \curlyvee, \sim, \odot_4, \hookrightarrow_4, \bot_\mathcal{C}, \top_\mathcal{C}),$$

where $\mathcal{C}$ is a complete refinement sequence of a finite universe U.

4.5 Operations Between Sequences of Orthopairs

In this section, we focus on operations between sequences of orthopairs. In particular, we show how they can be obtained starting from the operations between orthopairs of an individual covering. The latter are listed in Sect. 2.2.

Theorem 30. *Let $\mathcal{C} = (C_1, \ldots, C_n)$ be a safe and complete refinement sequence of U and let $X, Y \subseteq U$, then*

1. $\mathcal{O}_\mathcal{C}(X) \curlywedge \mathcal{O}_\mathcal{C}(Y) = ((A_1, B_1), \ldots, (A_n, B_n))$,
2. $\mathcal{O}_\mathcal{C}(X) \curlyvee \mathcal{O}_\mathcal{C}(Y) = ((D_1, E_1), \ldots, (D_n, E_n))$,
3. $\sim \mathcal{O}_\mathcal{C}(X) = ((F_1, G_1), \ldots, (F_n, G_n))$,

where

1. $(A_i, B_i) = (\mathcal{L}_i(X), \mathcal{E}_i(X)) \wedge_K (\mathcal{L}_i(Y), \mathcal{E}_i(Y))$
2. $(D_i, E_i) = (\mathcal{L}_i(X), \mathcal{E}_i(X)) \vee_K (\mathcal{L}_i(Y), \mathcal{E}_i(Y))$
3. $(F_i, G_i) = \neg(\mathcal{L}_i(X), \mathcal{E}_i(X))$,

for each $i \in \{1, \ldots, n\}$. The operations $\wedge_\mathcal{K}$ and $\vee_\mathcal{K}$ are given in Definition 8, and $\neg(A, B) = (B, A)$.

Proof. We only provide the proof of point 1, since we can demonstrate the remaining cases in a similar way. Then, we suppose that Z is the subset of U such that $\mathcal{O}_\mathcal{C}(X) \curlywedge \mathcal{O}_\mathcal{C}(Y) = \mathcal{O}_\mathcal{C}(Z)$. Since $\mathcal{C}$ is safe, $\mathcal{O}_\mathcal{C}(X) \curlywedge \mathcal{O}_\mathcal{C}(Y) = \alpha^{-1}((X^1, X^2) \sqcap (Y^1, Y^2)) = \alpha^{-1}((X^1 \cap Y^1, X^2 \cup Y^2))$. Then, $Z^1 = X^1 \cap Y^1$ and $Z^2 = X^2 \cup Y^2$. On the other hand, we recall that

$$(\mathcal{L}_i(X), \mathcal{E}_i(X)) \wedge_\mathcal{K} (\mathcal{L}_i(Y), \mathcal{E}_i(Y)) = (\mathcal{L}_i(X) \cap \mathcal{L}_i(Y), \mathcal{E}_i(X) \cup \mathcal{E}_i(Y)).$$

So, fixed $i \in \{1, \ldots, n\}$, $x \in \mathcal{L}_i(Z)$ if and only if there exists $N \in P_\mathcal{C}$ such that $N \subseteq Z$. Therefore, there exists $N \in P_\mathcal{C}$ such that $N \in X^1 \cap Y^1$, and so $N \subseteq X \cap Y$. This is equivalent to say that $x \in \mathcal{L}_i(X) \cap \mathcal{E}_i(Y)$. Similarly, we can prove that $x \in \mathcal{E}_i(Z)$ if and only if $\mathcal{E}_i(X) \cup \mathcal{E}_i(Y)$.

Example 36. Let $\mathcal{C} = (C_1, C_2)$ be the refinement sequence of $\{a, b, c, d, e\}$, such that $C_1 = \{\{a, b, c, d, e\}\}$ and $C_2 = \{\{a, b\}, \{c, d\}\}$. Since $\mathcal{C}$ is safe and complete, the previous theorem holds. Then,

$$\mathcal{O}_\mathcal{C}(\{a, b\}) \curlywedge \mathcal{O}_\mathcal{C}(\{a, b, c\}) = ((\emptyset, \emptyset), (\{a, b\}, \{c, d\})),$$

where

$(\mathcal{L}_1(\{a, b\}), \mathcal{E}_1(\{a, b\})) \wedge_\mathcal{K} (\mathcal{L}_1(\{a, b, c\}), \mathcal{E}_1(\{a, b, c\})) = (\emptyset, \emptyset) \wedge_\mathcal{K} (\emptyset, \emptyset) = (\emptyset, \emptyset)$.
$(\mathcal{L}_2(\{a, b\}), \mathcal{E}_2(\{a, b\})) \wedge_\mathcal{K} (\mathcal{L}_2(\{a, b, c\}), \mathcal{E}_2(\{a, b, c\})) = (\{a, b\}, \{c, d\}) \wedge_\mathcal{K} (\{a, b\}, \emptyset) = (\{a, b\}, \{c, d\}))$.

Moreover,

$$\mathcal{O}_\mathcal{C}(\{a, b\}) \curlyvee \mathcal{O}_\mathcal{C}(\{a, b, c\}) = ((\emptyset, \emptyset), (\{a, b\}, \emptyset)),$$

where

$(\mathcal{L}_1(\{a, b\}), \mathcal{E}_1(\{a, b\})) \vee_\mathcal{K} (\mathcal{L}_1(\{a, b, c\}), \mathcal{E}_1(\{a, b, c\})) = (\emptyset, \emptyset) \vee_\mathcal{K} (\emptyset, \emptyset) = (\emptyset, \emptyset)$.
$(\mathcal{L}_2(\{a, b\}), \mathcal{E}_2(\{a, b\})) \vee_\mathcal{K} (\mathcal{L}_2(\{a, b, c\}), \mathcal{E}_2(\{a, b, c\})) = (\{a, b\}, \{c, d\}) \vee_\mathcal{K} (\{a, b\}, \emptyset) = (\{a, b\}, \emptyset))$.

Moreover,

$$\sim \mathcal{O}_\mathcal{C}(\{a, b\}) = ((\emptyset, \emptyset), (\{c, d\}, \{a, b\})),$$

where

$(\mathcal{L}_1(\{a, b\}), \mathcal{E}_1(\{a, b\})) = \neg(\emptyset, \emptyset) = (\emptyset, \emptyset)$;
$(\mathcal{L}_2(\{a, b\}), \mathcal{E}_2(\{a, b\})) = \neg(\{a, b\}, \{c, d\}) = (\{c, d\}, \{a, b\})$.

The following theorems allow us to express the operations $\hookrightarrow_1$, $\star_2$, $\hookrightarrow_2$, $\star_3$ and $\hookrightarrow_3$ through the operations between orthopairs of an individual covering (see Definition 11 and Definition 12). We present the proof only for the operation $\odot_3$ of Theorem 33, because it is possible to give the proof for the other operations with similar procedures. We recall that, given a refinement sequence $\mathcal{C} = (C_1, \ldots, C_n)$, in Remark 10, we denote the union of all blocks of C_i with U_i, for each $i \in \{1, \ldots, n\}$.

Theorem 31. *Let $\mathcal{C} = (C_1, \ldots, C_n)$ be a safe and complete refinement sequence of U. Then,*

$$\mathcal{O}_{\mathcal{C}}(X) \hookrightarrow_1 \mathcal{O}_{\mathcal{C}}(Y)$$

is the sequence $((A_1, B_1), \ldots, (A_n, B_n))$ defined as follows. Firstly, we set

$$(A_i', B_i') = (\mathcal{L}_i(X), \mathcal{E}_i(X)) \to_{\mathcal{N}} (\mathcal{L}_i(Y), \mathcal{E}_i(Y)),$$

for each i from 1 to n. Then, we set $(A_n, B_n) = (A_n', B_n')$ and

$$A_i = A_i' \setminus \cup \{N \in C_i \mid N' \subseteq N \text{ with } N' \in C_{i+1} \text{ and } N' \subseteq U_{i+1} \setminus A_{i+1}\},$$

and $B_i = B_i'$ for each $i < n$.

Theorem 32. *Let $\mathcal{C} = (C_1, \ldots, C_n)$ be a safe and complete refinement sequence of U. Then,*

$$\mathcal{O}_{\mathcal{C}}(X) \odot_2 \mathcal{O}_{\mathcal{C}}(Y)$$

is the sequence $((A_1, B_1), \ldots, (A_n, B_n))$ defined as follows. Firstly, we set

$$(A_i', B_i') = (\mathcal{L}_i(X), \mathcal{E}_i(X)) *_{\mathcal{L}} (\mathcal{L}_i(Y), \mathcal{E}_i(Y)),$$

for each i from 1 to n. Then, we set $(A_n, B_n) = (A_n', B_n')$, $A_i = A_i'$, and

$$B_i = B_i' \setminus \cup \{N \in C_i \mid N' \subseteq N \text{ with } N' \in C_{i+1} \text{ and } N' \subseteq U_{i+1} \setminus B_{i+1}\}$$

for each $i < n$. Moreover,

$$\mathcal{O}_{\mathcal{C}}(X) \hookrightarrow_2 \mathcal{O}_{\mathcal{C}}(Y)$$

is the sequence defined as follows. Firstly, we set

$$(A_i', B_i') = (\mathcal{L}_i(X), \mathcal{E}_i(X)) \to_{\mathcal{L}} (\mathcal{L}_i(Y), \mathcal{E}_i(Y)),$$

for each i from 1 to n. Then, we set $(A_n, B_n) = (A_n', B_n')$,

$$A_i = A_i' \setminus \cup \{N \in C_i \mid N' \subseteq N \text{ with } N' \in C_{i+1} \text{ and } N' \subseteq U_{i+1} \setminus A_{i+1}\},$$

and $B_i = B_i'$, for each $i < n$.

Theorem 33. *Let $\mathcal{C} = (C_1, \ldots, C_n)$ be a safe refinement sequence of partial partitions of U, then*

$$\mathcal{O}_{\mathcal{C}}(X) \odot_3 \mathcal{O}_{\mathcal{C}}(Y)$$

is the sequence of orthopairs $((A_1, B_1), \ldots, (A_n, B_n))$ defined as follows. Firstly we set

$$(A_i', B_i') = (\mathcal{L}_i(X), \mathcal{E}_i(X)) *_{\mathcal{S}} (\mathcal{L}_i(Y), \mathcal{E}_i(Y))$$

for each i from 2 to n. Then, we set $(A_1, B_1) = (A_1', B_1')$,

$$A_i = A_i' \cup \{N \in C_i \mid N \subseteq A_{i-1}\}, \text{ and } B_i = B_i' \setminus A_i,$$

for each $i > 0$.

Moreover,

$$\mathcal{O}_\mathcal{C}(X) \hookrightarrow_3 \mathcal{O}_\mathcal{C}(Y)$$

is the sequence of orthopairs $((A_1, B_1), \ldots, (A_n, B_n))$ *defined as follows. Firstly, we set*

$$(A_i', B_i') = (\mathcal{L}_i(X), \mathcal{E}_i(X)) \rightarrow_S (\mathcal{L}_i(Y), \mathcal{E}_i(Y))$$

for each $i > 2$. *Then, we set*

$$(A_1, B_1) = (A_1', B_1'), B_i = B_i' \cup \{N \in P_i \mid N \subseteq B_{i-1}\}, \text{ and } A_i = A_i' \setminus B_i,$$

for each $i > 0$.

In order to prove Theorem 33, we need to move from sequences of orthopairs to pairs of disjoint upsets. Let $\mathcal{C}$ be a refinement sequence of U such that $\mathcal{C} = \mathcal{C}'$. Then, the operation $\star_{K_O}^3$ coincides with $\star_3$ on $K(\mathcal{C})$. Indeed, $\mathcal{C} = \mathcal{C}'$ implies that β is the identity function (β is defined in Theorem 22). Consequently, for any $X, Y \subseteq U$, we have $(X^1, X^2) \star_{K_O}^3 (Y^1, Y^2) = \beta^{-1}((X^1, X^2) \star_3 (Y^1, Y^2)) = (X^1, X^2) \star_3 (Y^1, Y^2)$.

On the other hand, if $\mathcal{C} \neq \mathcal{C}'$ the IUML-algebras $K_O(\mathcal{C})$ and $K_O(\mathcal{C}')$ are not isomorphic. In any case, we can find a relationship between operations in $K_O(\mathcal{C}')$ and Sobociński conjunction, as follows.

Proposition 13. *Let* $\mathcal{C}$ *be a refinement sequence of partial partitions of* U, *let* $X, Y \subseteq U$, *and let* $F_X^\mathcal{C}$ *be the function defined by Eq. 1. Then,*

$$(X_\mathcal{C}^1, X_\mathcal{C}^2) \star_{K_O}^3 (Y_\mathcal{C}^1, Y_\mathcal{C}^2) = \beta^{-1}((Z_{\mathcal{C}'}^1, Z_{\mathcal{C}'}^2)),$$

where

$$Z_{\mathcal{C}'}^1 = \uparrow \{N \in P_{\mathcal{C}'} \mid F_X^{\mathcal{C}'}(N) \circledast_S F_Y^{\mathcal{C}'}(N) = 1\}$$

and

$$Z_{\mathcal{C}'}^2 = \{N \in P_{\mathcal{C}'} \mid F_X^{\mathcal{C}'}(N) \circledast_S F_Y^{\mathcal{C}'}(N) = 0\} \setminus Z_{\mathcal{C}'}^1.$$

Proof. By Definition 49, we must prove that $Z_{\mathcal{C}'}^1 = (X_{\mathcal{C}'}^1 \cap Y_{\mathcal{C}'}^1) \cup (X \diamond Y)$ and $Z_{\mathcal{C}'}^2 = (X_{\mathcal{C}'}^2 \cup Y_{\mathcal{C}'}^2) \setminus (X \diamond Y)$, where $X \diamond Y$ is related to $\mathcal{C}'$.

A node N belongs to $(X_{\mathcal{C}'}^1 \cap Y_{\mathcal{C}'}^1) \cup (X \diamond Y)$ if and only if $F_X(N) = 1$ and $F_Y(N) = 1$, or there exists $M \in P_{\mathcal{C}'}$ such that $N \subseteq M$ and $F_X(M) = 1$ and $F_Y(M) = 1 \backslash 2$, or $F_X(M) = 1 \backslash 2$ and $F_Y(M) = 1$. This is equivalent to affirm that $F_X(N) \circledast_S F_Y(N) = 1$ or there exists $M \in P_{\mathcal{C}'}$ such that $N \subseteq M$ and $F_X(M) \circledast_S F_Y(M) = 1$, since $\circledast_S$ is the Sobociński conjunction.

Similarly, N belongs to $(X_{\mathcal{C}'}^2 \cup Y_{\mathcal{C}'}^2) \setminus (X \diamond Y)$ if and only if $F_X(N) = 0$ or $F_Y(N) = 0$ and there does not exist $M \in P_{\mathcal{C}'}$ such that $N \subseteq M$ and $F_X(M) \circledast_S F_Y(M) = 1$. Then, $N \in \{N \in P_{\mathcal{C}'} \mid F_X(N) \circledast_S F_Y(N) = 0\} \setminus Z^1$.

Proof (Theorem 33). By definition of α (see Theorem 17), we have $(X^1, X^2) = \alpha(\mathcal{O}_C(X))$, $(Y^1, Y^2) = \alpha(\mathcal{O}_C(Y))$. Let Z be the subset of U such that

$$(Z^1, Z^2) = \alpha(\mathcal{O}_C(X)) \odot_3 \alpha(\mathcal{O}_C(Y)).$$

By induction on i we prove that $(\mathcal{L}_i(Z), \mathcal{E}_i(Z)) = (A_i, B_i)$.

Let $i = 1$. By definition and recalling that $Z^1 = \{N \in P_C \mid N \subseteq Z\}$, we have

$$\mathcal{L}_1(Z) = \bigcup \{N \in C_1 \mid N \subseteq Z\} = \bigcup \{N \in C_1 \cap Z^1\}.$$

By Proposition 13, $Z^1 = \uparrow \{N \in P_C \mid F_X(N) \circledast_S F_Y(N) = 1\}$, hence $Z^1 \cap C_1 = \{N \in C_1 \mid F_X(N) \circledast_S F_Y(N) = 1\}$. We have, by Proposition 4:

$$\mathcal{L}_1(Z) = \bigcup \{N \in C_1 \mid F_X(N) \circledast_S F_Y(N) = 1\} = A_1.$$

Now, we fix $i > 1$ and suppose by induction hypothesis that $A_{i-1} = \mathcal{L}_{i-1}(Z)$. Then by Proposition 4 and 13,

$$\mathcal{L}_i(Z) = \bigcup_{N \in Z^1 \cap C_i} N$$

$$= \bigcup \{N \in C_i \mid F_X(N) \circledast_S F_Y(N) = 1\} \cup \bigcup \{N \in C_i \mid N \subseteq M \text{ with } M \in Z^1 \cap C_{i-1}\}.$$

We notice that $A_i' = \cup \{N \in C_i \mid F_X(N) \circledast_S F_Y(N) = 1\}$ and $A_{i-1} = \mathcal{L}_{i-1}(Z) = \cup \{M \mid M \in Z^1 \cap C_{i-1}\}$. Consequently,

$$\mathcal{L}_i(Z) = A_i' \cup \{N \in C_i \mid N \subseteq A_{i-1}\}.$$

Similarly, by Propositions 4 and 13, we can prove that $B_i = B_i' \setminus A_i$, for each $i \in \{1, \ldots, n\}$. $\qquad \square$

In other words, the operation $\odot_3$ maps each pair of sequences of orthopairs to the sequence of orthopairs given by applying the Sobociński conjunction between orthopairs relative to the same partition and then closing with respect to the inclusion in the first component.

Hence, we can say that if we apply $\odot_3$ to sequences of orthopairs, the indeterminate value is always overcome by the determined ones, and in addition, as soon as a determined value is reached with respect to a given level of partial partitions, it is automatically given to all the blocks in the next refinements.

Example 37. Let C' be the refinement sequence of U of Example 16. We consider $X, Y \subseteq U$ such that $\mathcal{O}_{C'}(X)$ is equal to $\mathcal{O}_C(X)$ defined in Example 24 and $\mathcal{O}_{C'}(Y) = (\mathcal{O}_{C_1'}(Y), \mathcal{O}_{C_2'}(Y), \mathcal{O}_{C_3'}(Y))$, where

$\mathcal{O}_{C_1'}(Y) = (\emptyset, \emptyset)$,
$\mathcal{O}_{C_2'}(Y) = (\{u_3, u_4\}, \{u_5, u_6, u_{15}, \ldots, u_{20}\})$ and
$\mathcal{O}_{C_3'}(Y) = (\{u_3, u_4, u_7, u_8\}, \{u_5, u_6, u_{11}, u_{12}, u_{15}, \ldots, u_{18}\})$.

Hence,

$$\mathcal{O}_{C_1'}(X) *_S \mathcal{O}_{C_1'}(Y) = (\emptyset, \emptyset),$$
$$\mathcal{O}_{C_2'}(X) *_S \mathcal{O}_{C_2'}(Y) = (\{u_7, \ldots, u_{14}\}, \{u_1, \ldots, u_6, u_{15}, \ldots, u_{20}\}),$$
$$\mathcal{O}_{C_3'}(X) *_S \mathcal{O}_{C_3'}(Y) = (\{u_7, u_8\}, \{u_1, \ldots, u_6, u_{11}, u_{12}, u_{15}, \ldots, u_{18}\}).$$

Then, in order to close with respect to the inclusion in the first component, we add the elements of block $\{u_{11}, u_{12}\}$ to the first component of $\mathcal{O}_{C_3'}(X) *_S \mathcal{O}_{C_3'}(Y)$ and we subtract them from the second component of $\mathcal{O}_{C_3'}(X) *_S \mathcal{O}_{C_3'}(Y)$.

Finally, we obtain that $\mathcal{O}_{\mathcal{C}'}(X) \odot_3 \mathcal{O}_{\mathcal{C}'}(Y)$ is the sequence of $\mathsf{SO}(\mathcal{C}')$ made of the following pairs.

$$(\emptyset, \emptyset),$$
$$(\{u_7, \ldots, u_{14}\}, \{u_1, \ldots, u_6, u_{15}, \ldots, u_{20}\}) \text{ and}$$
$$(\{u_7, u_8, u_{11}, u_{12}\}, \{u_1, \ldots, u_6, u_{15}, \ldots, u_{18}\}).$$

We observe that $\mathcal{O}_{\mathcal{C}'}(X) \odot_3 \mathcal{O}_{\mathcal{C}'}(Y)$ provides precise information about the blocks $\{u_{15}, \ldots, u_{20}\}$, $\{u_1, u_2\}$, $\{u_7, \ldots, u_{10}\}$ and $\{u_{11}, \ldots, u_{14}\}$, while we do not know what happens to the elements u_{19} and u_{20} in $\mathcal{O}_{\mathcal{C}'}(X)$ and to the elements $u_1, u_2, u_9, u_{10}, u_{13}$ and u_{14} in $\mathcal{O}_{\mathcal{C}'}(Y)$. Hence, the uncertainty represented by the sequence $\mathcal{O}_{\mathcal{C}'}(X) \odot_3 \mathcal{O}_{\mathcal{C}'}(Y)$ is smaller than uncertainty that is in $\mathcal{O}_{\mathcal{C}'}(X)$ and $\mathcal{O}_{\mathcal{C}'}(Y)$.

Remark 20. The operations $\odot_4$ and $\hookrightarrow_4$ are not obtained by the generalization of some three-valued connectives. On the other hand, they allow us to define a new pair of operations between orthopairs that is following.

Let C be a covering of U, and let $X, Y \subseteq U$. Then,

$$(\mathcal{L}(X), \mathcal{E}(X)) \odot_4 (\mathcal{L}(Y), \mathcal{E}(Y)) = \begin{cases} (\emptyset, U), & \text{if } \mathcal{L}(X) = \emptyset \text{ and } \mathcal{L}(Y) = \emptyset; \\ (\mathcal{L}(X), \mathcal{E}(X)), & \text{if } \mathcal{L}(X) = \emptyset \text{ and } \mathcal{L}(Y) \neq \emptyset; \\ (\mathcal{L}(Y), \mathcal{E}(Y)), & \text{if } \mathcal{L}(X) \neq \emptyset \text{ and } \mathcal{L}(Y) = \emptyset; \\ (\mathcal{L}(X) \cap \mathcal{L}(Y), \mathcal{E}(X) \cap \mathcal{E}(Y)), & \text{if } \mathcal{L}(X) \neq \emptyset \text{ and } \mathcal{L}(Y) \neq \emptyset. \end{cases}$$
$$(36)$$

and

$$(\mathcal{L}(X), \mathcal{E}(X)) \hookrightarrow_4 (\mathcal{L}(Y), \mathcal{E}(Y)) = \begin{cases} (U, \emptyset), & \text{if } \mathcal{L}(X) = \emptyset \text{ and } \mathcal{E}(Y) = \emptyset; \\ (\mathcal{E}(X), \mathcal{L}(X)), & \text{if } \mathcal{L}(X) = \emptyset \text{ and } \mathcal{E}(Y) \neq \emptyset; \\ (\mathcal{L}(Y), \mathcal{E}(Y)), & \text{if } \mathcal{L}(X) \neq \emptyset \text{ and } \mathcal{E}(Y) = \emptyset; \\ (\mathcal{E}(X) \cap \mathcal{L}(Y), \mathcal{L}(X) \cap \mathcal{E}(Y)), & \text{if } \mathcal{L}(X) \neq \emptyset \text{ and } \mathcal{E}(Y) \neq \emptyset. \end{cases}$$
$$(37)$$

4.6 Application Scenario

In this section, we explain how an examiner's opinion on a number of candidates applying for a job can be represented by a sequence of orthopairs. Also, we show how opinions of two or more examiners can be combined by employing the operations $\curlywedge$, $\curlyvee$, $\odot_2$, $\odot_3$ and $\odot_4$ in order to get a final decision on each candidate. Moreover, such results are found in [21].

Imagine that a food company needs to recruit staff through a commission composed of several examiners, and managed by a committee chair. We indicate the set of twenty-four candidates with $\{c_1, \ldots, c_{24}\}$. The first selection will be to investigate the curriculum vitae of each candidate, after that all shortlisted applicants will be called for the first job interview. We suppose that the chair identifies some groups of applicants of $\{c_1, \ldots, c_{24}\}$ that have some specific characteristics which in his/her opinion are useful to work for the given company. Step by step, as it will be explained, the chair continues to refine each of these groups by identifying other suitable characteristics to work for the company. We underline that the chair selects sets made of applicants that have a specific characteristic in order to allow to each examiner to express his/her opinion on groups of candidates and not on every individual candidate. In this way, the first selection process is simplified.

In detail, the refinement process is made as follows. Initially, the chair identifies two characteristics: "to have a master degree in chemistry" and "to have a master degree in biology". Consequently, the covering $C_1 = \{b_1, b_2\}$ of $\{c_1, \ldots, c_{24}\}$ is determined, where $b_1 = \{c_1, \ldots, c_{12}\}$ is made of candidates with a master degree in chemistry and $b_2 = \{c_{13}, \ldots, c_{23}\}$ is made of candidates with a master degree in biology. Successively, the chair decides that the best candidates of b_1 are those specialized in "industrial chemistry", namely those of the set $b_3 = \{c_1, \ldots, c_5\}$ or in "pharmaceutical technology", namely the candidates of the set $b_4 = \{c_6, \ldots, c_{11}\}$. Moreover, the chair thinks that the best candidates of b_2 are those of $b_5 = \{c_{13}, \ldots, c_{17}\}$ that are specialized in "Biology of immunology" and those of $b_6 = \{c_{18}, \ldots, c_{22}\}$ that are specialized in "Food biology". In this way, the partial covering $C_2 = \{b_3, b_4, b_5, b_6\}$ of $\{c_1, \ldots, c_{24}\}$ is determined. Eventually, the chair considers $b_7 = \{c_1, c_2\}, b_8 = \{c_3, c_4\}, b_9 = \{c_6, c_7\}, b_{10} = \{c_8, c_9\}, b_{11} = \{c_{13}, c_{14}\}, b_{12} = \{c_{15}, c_{16}\}$ and $b_{13} = \{c_{18}, c_{19}\}$ and $b_{14} = \{c_{20}, c_{21}\}$, where b_7, b_9, b_{11} and b_{13} are respectively the subsets of b_3, b_4, b_5 and b_6 of candidates that have a certificate of Spanish language, instead b_8, b_{10}, b_{12} and b_{14} are respectively the subsets of b_3, b_4, b_5, b_6 of candidates that have a certificate of French language. Trivially, $C_3 = \{b_7, \ldots, b_{14}\}$ is also a partial covering of $\{c_1, \ldots, c_{24}\}$, and $\mathcal{C} = (C_1, C_2, C_3)$ is a refinement sequence of $\{c_1, \ldots, c_{24}\}$. More precisely, C_1, C_2 and C_3 are partial partitions of $\{c_1, \ldots, c_{24}\}$. The data used for the chair's classification are contained in the incomplete information table as Table 12, where $\{c_1, \ldots, c_{24}\}$ is the universe and {Master degree, Specialization, Language certification} is the set of attributes. The poset assigned to $\mathcal{C}$ is a forest, and it is shown in Fig. 24.

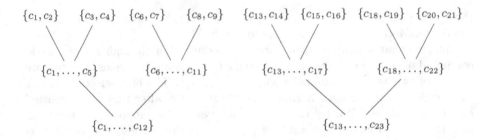

Fig. 24. Forest of the candidates

Table 12. Information table of the candidates

	Master degree	Specialization	Language certification
c_1	Chemistry	Industrial Chemistry	Spanish
c_2	Chemistry	Industrial Chemistry	Spanish
c_3	Chemistry	Industrial Chemistry	French
c_4	Chemistry	Industrial Chemistry	French
c_5	Chemistry	Industrial Chemistry	×
c_6	Chemistry	Pharmaceutical Technology	Spanish
c_7	Chemistry	Pharmaceutical Technology	Spanish
c_8	Chemistry	Pharmaceutical Technology	French
c_9	Chemistry	Pharmaceutical Technology	French
c_{10}	Chemistry	Pharmaceutical Technology	×
c_{11}	Chemistry	Pharmaceutical Technology	×
c_{12}	Chemistry	×	×
c_{13}	Biology	Immunology	Spanish
c_{14}	Biology	Immunology	Spanish
c_{15}	Biology	Immunology	Spanish
c_{16}	Biology	Immunology	French
c_{17}	Biology	Immunology	×
c_{18}	Biology	Food Biology	Spanish
c_{19}	Biology	Food Biology	Spanish
c_{20}	Biology	Food Biology	French
c_{21}	Biology	Food Biology	French
c_{22}	Biology	Food Biology	×
c_{23}	Biology	×	×
c_{24}	×	×	×

It is easy to notice that $\mathcal{C}$ is safe and complete.
Clearly, $P_{\mathcal{C}}$ is isomorphic to the forest of Fig. 25.

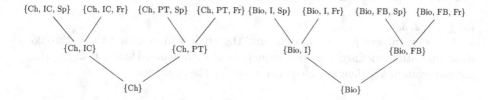

Fig. 25. Forest of the values of the candidates

Each node of Fig. 25 is the set of all values contained in Table 12 that characterizes the block of candidates of the respective node in $P_{\mathcal{C}}$ (we set Ch = Chemistry, IC = Industrial Chemistry, PT = Pharmaceutical Technology, Bio = Biology, I = Immunology, FB = Pharmaceutical Technology, Sp = Spanish, Fr = French). As an example, $\{Ch, IC, Fr\}$ is the set of the values that characterize the block $\{c_3, c_4\}$.

Once the classification process is completed, the chair invites every examiner to express his / her opinion about every block of $P_{\mathcal{C}}$, starting from the blocks that are minimal elements of $P_{\mathcal{C}}$ to those that are maximal elements of $P_{\mathcal{C}}$. Namely, examiners must first reveal their point of view on the nodes of level 0 of $P_{\mathcal{C}}$, then on those of level 1 of $P_{\mathcal{C}}$, and finally on those of level 2 of $P_{\mathcal{C}}$. For example, they can evaluate the blocks of $P_{\mathcal{C}}$ by following this order: $\{c_1, \ldots, c_{12}\}$, $\{c_{13}, \ldots, c_{23}\}$, $\{c_1, \ldots, c_5\}$, $\{c_6, \ldots, c_{11}\}$, $\{c_{13}, \ldots, c_{17}\}$, $\{c_{18}, \ldots, c_{22}\}$, $\{c_1, c_2\}$, $\{c_3, c_4\}$, $\{c_6, c_7\}$, $\{c_8, c_9\}$, $\{c_{13}, c_{14}\}$, $\{c_{15}, c_{16}\}$, $\{c_{18}, c_{19}\}$, $\{c_{20}, c_{21}\}$. Moreover, given a block b of $P_{\mathcal{C}}$ and an examiner E, we assume that three possibilities can occur: E could be in favour of the recruitment of all candidates in b, or E could not want to hire them, or E could be doubtful about them. Trivially, if E is in favour of the applicants of b, then E is also in favour of the candidates of all blocks included in b. For example, if E wants to recruit all candidates having a master degree in Chemistry, namely those of $\{c_1, \ldots, c_{12}\}$, then E is also in favour of hiring the candidates of $\{c_1, \ldots, c_5\}$ and $\{c_6, \ldots, c_{11}\}$, regardless of their specialization, and consequently also all candidates of $\{c_1, c_2\}$, $\{c_3, c_4\}$, $\{c_6, c_7\}$, and $\{c_8, c_9\}$, regardless of their language certification. Similarly, if E is not in favour of the applicants of b, then E is against hiring all candidates of every block included in b. Therefore, the opinion of E about all blocks of candidates in $P_{\mathcal{C}}$ is represented by the sequence of orthopairs $\mathcal{O}_{\mathcal{C}}(E)$ belonging to $SO(\mathcal{C})$, that is

$$\mathcal{O}_{\mathcal{C}}(E) = ((\mathcal{L}_1(E), \mathcal{E}_1(E)), (\mathcal{L}_2(E), \mathcal{E}_2(E)), (\mathcal{L}_3(E), \mathcal{E}_3(E))),$$

such that

$$\mathcal{L}_j(\mathsf{E}) = \cup\{b \in C_j \mid \mathsf{E} \text{ is in favour of hiring the candidates of } b\} \text{ and}$$
$$\mathcal{E}_j(\mathsf{E}) = \cup\{b \in C_j \mid \mathsf{E} \text{ is not in favour of hiring the candidates of } b\},$$

for $j = 1, 2, 3$.

Once examiners give their opinions, the chair can combine these through some operations defined between sequences of orthopairs. Hence, if $\mathsf{E}_1, \ldots, \mathsf{E}_m$ are our examiners, then the chair can consider the sequence

$$\mathcal{O}_\mathcal{C}(\mathsf{E}_1) \star \ldots \star \mathcal{O}_\mathcal{C}(\mathsf{E}_m),$$

where $\star \in \{\curlywedge, \curlyvee, \odot_2, \odot_3, \odot_4\}$ (these operations are defined in Sect. 4.5).

So, if a candidate belongs at least to one of the first components of the pairs in $\mathcal{O}_\mathcal{C}(\mathsf{E}_1) \star \ldots \star \mathcal{O}_\mathcal{C}(\mathsf{E}_m)$, then he / she will pass the first selection; if he / she belongs to at least one of the second components of the pairs in $\mathcal{O}_\mathcal{C}(\mathsf{E}_1) \star \ldots \star \mathcal{O}_\mathcal{C}(\mathsf{E}_m)$, then he / she will be excluded; otherwise, the chair will decide about him / her.

In order to provide the reader with a more intuitive representation of the examiners opinion and their combinations through our operations, we can describe sequences of orthopairs as labelled graphs defined in Remark 15. Thus, the labelled poset assigned to the sequence $\mathcal{O}_\mathcal{C}(X)$ of $\mathsf{SO}(\mathcal{C})$ is determined by the function

$$l_X : P_\mathcal{C} \mapsto \{\bullet, \circ, ?\}$$

such that

$$l_X(b) = \begin{cases} \bullet & \text{if } b \subseteq \mathcal{L}_i(X) \text{ for some } i \in \{1, 2, 3\}, \\ \circ & \text{if } b \subseteq \mathcal{E}_i(X) \text{ for some } i \in \{1, 2, 3\}, \\ ? & \text{otherwise}, \end{cases}$$

where $(\mathcal{L}_i(X), \mathcal{E}_i(X))$ denotes the i-th orthopair of $\mathcal{O}_\mathcal{C}(X)$.

Now, we assume that the examiners of the commission are two: E_1 and E_2. Moreover, the opinions of E_1 and E_2 are respectively expressed by the following labelled posets.

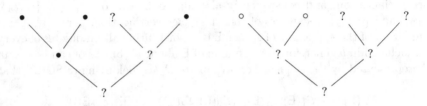

Fig. 26. Labelled forest of $\mathcal{O}_\mathcal{C}(\mathsf{E}_1)$

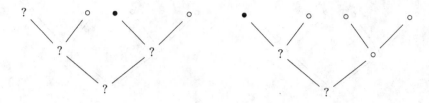

Fig. 27. Labelled forest of $\mathcal{O}_C(E_2)$

The labelled posets assigned to $\mathcal{O}_C(E_1) \curlywedge \mathcal{O}_C(E_2)$, $\mathcal{O}_C(E_1) \curlyvee \mathcal{O}_C(E_2)$, $\mathcal{O}_C(E_1) \odot_2$ $\mathcal{O}_C(E_2)$, $\mathcal{O}_C(E_1) \odot_3 \mathcal{O}_C(E_2)$ and $\mathcal{O}_C(E_1) \odot_4 \mathcal{O}_C(E_2)$ are respectively the following.

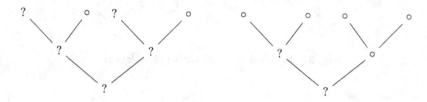

Fig. 28. Labelled forest of $\mathcal{O}_C(E_1) \curlywedge \mathcal{O}_C(E_2)$

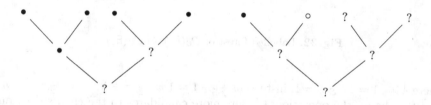

Fig. 29. Labelled forest of $\mathcal{O}_C(E_1) \curlyvee \mathcal{O}_C(E_2)$

We can observe that each of the previous operation determines the choice or the exclusion of some candidates of $\{c_1, \ldots, c_{24}\}$ with respect to the first selection. For example, $\odot_2$ involves the exclusion of candidates $c_3, c_4, c_8, c_9, c_{13}, \ldots, c_{23}$, and it does not allow any candidate to be admitted.

We can make the following remarks, in order to compare the results generated with $\curlywedge$, $\curlyvee$, $\odot_2$ and $\odot_3$. By theorems proved in Sect. 4.5, by Theorem 1, and by Theorem 2, we can affirm that $\curlywedge$, $\curlyvee$, $\odot_2$ and $\odot_3$ are respectively obtained starting from the three-valued operations $\wedge$, $\vee$, $\circledast_{\mathcal{L}}$ and $\circledast_{\mathcal{S}}$. Therefore, we obtain more excluded candidates with $\odot_2$ than with $\curlywedge$, $\curlyvee$ and $\odot_3$; indeed, $\odot_2$ is determined starting from the Łukasiewicz conjunction $\circledast_{\mathcal{L}}$, where $\frac{1}{2} \circledast_{\mathcal{L}} \frac{1}{2} = 0$, instead of $\frac{1}{2} \vee \frac{1}{2} = \frac{1}{2} \circledast_{\mathcal{S}} \frac{1}{2} = \frac{1}{2} \wedge \frac{1}{2} = \frac{1}{2}$. More candidates pass the first selection with $\odot_3$ than with $\curlywedge$ and $\odot_2$, since $\odot_3$ is obtained from the Sobociński conjunction $\circledast_{\mathcal{S}}$,

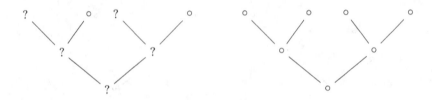

Fig. 30. Labelled forest of $\mathcal{O}_C(\mathsf{E}_1) \odot_2 \mathcal{O}_C(\mathsf{E}_2)$

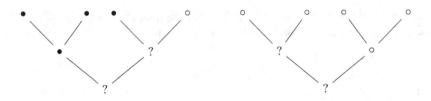

Fig. 31. Labelled forest of $\mathcal{O}_C(\mathsf{E}_1) \odot_3 \mathcal{O}_C(\mathsf{E}_2)$

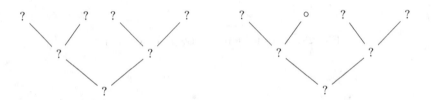

Fig. 32. Labelled forest of $\mathcal{O}_C(\mathsf{E}_1) \odot_4 \mathcal{O}_C(\mathsf{E}_2)$

where $\frac{1}{2} \circledast_S 1 = 1 \circledast_S \frac{1}{2} = 1$, instead of $\frac{1}{2} \circledast_\mathcal{L} 1 = 1 \circledast_\mathcal{L} \frac{1}{2} = \frac{1}{2} \wedge 1 = 1 \wedge \frac{1}{2} = \frac{1}{2}$. On the other hand, the operation $\curlywedge$ refers more candidates to the chair's decision than $\odot_2$ and $\odot_3$, since it is defined starting from the Kleene conjunction $\wedge$, where $\frac{1}{2} \wedge \frac{1}{2} = \frac{1}{2} \wedge 1 = 1 \wedge \frac{1}{2} = \frac{1}{2}$.

In this context, the operation $\odot_4$ can be interpreted as follows. Given $j \in \{1, 2\}$, we say that the opinion of E_j is *overall positive*, when E_j is in favour of recruiting of at least one block of candidates of P_C, otherwise E_j's opinion is *overall negative*. If the opinions of E_1 and E_2 are both overall negative, then all candidates of $\{c_1, \ldots, c_{24}\}$ are excluded. If only the E_1's opinion (or the E_2's opinion) is overall positive, then the candidates that are negative for E_2 (or E_1) are excluded (by negative candidates for E_2 (or E_1), we mean those belonging to each block b such that $l_{\mathsf{E}_2}(b) = \circ$ (or $l_{\mathsf{E}_1}(b) = \circ$)), and the chairman decides for the remaining applicants. If the opinions of E_1 and E_2 are both overall positive, then the candidates of each block b in P_C such that $l_{\mathsf{E}_1}(b) = l_{\mathsf{E}_2}(b) = \bullet$ pass the first selection, the candidate of each block b in P_C such that $l_{\mathsf{E}_1}(b) = l_{\mathsf{E}_2}(b) = \circ$ are excluded, and the chairman decides for the remaining applicants.

We can notice that each operation belonging to $\{\curlywedge, \curlyvee, \odot_2, \odot_3, \odot_4\}$ represents a way to repartition the universe $\{c_1, \ldots, c_{24}\}$ in three sets of candidates: the selected candidates (those belonging to some blocks with label $\bullet$), the excluded candidates (those belonging to some blocks with label $\circ$), and the remaining candidates on which the evaluation is uncertain (those belonging to blocks that all with label ?). More generally, each sequence of orthopairs of $SO(\mathcal{C})$ determines a tri-partition (i.e. partition made of three elements) of $\{c_1, \ldots, c_{24}\}$. For example, $\mathcal{O}_\mathcal{C}(\mathsf{E}_1)$ and $\mathcal{O}_\mathcal{C}(\mathsf{E}_2)$ generate respectively the following partitions of $\{c_1, \ldots, c_{24}\}$.

$$P_{\mathsf{E}_1} = \{\{c_1, \ldots, c_5, c_8, c_9\}, \{c_{13}, \ldots, c_{16}\}, \{c_6, c_7, c_{10}, c_{11}, c_{12}, c_{17}, \ldots, c_{24}\}\},$$
$$P_{\mathsf{E}_2} = \{\{c_6, c_7, c_{13}, c_{14}\}, \{c_3, c_4, c_8, c_9, c_{15}, c_{16}, c_{18}, \ldots, c_{22}\},$$
$$\{c_1, c_2, c_5, c_{10}, c_{11}, c_{12}, c_{17}, c_{23}, c_{24}\}\}.$$

Tri-partitions are at the basis of three-way decision (3WD) theory proposed by Yao [103]. A three-way decision procedure mainly consists in two steps: *dividing* the universe in three regions and then *acting*, i.e. taking a different strategy on objects belonging to different regions. In 3WD theory, the standard tools to trisect the universe are the classical rough sets and orthopairs, namely those generated by a partition [104]. Then, the lower approximation, the impossibility domain and the boundary region are called *acceptance region*, *rejection region* and *uncertain region*, respectively. On the other hand, a sequence of orthopairs divides the universe in a more precise way also starting from an incomplete information table, in which the data are missing. For example, if we focus on the labelled forest assigned to $\mathcal{O}_\mathcal{C}(\mathsf{E}_1)$, then we can observe that level 2 gives arise the tri-partition $\{\{c_1, c_2, c_3, c_4, c_8, c_9\}, \{c_{13}, c_{14}, c_{15}, c_{16}\}, \{c_6, c_7, c_{18}, c_{19}, c_{20}, c_{21}\}\}$, but level 1 allows us to put in the acceptance region also the element c_5.

Furthermore, operations between sequences of orthopairs represent several ways to aggregate different tri-partitions of the same universe. For example, if we consider $\curlyvee$, then the tri-partition made of $\{c_1, \ldots, c_9, c_{13}, c_{14}\}$, $\{c_{15}, c_{16}\}$ and $\{c_{10}, c_{11}, c_{12}, c_{17}, \ldots, c_{24}\}$ is generated starting from P_{E_1} and P_{E_2}.

Once the three regions have been obtained, one might need to expand or reduce one of them. For example, it could occur that the accepted candidates with $\curlyvee$ may be too many. Then, we can assign a weight to every object of the universe, by considering the labels of each block to which it belongs. Let $P_\mathcal{C}^j$ be the j-th level of $P_\mathcal{C}$ defined in Eq. 30 such that $j \in \{1, \ldots, n\}$, where n is the maximum number of elements of a chain in $P_\mathcal{C}$. For each $c \in \{c_1, \ldots, c_{24}\}$, we set

$$p_j(c) = \begin{cases} 1 & \text{if } c \in b \text{ where } b \in P_\mathcal{C}^k \text{ with } k \leq j \text{ and it is labelled with } \bullet; \\ 0 & \text{if } c \in b \text{ where } b \in P_\mathcal{C}^k \text{ with } k \leq j \text{ and it is labelled with } \circ; \\ \frac{1}{2} & \text{otherwise.} \end{cases}$$

Moreover, we assign to c, the following final weight.

$$w(c) = \frac{\sum_{j=1}^{n} p_j(c)}{n}.$$

If we focus on the sequences of orthopairs obtained starting from operation $\odot_3$, we have

- $w(c_1) = w(c_2) = w(c_3) = w(c_4) = w(c_5) = \frac{\frac{1}{2}+1+1}{3} = \frac{5}{6}$;

- $w(c_6) = w(c_7) = \frac{\frac{1}{2}+\frac{1}{2}+1}{3} = \frac{2}{3}$;

- $w(c_8) = w(c_9) = w(c_{13}) = w(c_{14}) = w(c_{15}) = w(c_{16}) = \frac{\frac{1}{2}+\frac{1}{2}+0}{3} = \frac{1}{3}$;

- $w(c_{18}) = w(c_{19}) = w(c_{20}) = w(c_{21}) = w(c_{22}) = \frac{\frac{1}{2}+0+0}{3} = \frac{1}{6}$;

- $w(c_{10}) = w(c_{11}) = w(c_{12}) = w(c_{17}) = w(c_{23}) = w(c_{24}) = \frac{\frac{1}{2}+\frac{1}{2}+\frac{1}{2}}{3} = \frac{1}{2}$.

Trivially, $w(c)$ belongs to the real interval $[0,1]$, and it expresses how much the candidate c must pass the first selection from 0 to 1.

The weights $w(c_1), \ldots, w(c_{24})$ can be used in several ways. For example, the chair could decide that the candidates with weight greater than $\frac{2}{3}$, and so c_1, c_2, c_3, c_4, c_5 pass the first selection, and that the remaining candidates are excluded. Moreover, he could choose two thresholds α and β in $[0,1]$ such that $\alpha \leq \beta$. Successively, he can redefine the following tri-partition of $\{c_1, \ldots, c_{24}\}$

- $\{c \in \{c_1, \ldots, c_{24}\} : w(c) \leq \alpha\}\}$ (*rejection region*),
- $\{c \in \{c_1, \ldots, c_{24}\} : \alpha < w(c) < \beta\}$ (*uncertain region*),
- $\{c \in \{c_1, \ldots, c_{24}\} : w(c) \geq \beta\}$ (*acceptance region*).

We observe that our procedure can be also applied for sequences of orthopairs generated by a sequence of equivalence relations that is not a refinement sequence. However, the advantage of considering sequences of refinements of orthopairs is that once we know that a block N is included in the acceptance region (or in the rejection region), we also know that all blocks included in N are included in the acceptance region (or in the rejection region). Similarly, if we know that $p_j(c) = 1$ (or $p_j(c) = 0$), we also know that $p_{j+1}(c) = 1$ (or $p_{j+1}(c) = 0$).

5 Modal Logic and Sequences of Orthopairs

> *"Then you should say what you mean," the March Hare went on. "I do," Alice hastily replied; "at least – at least I mean what I say–that's the same thing, you know." "Not the same thing a bit!" said the Hatter. "You might just as well say that 'I see what I eat' is the same thing as 'I eat what I see'!" "You might just as well say," added the March Hare, "that 'I like what I get' is the same thing as 'I get what I like'!" "You might just as well say," added the Dormouse, who seemed to be talking in his sleep, "that 'I breathe when I sleep' is the same thing as 'I sleep when I breathe'!"*
>
> Lewis Carroll *(Alice's Adventures in Wonderland)*

In this chapter, firstly, we recall some basic notions of modal logic and the existing connections between modal logic and rough sets (see Sect. 5.1). In Sect. 5.2, we develop the original modal logic SO_n, defining its language, introducing its Kripke models, and providing its axiomatization. Moreover, we investigate the properties of our logic system, such as the consistency, the soundness and the completeness with respect to Kripke semantics. In Sect. 5.3 we explore the relationships between modal logic SO_n and sequences of orthopairs. Also, we consider the operations between orthopairs and between sequences of orthopairs from the logical point of view. In the last section of this chapter, we employ modal logic SO_n to represent the knowledge of an agent that increases over time, as new information is provided.

5.1 Modal Logic $S5$ and Rough Sets

Modal logic is the logic of *necessity* and *possibility* [38]. It is characterized by the symbols $\Box$ and $\Diamond$, called *modal operators*, such that the formula $\Box\varphi$ means "it is necessary that φ" or, in other words, "φ is the case in every possible circumstance", and the formula $\Diamond\varphi$ means "it is possible that φ" or, in other words, "φ is the case in at least one possible circumstance". However, *necessity* and *possibility* are not the only modalities, since the term *modal logic* is used more broadly to cover a family of logics with similar rules and a variety of different symbols [51]. In this thesis, we are interested in propositional modal logic $S5$, that was proposed by Clarence Irving Lewis and Cooper Harold Langford in their book *Symbolic Logic* [68].

Now, we briefly describe the syntax and the semantics of modal logic $S5$ [29]. The $S5$-language contains all symbols of propositional logic, plus the modalities $\Box$ and $\Diamond$. In terms of semantics, the formulas of $S5$-language are interpreted with the *Kripke models*. A Kripke model of $S5$ is a triple consisting of a universe U (its element are named *possible worlds*), an equivalence relation R on U, and an evaluation function v, that assigns to a propositional variable p the set of all worlds of U in which p is true. We can extend v on the formulas of propositional logic as usual and on the modal formulas as following. Let p be a propositional variable, and let $u \in U$,

$\Box p$ *is true in* u if and only if "p is true in every world v of U such that uRv", and

$\Diamond p$ *is true in* u if and only if "p is true at least in a world v of U such that uRv".

The axiom schemas are obtained by adding the following schemas to those of propositional logic.

Definition 51 (Axioms of $S5$).

K. $\Box(\varphi \to \psi) \to (\Box\varphi \to \Box\psi)$ *(distribution axiom)*;
T. $\Box\varphi \to \varphi$ *(necessitation axiom)*;
5. $\Diamond\varphi \to \Box\Diamond\varphi$.

We notice that Axiom 5 it is equivalent to the set of axioms made of

B. $\varphi \to \Box\Diamond\varphi$ *and*
4. $\Box\varphi \to \Box\Box\varphi$.

The inference rules are the *modus ponens* and the *necessitation rule* $(\varphi/\Box\varphi)$. We stress that $S5$ belongs to the family of *normal modal logics*, that are characterized by adding the necessitation rule, and a list of axiom schemas Ax including **K** to the principles of propositional logic. The weakest normal modal logic is named K in honour of Saul Kripke, where $Ax=\{\mathbf{K}\}$. Thus, $S5$, as every normal modal logic, is an extension of K. A further example of normal modal logic is $S4$, that is obtained by adding to system K the axiom schemas **T** a and **4**.

The system $S5$ is sound and complete with respect to the class of all Kripke models of $S5$.

Moreover, propositional modal logic is also interpreted as an extension of classical propositional logic with two added operators expressing modality [55]. Since Pawlak rough set algebra is an extension of Boolean algebra (see Remark 3), the relationship between propositional modal logic and rough sets appears intuitive. In particular, modal logic $S5$ is connected with rough set theory, since the necessity and possibility can be interpreted as the lower and the upper approximation [79, 80]. Hence, let (U, R, v) be a Kripke model of $S5$, we have that

$$||\Box\varphi||_{\mathsf{v}} = \mathcal{L}_R(||\varphi||_{\mathsf{v}}) \quad \text{and} \quad ||\Diamond\varphi||_{\mathsf{v}} = \mathcal{U}_R(||\varphi||_{\mathsf{v}}),$$

where $||\varphi||_{\mathsf{v}}$, $||\Box\varphi||_{\mathsf{v}}$ and $||\Diamond\varphi||_{\mathsf{v}}$ are made of possible worlds in which φ, $\Box\varphi$ and $\Diamond\varphi$ are true, respectively.

It is important to recall that $S5$ can be considered as an epistemic logic in the sense that it is suitable for representing and reasoning about the knowledge of an individual agent [42, 46, 67]. Indeed, the formula $\Box\varphi$ can be read as "the agent knows φ". Moreover, the axioms of $S5$ express the properties of the knowledge. For instance, Schema **4** expresses the fact that if an agent knows φ, then she knows that she knows φ (*the positive introspection axiom*).

5.2   Modal Logic SO_n

In this section, the novel modal logic SO_n is developed.

From now, by refinement sequence, we mean a refinement sequence of *partial partitions* of the given universe, and we fix an integer $n > 0$.

Language of SO_n

We indicate the language of SO_n with L. Then, the alphabet of L consists of

- a set Var of propositional variables;
- the logical connectives $\wedge$ and $\neg$;
- the sequences of modal operators $(\square_1, \ldots, \square_n)$ and $(\bigcirc_1, \ldots, \bigcirc_n)$.

The propositional variables are typically denoted with $p, q, r, \ldots$ and refer to the statements that are considered basic, for example "the book is red". The symbols $\wedge$ and $\neg$ are respectively the *conjunction* and *negation* of classical propositional logic. Fixed $i \in \{1, \ldots, n\}$, we call *i-box* and *i-circle* the modal operators $\square_i$ and $\bigcirc_i$, respectively.

We denote the well formed formulas of L with Greek letters. As usual, the set Form of all well formed formulas of L is the smallest set that contains Var and satisfies the following conditions. Let $\varphi, \psi \in$ Form,

- if $\varphi \in$ Form, then $\neg\varphi$, $\square_i\varphi$, $\bigcirc_i\varphi \in$ Form, for each $i \in \{1, \ldots, n\}$;
- if $\varphi, \psi \in$ Form, then $\varphi \wedge \psi \in$ Form.

We simply call the elements of Form *formulas* or *sentences*. Moreover, the alphabet of L also contains the brackets "(" and ")" to establish the order wherewith the connectives work in the complex formulas. In this way, the language is clear and has no ambiguity.

The abbreviations introduced in the next definition, except the last one, are the standard abbreviations defined for the classical propositional logic [64].

Definition 52 (Abbreviations in L). *Let $\varphi, \psi \in$ Form and $p \in$ Var,*

1. $\bot := p \wedge \neg p$ *(false)*;
2. $\top := \neg\bot$ *(true)*;
3. $\varphi \vee \psi := \neg(\neg\varphi \wedge \neg\psi)$ *(disjunction)*;
4. $\varphi \rightarrow \psi := \neg\varphi \vee \psi$ *(implication)*;
5. $\varphi \equiv \psi := (\varphi \rightarrow \psi) \wedge (\psi \rightarrow \varphi)$ *(equivalence)*;
6. $\triangle_i\varphi := \square_i\neg\varphi$, *(i-triangle) with $i \in \{1, \ldots, n\}$.*

We employ the convention that $\leftrightarrow$ dominates $\rightarrow$, and $\rightarrow$ dominates the remaining symbols. For example, the formula $\square_i p \rightarrow q$ is understood as $(\square_i p) \rightarrow q$.

By *schema*, we mean a set of formulas all having the same form. For example, the schema $\varphi \wedge \psi$ is the set $\{\varphi \wedge \psi \mid \varphi, \psi \in$ Form$\}$.

Semantics of SO_n

We define the Kripke models of SO_n, which we also call *orthopaired Kripke models* or *SO_n-models*.

Definition 53. *A Kripke model of SO_n is a triple*

$$\mathcal{M} = (U, (R_1, \dots, R_n), v),$$

where

1. *U is a non-empty set of objects,*
2. *$(R_1, \dots, R_n)$ is a sequence of equivalence relations on U (i.e. for i from 1 to n, $R_i \subseteq (U \times U)$ and R_i is reflexive, symmetric and transitive) such that, let $u \in U$,*
   - *$R_1(u) \neq \{u\}$, and*
   - *$R_{i+1}(u) \subseteq R_i(u)$, for each $i < n$;*
3. *v is an evaluation function that assigns a subset of U to each element of Var (i.e. $v : \mathsf{Var} \mapsto 2^U$, where 2^U is the power set of U).*

We say that U is the *domain* or the *universe* of $\mathcal{M}$, the elements of U are the *states* or the *possible worlds* of $\mathcal{M}$, and $R_1, \dots, R_n$ are the *accessibility relations* of $\mathcal{M}$. The pair $(U, (R_1, \dots, R_n))$ is called Kripke frame of SO_n. Moreover, let $p \in \mathsf{Var}$, if $u \in v(p)$, then we can say that *p is true at u in $\mathcal{M}$*.

Remark 21. The domain of an orthopaired Kripke model has at least two elements.

Example 38. Let $\mathsf{Var} = \{p, q, r\}$, we suppose that

- $U = \{a, b, c, d\}$,
- $R_1 = \{(a, b), (b, a), (c, d), (d, c)\} \cup \{(u, u) \mid u \in U\}$,
- $R_2 = \{(a, b), (b, a)\} \cup \{(u, u) \mid u \in U\}$,
- v is a function from Var to 2^U such that $v(p) = \{a, b, c\}$, $v(q) = \{c, d\}$ and $v(r) = \{a, c\}$.

Then, $\mathcal{M} = (U, (R_1, R_2), v)$ is a Kripke model of SO_n.

Orthopaired Kripke models are also models of modal logic $S5^n$ developed in [46]. However, a Kripke model of $S5^n$ is not always a Kripke model of SO_n; in fact, the accessibility relations of each $S5^n$-model have only the property to be equivalence relations.

Definition 54 (Kripke models of SO_n as graphs). *A Kripke model $\mathcal{M} = (U, (R_1, \dots, R_n), v)$ of SO_n is represented by the graph $\mathcal{G}_\mathcal{M}$, where*

- *the set of the vertices is U,*
- *two vertices are connected with the labeled edge i if and only if*

$$i = \max\{j \in \{1, \dots, n\} \mid (a, b) \in R_j\}.$$

- *the label of $u \in U$ is the list of the propositional variables that are true at u in $\mathcal{M}$.*

Example 39. Suppose that $\mathsf{Var} = \{p\}$ and $\mathcal{M} = (U, (R_1, R_2), v)$ is a Kripke model of SO_n, where

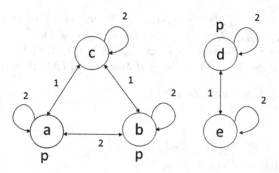

Fig. 33. Graph $\mathcal{G}_\mathcal{M}$

- $U = \{a, b, c, d, e\}$;
- $R_1 = \{(a, b), (b, a), (a, c), (c, a), (b, c), (c, b), (d, e), (e, d)\} \cup \{(u, u) \mid u \in U\}$,
- $R_2 = \{(a, b), (b, a)\} \cup \{(u, u) \mid u \in U\}$,
- $v(p) = \{a, b, d\}$.

The graph $\mathcal{G}_\mathcal{M}$ is as in Fig. 33.

The notion of *truth* of a formula in a Kripke model of SO_n is given by the next definition.

Definition 55. *Let* $\mathcal{M} = (U, (R_1, \ldots, R_n), v)$ *be a Kripke model of* SO_n. *The notion of* $(\mathcal{M}, u) \models \varphi$ *is inductively defined as follows.*

1. $(\mathcal{M}, u) \models p$, *with* $p \in$ Var *iff* "$u \in v(p) = \|p\|_v$";
2. $(\mathcal{M}, u) \models (\varphi \wedge \psi)$ *iff* "$(\mathcal{M}, u) \models \varphi$ *and* $(\mathcal{M}, u) \models \psi$";
3. $(\mathcal{M}, u) \models \neg\varphi$ *iff* "$(\mathcal{M}, u) \not\models \varphi$";
4. $(\mathcal{M}, u) \models \Box_i\varphi$ *iff* "$R_i(u) \subseteq \|\varphi\|_v$ *and* $R_i(u) \neq \{u\}$";
5. $(\mathcal{M}, u) \models \bigcirc_i\varphi$ *iff* "$u \models \varphi$ *and* $R_i(u) \neq \{u\}$";

where $\|\varphi\|_v$ *is the* truth set *of* φ, *that is*

$$\|\varphi\|_v = \{u \in U \mid (\mathcal{M}, u) \models \varphi\}.$$

$(\mathcal{M}, u) \models \varphi$ can be read as "φ is true at u in $\mathcal{M}$" or "φ holds at u in $\mathcal{M}$" or "$(\mathcal{M}, u)$ satisfies φ". Moreover, we say that "φ is false at u in $\mathcal{M}$" if and only if $(\mathcal{M}, u) \not\models \varphi$. We can write $u \models \varphi$, instead of $(\mathcal{M}, u) \models \varphi$, when $\mathcal{M}$ is clear from the context.

Remark 22. The points 1, 2 and 3 of Definition 55 are given for standard Kripke semantics too. Also, once fixed $i \in \{1, \ldots, n\}$, $u \models \Box_i\varphi$ differs from $u \models \Box\varphi$, where $\Box$ is the necessity operator of $S5$ logic interpreted by R_i, since the additional condition $R_i(u) \neq \{u\}$ is required.

The next proposition follows by Definition 52 and Definition 55.

Proposition 14. *Let* $\mathcal{M} = (U, (R_1, \ldots, R_n), v)$ *be a Kripke model of* SO_n. *Then,*

1. $(\mathcal{M}, u) \models (\varphi \vee \psi)$ *iff* "*either* $(\mathcal{M}, u) \models \varphi$ *or* $(\mathcal{M}, u) \models \psi$";
2. $(\mathcal{M}, u) \models \triangle_i \varphi$ *iff* "$R_i(u) \cap ||\varphi||_v = \emptyset$ *and* $R_i(u) \neq \{u\}$";
3. $(\mathcal{M}, u) \models \varphi \to \psi$ *iff* "$(\mathcal{M}, u) \models \varphi$ *implies that* $(\mathcal{M}, u) \models \psi$";
4. $(\mathcal{M}, u) \models \varphi \equiv \psi$ *iff* "$(\mathcal{M}, u) \models \varphi$ *if and only if* $(\mathcal{M}, u) \models \psi$";

for each $u \in U$, $\varphi, \psi \in$ *Form and* $i \in \{1, \dots, n\}$.

Remark 23. It is clear that

- $(\mathcal{M}, u) \models \square_1 \varphi$ iff $R_1(u) \subseteq ||\varphi||_v$;
- $(\mathcal{M}, u) \models \triangle_1 \varphi$ iff $R_1(u) \cap ||\varphi||_v = \emptyset$;
- $(\mathcal{M}, u) \models \varphi$ iff $(\mathcal{M}, u) \models \bigcirc_1 \varphi$;
- If $(\mathcal{M}, u) \models \bigcirc_i \varphi$, then $(\mathcal{M}, u) \models \varphi$;
- If $(\mathcal{M}, u) \models \square_i \varphi$, then $(\mathcal{M}, u) \models \bigcirc_i \varphi$;

for each i from 1 to n.

The following theorem expresses the connection between the logical connectives of L and the set-theoretic operations.

Theorem 34. *Let* $\mathcal{M} = (U, (R_1, \dots, R_n), v)$ *be a Kripke model of* SO_n. *Then,*

1. $||\perp||_v = \emptyset$;
2. $||\top||_v = U$;
3. $||\neg\varphi||_v = U \setminus ||\varphi||_v$;
4. $||\varphi \wedge \psi||_v = ||\varphi||_v \cap ||\psi||_v$;
5. $||\varphi \vee \psi||_v = ||\varphi||_v \cup ||\psi||_v$;
6. $||\varphi \to \psi||_v = (U \setminus ||\varphi||_v) \cup ||\psi||_v$;
7. $||\varphi \equiv \psi||_v = ((U \setminus ||\varphi||_v) \cup ||\psi||_v) \cap ((U \setminus ||\psi||_v) \cup ||\varphi||_v)$;
8. $||\square_i \varphi||_v = \{u \in U \mid R_i(u) \subseteq ||\varphi||_v$ *and* $R_i(u) \neq \{u\}\}$;
9. $||\triangle_i \varphi||_v = \{u \in U \mid R_i(u) \cap ||\varphi||_v = \emptyset$ *and* $R_i(u) \neq \{u\}\}$; *for* i *from 1 to* n.

Let Cl_n be the class of the Kripke models of SO_n, we define the notion of validity in the models that belong to Cl_n.

Definition 56. *Let* $\mathcal{M} \in \mathsf{Cl}_n$. *Then, for each* $\varphi \in$ *Form, we write*

- $\models^{\mathcal{M}} \varphi$ *iff* "$(\mathcal{M}, u) \models \varphi$, *for every world* u *in* $\mathcal{M}$", *and we say that* φ *is valid in* $\mathcal{M}$;
- $\models^{\mathsf{Cl}_n} \varphi$ *iff* "$\models^{\mathcal{M}} \varphi$, *for every model* $\mathcal{M}$ *in* Cl_n", *and we say that* φ *is valid in* Cl_n.

From the previous notions of validity, two logical consequence relations can be formally defined.

Definition 57. *For each* $\mathcal{M} \in \mathsf{Cl}_n$, $\varphi \in$ *Form and* $\Gamma \subseteq$ *Form, we write*

- $\Gamma \models^{\mathcal{M}} \varphi$ *iff* "*if* $\models^{\mathcal{M}} \Gamma$, *then* $\models^{\mathcal{M}} \varphi$", *and*
- $\Gamma \models^{\mathsf{Cl}_n} \varphi$ *iff* "*if* $\models^{\mathsf{Cl}_n} \Gamma$, *then* $\models^{\mathsf{Cl}_n} \varphi$".

Proposition 15. *Let* $i \in \{1, \ldots, n\}$*, the instances of the following schemes are* SO_n*-tautologies.*

Ab$_{\triangle_1}$. $\triangle_1 \bot$.
Dist$_{\square_i}$. $\square_i(\varphi \wedge \psi) \equiv \square_i\varphi \wedge \square_i\psi$.
Dist$_{\triangle_i}$. $\triangle_i(\varphi \vee \psi) \equiv \triangle_i\varphi \wedge \triangle_i\psi$.
P$_1$. $\neg \bigcirc_i \varphi \rightarrow (\neg\square_i\varphi \vee \neg\triangle_i\varphi)$.
P$_2$. $(\neg \bigcirc_i \varphi \wedge \varphi) \rightarrow (\neg\square_i\varphi \wedge \neg\triangle_i\varphi)$.

Proof. Let $\mathcal{M} = (U, (R_1, \ldots, R_n), \mathsf{v}) \in \mathsf{Cl}_n$, and let $u \in U$.

Ab$_{\triangle_1}$. By Definition 53, $R_1(u) \neq \{u\}$; moreover, by Theorem 34, $\|\bot\|_\mathsf{v} = \emptyset$. Then, $(\mathcal{M}, u) \models \triangle_1 \bot$.

Dist$_{\square_i}$. By Theorem 34, $\|\varphi \wedge \psi\|_\mathsf{v} = \|\varphi\|_\mathsf{v} \cap \|\psi\|_\mathsf{v}$. Trivially, $R_i(u) \subseteq \|\varphi \wedge \psi\|_\mathsf{v}$ if and only if $R_i(u) \subseteq \|\varphi\|_\mathsf{v}$ and $R_i(u) \subseteq \|\psi\|_\mathsf{v}$. Then, $(\mathcal{M}, u) \models \square_i(\varphi \wedge \psi)$ if and only if $(\mathcal{M}, u) \models \square_i\varphi \wedge \square_i\psi$.

Dist$_{\triangle_i}$. $(\mathcal{M}, u) \models \triangle_i(\varphi \vee \psi)$ if and only if $R_i(u) \subseteq \|\varphi \vee \psi\|_\mathsf{v}$ and $R_i(u) \neq \{u\}$. By Proposition 14, $R_i(u) \cap \|\varphi \vee \psi\|_\mathsf{v} = R_i(u) \cap (\|\varphi\|_\mathsf{v} \cup \|\psi\|_\mathsf{v})$. Since $R_i(u) \cap (\|\varphi\|_\mathsf{v} \cup \|\psi\|_\mathsf{v}) = (R_i(u) \cap \|\varphi\|_\mathsf{v}) \cup (R_i(u) \cap \|\psi\|_\mathsf{v})$, we have that $R_i(u) \cap \|\varphi \vee \psi\|_\mathsf{v} = \emptyset$ if and only if $R_i(u) \cap \|\varphi\|_\mathsf{v} = \emptyset$ and $R_i(u) \cap \|\psi\|_\mathsf{v} = \emptyset$. Then, $(\mathcal{M}, u) \models \triangle_i\varphi$ and $(\mathcal{M}, u) \models \triangle_i\psi$.

P$_1$. Suppose that $(\mathcal{M}, u) \models \neg \bigcirc_i \varphi$. Then, $(\mathcal{M}, u) \not\models \varphi$ or $R_i(u) = \{u\}$. If $(\mathcal{M}, u) \not\models \varphi$, then $\neg\square_i\varphi$ is true at u in $\mathcal{M}$. If $R_i(u) = \{u\}$, then both $\neg\square_i\varphi$ and $\neg\triangle_i\varphi$ are true at u in $\mathcal{M}$.

P$_2$. If $(\mathcal{M}, u) \models \neg \bigcirc_i \varphi \wedge \varphi$, then $R_i(u) = \{u\}$. Consequently, both $\neg\square_i\varphi$ and $\neg\triangle_i\varphi$ are true at u in $\mathcal{M}$.

Axiomatic System of SO_n

The orthopaired modal logic SO_n is the smallest set of sentences that contains the instances of the axiom schemes of propositional logic and the instances of the axiom schemes of Definition 58, and that is closed under the inference rules of Definition 59.

Definition 58 (Axioms of SO_n).

Z$_{\square_1}$. $\square_1 \top$.
Def$_1$. $\square_i\varphi \equiv \triangle_i\neg\varphi$.
Def$_2$. $\bigcirc_i\varphi \equiv \bigcirc_i \top \wedge \varphi$.
K$_{\square_i}$. $\square_i(\varphi \rightarrow \psi) \rightarrow (\square_i\varphi \rightarrow \square_i\psi)$.
T$_{\square_i}$. $\square_i\varphi \rightarrow \varphi$.
B$_{\square_i}$. $\bigcirc_i\varphi \rightarrow \square_i\neg\triangle_i\varphi$.
4$_{\square_i}$. $\square_i\varphi \rightarrow \square_i\square_i\varphi$.
Eq. $\bigcirc_i \top \equiv \square_i \top$.
R1$_{\bigcirc_i}$. $\bigcirc_i\varphi \rightarrow (\square_j\varphi \rightarrow \square_i\varphi)$, *with* $j \leq i$.
R2$_{\bigcirc_i}$. $\square_i\varphi \rightarrow \bigcirc_i\varphi$.
Nst$_{\bigcirc_i}$. $\bigcirc_i\varphi \rightarrow \bigcirc_j\varphi$, *with* $0 < j \leq i$.

Definition 59 (Inference rules of SO_n).

MP. $\dfrac{\varphi, \varphi \to \psi}{\psi}$ *(Modus Ponens).*

$\square_i$**Mn.** $\dfrac{\varphi \to \psi}{\square_i\varphi \to \square_i\psi}$, *for each $i \in I$.*

We notice that Schema $\mathbf{Z}_{\square_1}$ ensures that all equivalence classes of the first accessibility relation of the SO_n-models are not singletons. Furthermore, fixed $i \in \{1, \ldots, n\}$, Schema $\mathbf{Def}_1$ allows us to obtain $\square_i$ through the modal operator $\triangle_i$; vice-versa, we also have that $\triangle_i\varphi \equiv \square_i\neg\varphi$. Trivially, $\mathbf{Def}_2$ is introduced to individuate the possible worlds of which the i-th equivalence class is a singleton. Schemas $\mathbf{K}_{\square_i}$, $\mathbf{T}_{\square_i}$ and $\mathbf{4}_{\square_i}$ are respectively the schemas $\mathbf{K}$, $\mathbf{T}$, and $\mathbf{4}$ that characterized $S4$ (see Definition 51), where $\square = \square_i$ and $\Diamond = \neg\triangle_i$. Thus, $\mathbf{K}_{\square_i}$ states that the operator $\square_i$ distributes over the implication $\to$; $\mathbf{T}_{\square_i}$ and $\mathbf{4}_{\square_i}$ express respectively that the accessibility relations of all SO_n-models are reflexive and transitive relations. On the other hand, taking $\square_i = \square$, $\mathbf{B}_{\square_i}$ is not equal to $\mathbf{B}$; they are different because the hypothesis of $\mathbf{B}_{\square_i}$ ($\bigcirc_i\varphi$) is stronger than the hypothesis of $\mathbf{B}$ (φ); so, we can say that each relation of each Kripke model of SO_n is a *strongly symmetric* relation. Furthermore, $\mathbf{B}_{\square_1}$ is equal to $\mathbf{B}$, since $\mathbf{Z}_{\square_1}$ requires that the condition $R_1(u) \neq \{u\}$ is satisfied, for each possible world u, and for each accessibility relation R_1 of the SO_n-models. Moreover, by Schema $\mathbf{B}_{\square_i}$, we can observe that the accessibility relations of the SO_n-models satisfy the *euclidean* property. Also, we have to stress that the modal operator $\triangle_i$ corresponds to the negation of the *possibility operator* $\Diamond$ of every modal logic. In addition, the schemas $\mathbf{Eq}$, $\mathbf{R1}_{\bigcirc_i}$, $\mathbf{R2}_{\bigcirc_i}$ and $\mathbf{Nst}_{\bigcirc_i}$ provide some connections between the operators $\bigcirc_i$ and $\square_i$. More precisely, $\mathbf{Eq}$ affirms that both $(\mathcal{M}, u) \models \bigcirc_i\top$ and $(\mathcal{M}, u) \models \square_i\top$ mean that $R_i(u)$ is not a singleton. $\mathbf{R1}_{\bigcirc_i}$ guarantees that each relation is finer than the previous one, namely $R_{i+1}(u) \subseteq R_i(u)$ for each $i > 1$. By $\mathbf{R2}_{\bigcirc_i}$, we have that $\bigcirc_i$ follows from $\square_i$. On the other side, $\mathbf{Nst}_{\bigcirc_i}$ states that if $R_i(u)$ is not a singleton, then all equivalence classes of the previous relations to R_i containing u are not singletons. Finally, we can notice that $\mathbf{T}_{\square_i}$ is obtained from $\mathbf{Def}_2$ and $\mathbf{R2}_{\bigcirc_i}$.

Remark 24. Suppose that Schema $\mathbf{Z}_{\square_1}$ is substituted by the schemas $\neg \bigcirc_1 \top$, $\ldots$, $\neg \bigcirc_n \top$. Then, each equivalence class of each accessibility relation of the SO_n-models is a singleton. In this case, it is clear that all axiom schemas of Definition 58 are trivially satisfied by each SO_n-model. Moreover, if $n = 1$, then the axiom schemas $\mathbf{Eq}$, $\mathbf{R1}_{\bigcirc_1}$, $\mathbf{R2}_{\bigcirc_1}$ and $\mathbf{Nst}_{\bigcirc_1}$ are trivially satisfied by each SO_1-model. Thus, the axiom schemas of our logic is obtain by adding $\mathbf{Z}_{\square_1}$ to those of modal logic $S5$ and by setting $\square_1 = \square$ and $\triangle_1 = \neg\Diamond$. Clearly, in this case, the Kripke models of SO_1 are all Kripke models of $S5$ such that the equivalence classes of their accessibility relations are not singletons.

Soundness and Completeness of SO_n

Next, we prove the soundness of SO_n system with respect to the class of models Cl_n already defined.

Theorem 35. *The axiom schemes of SO_n are valid in the class Cl_n, and the rules preserve the validity in this class.*

Proof. Let $\mathcal{M} = (U, (R_1, \ldots, R_n), v)$ be a model of Cl_n. Fixed $u \in U$, we prove that each instance of the axiom schemas of SO_n is true at u in $\mathcal{M}$.

$Z_{\square_1}$. By Definition 53, $R_1(u) \neq \{u\}$, and by Theorem 34, $||T||_v = U$. Then, $(\mathcal{M}, u) \models \square_1 T$.

Def$_1$. $(\mathcal{M}, u) \models \square_i \varphi$ if and only if $R_i(u) \subseteq ||\varphi||_v$ and $R_i(u) \neq \{u\}$, by Definition 55. Moreover, $R_i(u) \subseteq ||\varphi||_v$ if and only if $R_i(u) \cap (U \setminus ||\varphi||_v) = \emptyset$. However, by Theorem 34, $U \setminus ||\varphi||_v = ||\neg\varphi||_v$, So, it is clear that $(\mathcal{M}, u) \models \triangle_i \neg\varphi$.

Def$_2$. It is trivial.

$K_{\square_i}$. Suppose that $(\mathcal{M}, u) \models \square_i(\varphi \to \psi)$ and $(\mathcal{M}, u) \models \square_i\varphi$. Then, $R_i(u) \neq \{u\}$, $R_i(u) \subseteq ||\varphi \to \psi||_v$ and $R_i(u) \subseteq ||\varphi||_v$. By Theorem 34, $||\varphi \to \psi||_v = (U \setminus ||\varphi||_v) \cup ||\psi||_v$. Therefore, it is obvious that $R_i(u) \subseteq ||\psi||_v$ and so $(\mathcal{M}, u) \models \square_i\psi$.

$T_{\square_i}$. Suppose that $(\mathcal{M}, u) \models \square_i\varphi$. Then, $R_i(u) \subseteq ||\varphi||_v$. By Definition 53, R_i is reflexive and so $u \in R_i(u)$. Consequently, $(\mathcal{M}, u) \models \varphi$.

$B_{\square_i}$. Suppose that $(\mathcal{M}, u) \models \bigcirc_i\varphi$. Then, $(\mathcal{M}, u) \models \varphi$ and $R_i(u) \neq \{u\}$. Since $u \in ||\varphi||_v$, we have that

$$R_i(u) \cap ||\varphi||_v \neq \emptyset. \tag{38}$$

On the other hand,

$$||\triangle_i\varphi||_v = \{v \in U \mid R_i(v) \neq \{v\} \text{ and } R_i(v) \cap ||\varphi||_v = \emptyset\}. \tag{39}$$

By Eq. 38 and 39, $R_i(u) \cap ||\triangle_i\varphi||_v = \emptyset$. Therefore, $R_i(u) \subseteq U \setminus ||\triangle_i\varphi||_v$ and so $R_i(u) \subseteq ||\neg\triangle_i\varphi||_v$. Consequently, $(\mathcal{M}, u) \models \neg\triangle_i\varphi$.

$4_{\square_i}$. If $(\mathcal{M}, u) \models \square_i\varphi$, then $R_i(u) \subseteq ||\varphi||_v$ and $R_i(u) \neq \{u\}$. On the other hand, $||\square_i\varphi||_v = \cup_{u \in U}\{R_i(u) \mid R_i(u) \neq \{u\}\}$. Then, $R_i(u) \subseteq ||\square_i\varphi||_v$. Therefore, $(\mathcal{M}, u) \models \square_i\square_i\varphi$.

Eq. By Theorem 34, we have that $||T||_v = U$. Then, both $\square_i T$ and $\bigcirc_i T$ are true at u in $\mathcal{M}$ if and only if $R_i(u) \neq \{u\}$.

$R1_{\bigcirc_i}$. Suppose that $(\mathcal{M}, u) \models \bigcirc_i\varphi$ and $(\mathcal{M}, u) \models \square_j\varphi$. Then $R_j(u) \subseteq ||\varphi||_v$. Since $j \leq i$, $R_i(u) \subseteq R_j(u)$. Therefore, $R_i(u) \subseteq ||\varphi||_v$. Since $(\mathcal{M}, u) \models \bigcirc_i\varphi$, we also have that $R_i(u) \neq \{u\}$. Then, $(\mathcal{M}, u) \models \square_i\varphi$.

$R2_{\bigcirc_i}$. Trivially, $R_i(u) \subseteq ||\varphi||_v$ implies that $u \in ||\varphi||_v$, since R_i is a reflexive relation.

Nest$_{\bigcirc_i}$. Let $j \leq i$, if $R_i(u) \neq \{u\}$ then $R_j(u) \neq \{u\}$, since $R_i(u) \subseteq R_j(u)$; indeed $(\mathcal{M}, u) \models \bigcirc_i\varphi \to \bigcirc_j\varphi$.

We prove that if the hypothesis of the inference rules are true at u in $\mathcal{M}$, then the thesis is also true at u in $\mathcal{M}$.

MP. It is trivial.

$\square_i$**Mn.** By Theorem 34, if $(\mathcal{M}, u) \models \varphi \to \psi$, then $||\varphi||_v \subseteq ||\psi||_v$. If $(\mathcal{M}, u) \models \square_i\varphi$, then $R_i(u) \subseteq ||\varphi||_v$ and $R_i(u) \neq \{u\}$. Then, it is clear that $(\mathcal{M}, u) \models \psi$.

Corollary 3. *The SO_n system is sound with respect to the class of models Cl_n (i.e. if $\vdash_{SO_n} \varphi$ then $\models_{Cl_n} \varphi$, for each $\varphi \in$ Form).*

We usually write "$\vdash_{SO_n} \varphi$" to mean that φ is a theorem of SO_n, this is $\vdash_{SO_n} \varphi$.

In terms of theoremhood, we can characterize notions of deducibility and consistency.

Definition 60. *A formula φ of Form is deductible or derivable from a set of sentences Γ in the system SO_n, written $\Gamma \vdash_{SO_n} \varphi$, if we have*

$$\vdash_{SO_n} (\varphi_1 \wedge \ldots \wedge \varphi_n) \rightarrow \varphi,$$

where $\varphi_1, \ldots, \varphi_n$ are formulas in Γ.

Definition 61. *A subset Γ of Form is consistent in SO_n, written $Cons_{SO_n}\Gamma$, if and only if the falsum is not deducible from Γ in SO_n, namely $\Gamma \nvdash_{SO_n} \bot$.*

Thus, Γ is inconsistent in SO_n just when $\Gamma \vdash_{SO_n} \bot$.

Next, we define the idea of a canonical model for axiomatic system SO_n, and we prove some fundamental theorems about completeness. Before of introducing the concept of canonical model, we need to define the concept of maximality. Intuitively, a set of formulas is maximal if it is consistent, and it contains as many formulas as it can without becoming inconsistent. We write $Max_{SO_n}\Gamma$ to indicate that Γ is SO_n-maximal, and we formally give the definition as follows.

Definition 62. *Let $\Gamma \subseteq$ Form, $Max_{SO_n}\Gamma$ if and only if*

1. *$Cons_{SO_n}\Gamma$, and*
2. *for each $\varphi \in$ Form, if $Cons_{SO_n}(\Gamma \bigcup \{\varphi\})$ then $\varphi \in \Gamma$.*

Now, we have to recall Theorem 36, the Lindenbaum's lemma and its two corollaries (found in [29]) for the maximal consistent sets of logical systems. By *logical system*, we mean be any set which contains certain initial axioms and which is closed under certain rules of inference. Moreover, we write $Max_\Sigma\Gamma$ to denote that Γ is Σ-maximal.

Theorem 36. *Let Σ be a logical system, and let $Max_\Sigma\Gamma$, then*

1. *$\neg\varphi \in \Gamma$ iff $\varphi \notin \Gamma$;*
2. *$\varphi \wedge \psi \in \Gamma$ iff $\varphi \in \Gamma$ and $\psi \in \Gamma$;*
3. *$\varphi \rightarrow \psi \in \Gamma$ iff if $\varphi \in \Gamma$, then $\psi \in \Gamma$.*

Theorem 37 (Lindenbaum's lemma). *[90] Let Σ be a logical system. If $Con_\Sigma\Gamma$, then there is a $Max_\Sigma\Delta$ such that $\Gamma \subseteq \Delta$*

Corollary 4. *Let Σ be a logical system. Then,*

$$\vdash_\Sigma \varphi \, if \, and \, only \, if \, \varphi \in \Delta,$$

for every $Max_\Sigma\Delta$.

Corollary 5. *Let Σ be a logical system. Then, $\Gamma \vdash_\Sigma \varphi$ if and only if φ is an element of every $Max_\Sigma \Delta$ such that $\Gamma \subseteq \Delta$.*

In terms of maximality we can define what we shall call the proof set of a formula. Relative to system SO_n, the proof set of a formula φ (denoted by $|\varphi|_{SO_n}$) is the set of SO_n-maximal sets containing φ.

Definition 63. *Let $\varphi \in$ Form, we set*

$$|\varphi|_{SO_n} = \{Max_{SO_n}\Gamma \mid \varphi \in \Gamma\}.$$

We can state that a formula is deducible from a set of formulas if and only if it belongs to every maximal extension of the set.

Theorem 38. *Let $\Gamma \subseteq$ Form, and let $\varphi \in$ Form. Then,*

$$\Gamma \vdash_{SO_n} \varphi \text{ if and only if } \varphi \in \Delta \text{ for every } \Delta \in |\Gamma|_{SO_n}$$

Proof. It follows from the Lindenbaum's Lemma.

Definition 64. *The* canonical model *of SO_n is the structure*

$$\mathcal{M}^* = (U^*, (R_1^*, \ldots, R_n^*), v^*)$$

that satisfies the following conditions.

1. $U^* = \{\Gamma \subseteq$ Form $: Max_{SO_n}\Gamma\};$
2. *For every $w', w \in U^*, w' \in R_i^*(w)$ iff $\{\varphi \mid \square_i\varphi \in w\} \subseteq w'$ (namely, wR_i^*w' if and only if every formula φ belongs to w', whenever $\square_i\varphi$ belongs to w), and $\bigcirc_i\top \in w;$*
3. $v^*(p) = |p|_{SO_n}$, *for each $p \in$ Var.*

The canonical model has this property: if $w \in U^*$, then the formulas that are true at w in $\mathcal{M}^*$ are all and only the formulas belonging to w. More precisely, the following theorem holds.

Theorem 39. *Let $\mathcal{M}^*$ be the canonical model of SO_n. Then, for every possible world w of $\mathcal{M}^*$ and for every formula φ of Form,*

$$(\mathcal{M}^*, w) \models \varphi \quad \text{if and only if} \quad \varphi \in w. \tag{40}$$

Proof. In order to prove the statement 40, we use the induction on the length of the formulas. By the definition of v^* and by Definition 63, the propositional variables satisfy 40 (case base). Suppose that the statement 40 holds for the formulas φ and ψ (induction hypothesis), we intend to prove that $\neg\varphi$, $\varphi \wedge \psi$, $\square_i\varphi$ and $\bigcirc_i\varphi$ satisfy 40 for each $i \in \{1, \ldots, n\}$ (induction step).

$(\neg\varphi)$. By Definition 55, $(\mathcal{M}^*, w) \models \neg\varphi$ if and only if $(\mathcal{M}^*, w) \not\models \varphi$. By induction hypothesis, we have that $\varphi \notin w$, namely $\neg\varphi \notin w$, since Theorem 36 holds.

$(\varphi \wedge \psi)$. By Definition 55, $(\mathcal{M}^*, w) \models \varphi \wedge \psi$ if and only if $(\mathcal{M}^*, w) \models \varphi$ and $(\mathcal{M}^*, w) \models \psi$. By induction hypothesis, we have that $\varphi \in w$ and $\psi \in w$, namely $\varphi \wedge \psi \in w$, since Theorem 36 holds.

$(\Box_i \varphi)$. Suppose that $(\mathcal{M}^*, w) \models \Box_i \varphi$. Then, by Definition 55, $R_i^*(u) \subseteq ||\varphi||_{v^*}$. Therefore, if $w' \in U^*$ and $\{\psi \mid \Box_i \psi \in w\} \subseteq w'$, then $(\mathcal{M}^*, w') \models \varphi$. By induction hypothesis, $\varphi \in w'$. Then, $w' \vdash_{SO_n} \varphi$, by Theorem 36. By Corollary 5, $\{\psi \mid \Box_i \psi \in w\} \vdash_{SO_n} \varphi$. So, by Definition 60, $\vdash_{SO_n} \psi_1 \wedge \ldots \wedge \psi_n \to \varphi$. By rule $\Box_i$**Mn**, $\vdash \Box_i \psi_1 \wedge \ldots \wedge \Box_n \psi \to \Box_i \varphi \in w$. Moreover, by modus ponens, $\Box_i \varphi \in w$.

Let $\Box_i \varphi \in w$, we intend to prove that $R_i^*(w) \subseteq ||\varphi||_{v^*}$ and $R_i^*(w) \neq \{w\}$. Firstly, suppose that $w' \in R_i^*(w)$, then $\{\psi \mid \Box_i \psi \in w\} \subseteq w'$. Thus, $\varphi \in w$, since $\Box_i \varphi \in w$. Then, $w \in ||\varphi||_{v^*}$.

By schema **R2**$_{\bigcirc_i}$, $\Box_i \varphi \to \bigcirc_i \varphi \in w$ and by hypothesis $\bigcirc_i \varphi \in w$. Then, by modus ponens, $\bigcirc_i \varphi \in w$, and so $R_i^*(w) \neq \{w\}$.

$(\bigcirc_i \varphi)$. $(\mathcal{M}^*, w) \models \bigcirc_i \varphi$ if and only if $(\mathcal{M}^*, w) \models \varphi$ and $(\mathcal{M}^*, w) \models \bigcirc_i \top$. Then, by induction hypothesis, $\varphi \in w$ and by definition of canonical model $\bigcirc_i \top \in w$. They are equivalent to say that $\varphi \wedge \bigcirc_i \top \in w$, namely $\bigcirc_i \varphi \in w$.

Theorem 40. *The canonical model* $\mathcal{M}^* = (U^*, (R_1^*, \ldots, R_n^*), v^*)$ *is a Kripke model of* SO_n.

Proof. (R_i^* *is reflexive*). Let $w \in U^*$ such that $\Box_i \varphi \in w$. By the schema $\mathbf{T}_i$ of Definition 58 ($\Box_i \varphi \to \varphi$) and by Theorem 36, we have that $\varphi \in w$. Then, $w R_i^* w$.

(R_i^* *is symmetric*). Suppose that $w R_i^* w'$, with $w \neq w'$. Therefore, $R_i^*(w) \neq \{w\}$ (consequently, $\bigcirc_i \top \in w$), and $\{\varphi \in \mathsf{Form} \mid \Box_i \varphi \in w\} \subseteq w'$. Let $\varphi \in \mathsf{Form}$ such that $\Box_i \varphi \in w'$. We have to prove that $\varphi \in w$. If $\varphi \notin w$, then $\neg \varphi \in w$. By Schema **Def**$_2$, $\bigcirc_i \neg \varphi \in w$. By Schema $\mathbf{B}_{\Box_i}$ and by Theorem 36, $\Box_i \neg \triangle_i \neg \varphi \in w$. By hypothesis, $\neg \triangle_i \neg \varphi \in w'$, namely $\triangle_i \neg \varphi \notin w'$. By Schema **Def**$_1$, $\Box_i \varphi \notin w'$. The latter is an absurd, since we have assumed that $\Box_i \varphi \in w'$.

(R_i^* *is transitive*). Suppose that $w R_i^* w'$ and $w' R_i^* w''$. Consequently, $\{\varphi \in \mathsf{Form} \mid \Box_i \varphi \in w\} \subseteq w'$ and $\{\varphi \in \mathsf{Form} \mid \Box_i \varphi \in w'\} \subseteq w''$. Let $\varphi \in \mathsf{Form}$ such that $\Box_i \varphi \in w$, we have to prove that $\varphi \in w''$. By schema $\mathbf{4}_{\Box_i}$ of Definition 58 and Theorem 36, if $\Box_i \varphi \in w$, then $\Box_i \Box_i \varphi \in w$. By hypothesis, $\Box_i \varphi \in w'$ and so $\varphi \in w''$.

($R_1^*(w) \neq \{w\}$, *for each* $w \in U^*$). We consider $w \in U^*$. By Definition 64, $\bigcirc_i \top \in w$. Then, $\bigcirc_1 \top \in w$ and so $R_1^*(w) \neq \{w\}$.

($R_{i+1}^*(w) \subseteq R_i^*(w)$, *for each* $i \in \{1, \ldots, n-1\}$). Let $w' \in R_{i+1}^*(w)$ and $\varphi \in \mathsf{Form}$ such that $\Box_i \varphi \in w$. We have to prove that $\varphi \in w'$. By Schema $\mathbf{T}_{\Box_i}$, the hypothesis that $\Box_i \varphi \in w$ implies that $\varphi \in w$. By Definition 64, $\bigcirc_{i+1} \top \in w$. Consequently, $\bigcirc_i \top \wedge \varphi \in w$ and so $\bigcirc_{i+1} \varphi \in w$.

Since $R_{i+1}(w) \neq \{w\}$, then $\bigcirc_{i+1} \top \in w$. By schema $\mathbf{R1}_{\bigcirc_i}$ of Definition 58 and Theorem 36, $\Box_{i+1} \varphi \in w$. Then, $\varphi \in w'$.

5.3 Orthopaired Kripke Model and Sequences of Orthopairs

In this section, we intend to investigate on the connections between sequences of orhopairs and modal logic SO_n. The relationships between rough sets and modal logic have been explored by several authors (see [69] for a list); the most studied one concerns Pawlak set theory and modal logic $S5$ [8,90]. As we have already said in Sect. 5.1, the intuitión behind this link is that the lower and the upper approximations can be regarded as two unary operations on subsets of the given universe. Thus, let U be a universe, and let R be an equivalence relation on U, the *Pawlak rough set algebra* $(2^U, \cap, \cup, \neg, \mathcal{L}_R, \mathcal{U}_R, \emptyset, U)$ is an extension of the Boolean algebra $(2^U, \cap, \cup, \neg, \emptyset, U)$ (see Remark 3), and then it may be interpreted in terms of the notions of topological space and topological Boolean algebra [8].

Firstly, we prove that there is a one-to-one correspondence between refinement sequences and Kripke frames of SO_n.

Without loss of generality, let be $\mathcal{C} = (C_1, \ldots, C_n)$ a refinement sequence of U, we suppose that its first partition C_1 covers U.

Let n be a positive integer. We denote the set of all refinement sequences made of n partial partitions with RS_n, and the set of all Kripke frames of SO_n made of n equivalence relations with F_n.

Definition 65. *We consider the map* $f : \mathsf{RS}_n \mapsto \mathsf{F}_n$, *where, let* $\mathcal{C} \in \mathsf{RS}_n$, $f(\mathcal{C}) = (U, (R_1, \ldots, R_n)) \in \mathsf{F}_n$ *such that*

1. $U = \cup\{b \mid b \in C_1\}$,
2. uR_iv *if and only if* $u = v$ *or* $\{u, v\} \subseteq b$, *with* $b \in C_i$; *for each* $u, v \in U$ *and* $i \in \{1, \ldots, n\}$.

Clearly, let $(U, (R_1, \ldots, R_n)) \in \mathsf{F}_n$, *then* $f^{-1}((U, (R_1, \ldots, R_n)))$ *is the refinement sequence* $(C_1, \ldots, C_n)$ *of* U *such that*

$$C_i = \{R_i(u) \mid u \in U \text{ and } R_i(u) \neq \{u\}\}.$$

Proposition 16. *The function* f *is a bijection.*

Proof. It is trivial.

Let $\mathcal{C} \in \mathsf{RS}_n$, we denote $f(\mathcal{C})$ with $\mathcal{F}_\mathcal{C}$. vice versa, let $\mathcal{F} \in \mathsf{F}_n$, we denote $f^{-1}(\mathcal{C})$ with $\mathcal{C}_\mathcal{F}$.

Example 40. Let $\mathcal{C} = (C_1 = \{\{a, b, c\}, \{d, e\}\}, C_2 = \{\{a, b\}\})$ be a refinement sequence of $\{a, b, c, d, e\}$. Then, $f(\mathcal{C}) = (\{a, b, c, d, e\}, (R_1, R_2))$, where

1. $R_1 = \{(a, b), (b, a), (a, c), (c, a), (b, c), (c, b), (d, e), (e, d)\} \cup \{(u, u) \mid u \in \{a, b, c, d, e\}\}$ and
2. $R_2 = \{(a, b), (b, a)\} \cup \{(u, u) \mid u \in \{a, b, c, d, e\}\}$.

Vice versa, $f^{-1}((\{a, b, c, d, e\}, (R_1, R_2)) = \mathcal{C}$.

Therefore, function f allows us to identify Kripke frames of SO_n logic having U as universe with refinement sequences of partial partitions of U. Furthermore, we can observe that Kripke frame $(U, (R_1, \ldots, R_n))$ corresponds to the sequences of Pawlak spaces $((U, R_1), \ldots, (U, R_n))$.

The following theorem establishes a connection between sequences of orthopairs and the modal operators $(\Box_1, \ldots, \Box_n)$ and $(\triangle_1, \ldots, \triangle_n)$ of SO_n logic.

Theorem 41. *Let* $\mathcal{F} = (U, (R_1, \ldots, R_n)) \in F_n$ *and* $(\mathcal{F}, v) \in C_n$. *Then,* $(||\Box_i\varphi||_v, ||\triangle_i\varphi||_v)$ *is the orthopair of* $||\varphi||_v$ *generated by the i-th partition of* $C_\mathcal{F}$. *Therefore,*

$$(\, (||\Box_1\varphi||_v, ||\triangle_1\varphi||_v), \ldots, (||\Box_n\varphi||_v, ||\triangle_n\varphi||_v) \,)$$

is the sequence of orthopairs of $||\varphi||_v$ *generated by* $C_\mathcal{F}$.

Proof. The proof follows by Definition 55 (point 4), Proposition 14 (point 2) and Definition 65.

Example 41. Let $\mathcal{F}$ be the Kripke frame of Example 40. We suppose that $\mathsf{Var} = \{p, q\}$ and we consider the Kripke model $(\mathcal{F}, v)$ such that $v(p) = \{a, b, c\}$, and $v(q) = \{a, b, d\}$. Then, $||p \wedge q||_v = \{a, b\}$. Moreover,

$$(\, (||\Box_1 \, p \wedge q||_v, ||\triangle_1 \, p \wedge q||_v), (||\Box_2 \, p \wedge q||_v, ||\triangle_2 \, p \wedge q||_v) \,) = ((\emptyset, \{d, e\}), (\{a, b\}, \emptyset)),$$

that is the sequence $\mathcal{O}_{C_\mathcal{F}}(||\varphi||_v)$.

Trivially, let v and v' be two evaluation functions such that $v \neq v'$, then the sequence $\mathcal{O}_{C_\mathcal{F}}(||\varphi||_v)$ is not usually equal to $\mathcal{O}_{C_\mathcal{F}}(||\varphi||_{v'})$.

Example 42. We consider the Kripke model $(\mathcal{F}, v)$ of Example 41 and the Kripke model $(\mathcal{F}, v')$ such that $v'(p) = \{a, d, e\}$ and $v' = \{d, e\}$. Then, $||p \wedge q||_{v'} = \{d, e\}$ and so $\mathcal{O}_{C_\mathcal{F}}(||\varphi||_{v'}) = ((\{d, e\}, \{a, b, c\}), (\emptyset, \{a, b\}))$, that is not equal to the sequence $\mathcal{O}_{C_\mathcal{F}}(||\varphi||_v)$.

Given a Kripke model $(\mathcal{F}, v)$ of SO_n and two formulas φ and ψ, there exists a formula obtained from φ and ψ that is valid in $(\mathcal{F}_C, v)$ if and only if the sequences of orthopairs of $||\varphi||_v$ and $||\psi||_v$ generated by $C_\mathcal{F}$ are equal to each other. More precisely, the following theorem holds.

Theorem 42. *Let* $\varphi, \psi \in \mathsf{Form}$ *and* $(\mathcal{F}, v) \in C_n$, *then*

$$\mathcal{O}_{C_\mathcal{F}}(||\varphi||_v) = \mathcal{O}_{C_\mathcal{F}}(||\psi||_v) \quad \textit{iff} \quad \models^{(\mathcal{F}, v)} \bigwedge_{i=1}^{n} (\Box_i\varphi \equiv \Box_i\psi) \wedge (\triangle_i\varphi \equiv \triangle_i\psi).$$

Proof. Notice that, by Proposition 14, $\models^{(\mathcal{F}, v)} (\Box_i\varphi \equiv \Box_i\psi)$ if and only if $||\Box_i\varphi||_v = ||\Box_i\psi||_v$, for each $i \in \{1, \ldots, n\}$. Then, the thesis clearly follows.

The following remark shows that the modal operators $\bigcirc_1, \ldots, \bigcirc_n$ allow us to understand what are the elements that are lost during the refinement process.

Remark 25. Let $\mathcal{C} = (C_1, \ldots, C_n)$ be a refinement sequence of U, through the modal operator $\bigcirc_i$, it is easy to check whether an element of U belongs to a block of the C_i; thus, let $u \in U$ and $i \in \{1, \ldots, n\}$, we have that

$$u \in \bigcup_{b \in C_i} b \text{ if and only if } ((\mathcal{F}_C, \mathsf{v}), u) \models \bigcirc_i \top,$$

for each evaluation function v.

Furthermore, we can express the property of safety of refinement sequences of partial partitions by using the modal operators $(\square_1, \ldots, \square_n)$ and $(\bigcirc_1, \ldots, \bigcirc_n)$ (the meaning of safe refinement sequence is given in Definition 44).

Theorem 43. *Let $\mathcal{C}$ be a refinement sequence of U. Then, $\mathcal{C}$ is safe if and only if the following condition holds:*

"if $(\mathcal{M}, u) \models \square_i \varphi$ and $i \leq j$, then $R_i(u) = R_j(u)$ or there exists $u' \in R_i(u)$ such that $(\mathcal{M}, u') \models \neg \bigcirc_j \varphi$" (or "if $(\mathcal{M}, u) \models \triangle_i \varphi$, then $R_i(u) = R_j(u)$ or there exists $u' \in R_i(u)$ such that $(\mathcal{M}, u') \models \neg \bigcirc_j \neg \varphi$"), for each $\varphi \in \mathsf{Form}$, $\mathcal{M} = (\mathcal{F}_C, \mathsf{v}) \in \mathcal{C}_n$, $u \in U$ and $i \in \{1, \ldots, n-1\}$.

Proof. ($\Rightarrow$). We suppose that $(\mathcal{M}, u) \models \square_i \varphi$ and $R_i(u) \neq R_j(u)$, with $j > i$. We notice that $R_i(u) \in C_i$, since $R_i(u) \neq \{u\}$. On the other hand, $R_i(u) \notin C_j$, since $R_i(u) \neq R_j(u)$. So, we call $N_1, \ldots, N_m$ the blocks of C_j that are included in $R_i(u)$. By Remark 13, the successors $N_1', \ldots, N_l'$ of $R_i(u)$ belong to C_k, where $i < k \leq j$. Since $\mathcal{C}$ is safe, there exists $u' \in R_i(u)$ such that $u' \notin N_1' \cup \ldots \cup N_l'$ (see Definition 44). Then, $u' \notin \cup\{b \mid b \in C_k\}$ and so $u' \notin \cup\{b \mid b \in C_j\}$. Then, $R_j(u') = \{u'\}$ and this means that $(\mathcal{M}, u') \models \neg \bigcirc_j \varphi$.

($\Leftarrow$). Let $N \in P_\mathcal{C}$. Suppose that $N_1, \ldots, N_m$ are the successors of N in $P_\mathcal{C}$. We intend to prove that $N_1 \cup \ldots \cup N_m \subset N$. We consider the evaluation function v such that $\mathsf{v}(p) = N$, where $p \in \mathsf{Var}$. If $N \in C_i$, then there exists $u \in U$ such that $N = R_i(u)$. Trivially, we have that $((\mathcal{F}_C, \mathsf{v}), u) \models \square_i p$. We notice that $N_1, \ldots, N_m$ belong to C_j, with $j > i$. By hypothesis, there exists $u' \in R_i(u) (= N)$ such that $((\mathcal{F}_C, \mathsf{v}), u) \models \neg \bigcirc_i p$. Then $R_j(u') \neq \{u'\}$ and so u' does not belong to some nodes of C_j. Therefore, $u' \in N$, but $u' \notin N_1 \cup \ldots \cup N_m$ and so by Definition 44, $\mathcal{C}$ is safe.

As a consequence of the previous theorem, we can express the results of Corollary 2 for refinement sequences of partial partitions by using the modal operators $(\square_1, \ldots, \square_n)$ and $(\bigcirc_1, \ldots, \bigcirc_n)$ as follows.

Theorem 44. *Let $\mathcal{C} = (C_1, \ldots, C_n)$ be a refinement sequence of U. Then, $\mathbb{K}_\mathcal{C}^3$ is a finite IUML-algebra if and only if the following condition holds:*

"if $(\mathcal{M}, u) \models \square_i \varphi$ and $i \leq j$, then $R_i(u) = R_j(u)$ or there exists $u' \in R_i(u)$ such that $(\mathcal{M}, u') \models \neg \bigcirc_j \varphi$" (or "if $(\mathcal{M}, u) \models \triangle_i \varphi$, then $R_i(u) = R_j(u)$ or there exists $u' \in R_i(u)$ such that $(\mathcal{M}, u') \models \neg \bigcirc_j \neg \varphi$"), for each $\varphi \in \mathsf{Form}$, $\mathcal{M} = (\mathcal{F}_C, \mathsf{v}) \in \mathcal{C}_n$, $u \in U$ and $i \in \{1, \ldots, n-1\}$.

However, by using modal logic, we can also express the results obtained for the structures $\mathbb{K}_\mathcal{C}^1$, $\mathbb{K}_\mathcal{C}^2$ and $\mathbb{K}_\mathcal{C}^4$ in Sect. 4, but only when $\mathcal{C}$ is a refinement sequence of partial partitions (we recall that such algebraic structures, except $\mathbb{K}_C^3$, are generated by refinement sequences of partial coverings of the given universe).

At the end of this section, we intend to include the operations $\curlywedge$, $\curlyvee$, $\hookrightarrow_1$, $\odot_2$, $\hookrightarrow_2$, $\odot_3$ and $\hookrightarrow_3$ defined on sequences of orthopairs of partial partitions (see Definition 50) in our modal logic.[2]

Theorem 45. *Let* $\varphi, \psi \in$ *Form and* $(\mathcal{F}, v) \in Cl_n$. *If* $C_\mathcal{F}$ *is safe, then*

$$\mathcal{O}_{C_\mathcal{F}}(||\varphi||_v) \curlywedge \mathcal{O}_{C_\mathcal{F}}(||\psi||_v) = ((A_1, B_1), \ldots, (A_n, B_n)),$$

where $(A_i, B_i) = (||\Box_i\varphi \wedge \Box_i\psi||_v{}^3, ||\triangle_i\varphi \vee \triangle_i\psi||_v)$, *for each* $i \in \{1, \ldots, n\}$, *and*

$$\mathcal{O}_{C_\mathcal{F}}(||\varphi||_v) \curlyvee \mathcal{O}_{C_\mathcal{F}}(||\psi||_v) = ((C_1, D_1), \ldots, (C_n, D_n)),$$

where $(C_i, D_i) = (||\Box_i\varphi \vee \Box_i\psi||_v, ||\triangle_i\varphi \wedge \triangle_i\psi||_v{}^4)$, *for each* $i \in \{1, \ldots, n\}$.

Proof. By Theorem 30, $\mathcal{O}_{C_\mathcal{F}}(||\varphi||_v) \curlywedge \mathcal{O}_{C_\mathcal{F}}(||\psi||_v) = ((A_1, B_1), \ldots, (A_n, B_n))$, such that $(A_i, B_i) = (\mathcal{L}_i(||\varphi||_v), \mathcal{E}_i(||\varphi||_v)) \wedge_K (\mathcal{L}_i(||\psi||_v), \mathcal{E}_i(||\psi||_v)) = (\mathcal{L}_i(||\varphi||_v) \cap \mathcal{L}_i(||\psi||_v), \mathcal{E}_i(||\varphi||_v) \cup \mathcal{E}_i(||\psi||_v))$. Suppose that $u \in U$, we have that $u \in \mathcal{L}_i(||\varphi||_v) \cap \mathcal{L}_i(||\psi||_v)$ if and only if $R_i(u) \subseteq ||\varphi||_v$, $R_i(u) \subseteq ||\psi||_v$ and $R_i(u) \neq \{u\}$, namely $u \models \Box_i\varphi \wedge \Box_i\psi$. Moreover, $u \in \mathcal{E}_i(||\varphi||_v) \cup \mathcal{E}_i(||\psi||_v)$ if and only if $R_i(u) \neq \{u\}$ and either $R_i(u) \subseteq ||\varphi||_v$ or $R_i(u) \subseteq ||\psi||_v$, namely $u \models \triangle_i\varphi \vee \triangle_i\psi$. The proof for the operation $\curlyvee$ is analogous.

Definition 66. *Let* $\varphi, \psi \in$ *Form, we recursively define the sequences of formulas* $(\alpha_1(\varphi, \psi), \ldots, \alpha_n(\varphi, \psi))$, $(\beta_1(\varphi, \psi), \ldots, \beta_n(\varphi, \psi))$, $(\gamma_1(\varphi, \psi), \ldots, \gamma_n(\varphi, \psi))$, $(\delta_1(\varphi, \psi), \ldots, \delta_n(\varphi, \psi))$, $(\epsilon_1(\varphi, \psi), \ldots, \epsilon_n(\varphi, \psi))$, $(\zeta_1(\varphi, \psi), \ldots, \zeta_n(\varphi, \psi))$, $(\eta_1(\varphi, \psi), \ldots, \eta_n(\varphi, \psi))$, $(\theta_1(\varphi, \psi), \ldots, \theta_n(\varphi, \psi))$, $(\iota_1(\varphi, \psi), \ldots, \iota_n(\varphi, \psi))$ *and* $(\kappa_1(\varphi, \psi), \ldots, \kappa_n(\varphi, \psi))$ *as follows.*

- $\alpha_n(\varphi, \psi) := \neg\Box_n\varphi \vee \Box_n\psi$;
- $\alpha_i(\varphi, \psi) := (\neg\Box_i\varphi \vee \Box_i\psi) \wedge \neg\alpha_{i+1}(\varphi, \psi)$, with $i \in \{1, \ldots, n-1\}$;
- $\beta_i(\varphi, \psi) := \Box_i\varphi \wedge \Box_i\psi$, with $i \in \{1, \ldots, n\}$;
- $\gamma_i(\varphi, \psi) := \Box_i\varphi \wedge \Box_i\psi$, with $i \in \{1, \ldots, n\}$;
- $\delta_n(\varphi, \psi) := \lambda_n(\varphi, \psi)$;
- $\delta_i(\varphi, \psi) := \lambda_i(\varphi, \psi) \wedge \neg\delta_{i+1}(\varphi, \psi)$, with $i \in \{1, \ldots, n-1\}$, where

$$\lambda_i(\varphi, \psi) := \neg(\Box_i\varphi \vee \Box_i\psi) \vee \Box_i\varphi \vee \Box_i\psi.$$

- $\epsilon_n(\varphi, \psi) := \mu_n(\varphi, \psi)$;
- $\epsilon_i(\varphi, \psi) := \mu_i(\varphi, \psi) \wedge \neg\epsilon_{i+1}(\varphi, \psi)$, with $i \in \{1, \ldots, n-1\}$, where

$$\mu_i(\varphi, \psi) := (\neg\Box_i\varphi \vee \Box_i\psi) \wedge (\triangle_i\varphi \vee \neg\triangle_i\psi).$$

- $\zeta_i(\varphi, \psi) := \Box_i\varphi \wedge \triangle_i\psi$, with $i \in \{1, \ldots, n\}$;

---

[2] We exclude the operations $\odot_4$ and $\hookrightarrow_4$, since they can not be obtained starting from operations between the orthopairs.

[3] By Preposition 15, $\Box_i\varphi \wedge \Box_i\psi = \Box_i(\varphi \wedge \psi)$.

[4] By Preposition 15, $\triangle_i\varphi \wedge \triangle_i\psi = \triangle_i(\varphi \vee \psi)$.

- $\eta_1(\varphi, \psi) := \nu_1(\varphi, \psi);$
- $\eta_i(\varphi, \psi) := \nu_i(\varphi, \psi) \vee \Box_i \eta_{i-1}(\varphi, \psi),$ with $i > 1$ and

$$\nu_i(\varphi, \psi) = (\Box_i \varphi \wedge \neg \triangle_i \psi) \vee (\Box_i \psi \wedge \neg \triangle_i \varphi).$$

- $\theta_i(\varphi, \psi) := (\triangle_i \varphi \vee \triangle_i \psi) \wedge \neg \eta_i(\varphi, \psi),$ with $i \in \{1, \ldots, n\}$;[5]
- $\iota_i(\varphi, \psi) := ((\neg \Box_i \varphi \vee \Box_i \psi) \wedge (\triangle_i \varphi \vee \neg \triangle_i \psi)) \wedge \kappa_i(\varphi, \psi),$ for each $i \in \{1, \ldots, n\}$;
- $\kappa_1(\varphi, \psi) := \Box_1 \varphi \wedge \triangle_1 \psi;$
- $\kappa_i(\varphi, \psi) := (\Box_i \varphi \wedge \triangle_i \psi) \vee \kappa_{i-1}(\varphi, \psi),$ for each $i \in \{2, \ldots, n\}.$

Theorem 46. *Let $\varphi, \psi \in$ Form and $(\mathcal{F}, v) \in \mathcal{C}_n$. If $\mathcal{C}_{\mathcal{F}}$ is safe, then*

$$\mathcal{O}_{\mathcal{C}_{\mathcal{F}}}(||\varphi||_v) \hookrightarrow_1 \mathcal{O}_{\mathcal{C}_{\mathcal{F}}}(||\psi||_v) = ((E_1, F_1), \ldots, (E_n, F_n)),$$

where $(E_i, F_i) = (||\alpha_i(\varphi, \psi)||_v, ||\beta_i(\varphi, \psi)||_v),$ for each $i \in \{1, \ldots, n\}.$

$$\mathcal{O}_{\mathcal{C}_{\mathcal{F}}}(||\varphi||_v) \odot_2 \mathcal{O}_{\mathcal{C}_{\mathcal{F}}}(||\psi||_v) = ((G_1, H_1), \ldots, (G_n, H_n)),$$

where $(G_i, H_i) = (||\gamma_i(\varphi, \psi)||_v, ||\delta_i(\varphi, \psi)||_v),$ for each $i \in \{1, \ldots, n\}.$

$$\mathcal{O}_{\mathcal{C}_{\mathcal{F}}}(||\varphi||_v) \hookrightarrow_2 \mathcal{O}_{\mathcal{C}_{\mathcal{F}}}(||\psi||_v) = ((I_1, J_1), \ldots, (I_n, J_n)),$$

where $(I_i, J_i) = (||\epsilon_i(\varphi, \psi)||_v, ||\zeta_i(\varphi, \psi)||_v),$ for each $i \in \{1, \ldots, n\}.$

$$\mathcal{O}_{\mathcal{C}_{\mathcal{F}}}(||\varphi||_v) \odot_3 \mathcal{O}_{\mathcal{C}_{\mathcal{F}}}(||\psi||_v) = ((K_1, L_1), \ldots, (K_n, L_n)),$$

where $(K_i, L_i) = (||\eta_i(\varphi, \psi)||_v, ||\theta_i(\varphi, \psi)||_v),$ for each $i \in \{1, \ldots, n\}.$

$$\mathcal{O}_{\mathcal{C}_{\mathcal{F}}}(||\varphi||_v) \hookrightarrow_3 \mathcal{O}_{\mathcal{C}_{\mathcal{F}}}(||\psi||_v) = ((M_1, N_1), \ldots, (M_n, N_n)),$$

where $(M_i, N_i) = (||\iota_i(\varphi, \psi)||_v, ||\kappa_i(\varphi, \psi)||_v),$ for each $i \in \{1, \ldots, n\}.$

Proof. We only provide the proof for the operation $\odot_3$, since those of the remaining cases are analogous.

Let $u \in U$,

$$((\mathcal{F}, v), u) \models \nu_i \text{ iff } ((\mathcal{F}, v), u) \models \Box_1 \varphi \wedge \neg \triangle_1 \psi \text{ or } ((\mathcal{F}, v), u) \models \Box_1 \psi \wedge \neg \triangle_1 \varphi,$$

that is

- $R_i(u) \subseteq ||\varphi||_v, R_i(u) \neq \{u\}$ and $R_i(u) \cap ||\psi||_v \neq \emptyset,$ or
- $R_i(u) \subseteq ||\psi||_v, R_i(u) \neq \{u\}$ and $R_i(u) \cap ||\varphi||_v \neq \emptyset.$

Consequently, we obtain that

$$((\mathcal{F}, v), u) \models \nu_i \text{ if and only if } u \in (\mathcal{L}_i(\varphi) \setminus \mathcal{E}_i(\psi)) \cup (\mathcal{L}_i(\psi) \setminus \mathcal{E}_i(\varphi)).$$

---

[5] Observe that this expression is equivalent to $(\Box_i \varphi \setminus \triangle_i \psi \wedge \Box_i \psi \setminus \triangle_i \varphi).$

Trivially, we can observe that

$$((\mathcal{F}, \mathsf{v}), u) \models \Box_i \eta_{i-1}(\varphi, \psi) \text{ iff } R_i(u) \subseteq ||\eta_{i-1}(\varphi, \psi)||_{\mathsf{v}} \text{ and } R_i(u) \neq \{u\},$$

and

$$((\mathcal{F}, \mathsf{v}), u) \models \theta_i(\varphi, \psi) \text{ iff } u \in \mathcal{E}_i(||\varphi||_{\mathsf{v}}) \cup \mathcal{E}_i(||\psi||_{\mathsf{v}}).$$

By Theorem 33 and by $(X, Y) *_{\mathcal{S}} (Z, W) = ((X \setminus W) \cup (Z \setminus Y), Y \cup W)$ (see Definition 11), we obtain that the i-th component of the sequence $\mathcal{O}_{\mathcal{C}_{\mathcal{F}}}(||\varphi||_{\mathsf{v}}) \odot_3 \mathcal{O}_{\mathcal{C}_{\mathcal{F}}}(||\psi||_{\mathsf{v}})$ is $(||\eta_i(\varphi, \psi)||_{\mathsf{v}}, ||\theta_i(\varphi, \psi)||_{\mathsf{v}})$.

5.4   Epistemic Logic SO_n

In this section, we employ modal logic SO_n and describe the knowledge of an agent during a sequence $(t_1, \ldots, t_n)$ of consecutive instants of time. Also, we intend to establish whether the given agent is *interested in knowing* the truth or falsity of the sentences at every instant of $(t_1, \ldots, t_n)$. In detail, we represent situations in which, given an agent $\mathcal{A}$ and a sequence $(t_1, \ldots, t_n)$,

- $\mathcal{A}$ knows more information at time t_{i+1} than at time t_i, and
- $\mathcal{A}$ is less interested in knowing at time t_{i+1} than at time t_i.

Example 43. We suppose that a restaurant owner manages seven restaurants in seven Italian cities: Viterbo, Rieti, Rome, Latina, Frosinone, Potenza and Matera. He needs to know the weather report for tomorrow in order to decide whether to set up the gardens of his restaurants. At time t_1, he knows by speaking with a friend, that it is cloudy throughout Lazio, consequently it is cloudy in Viterbo, Rieti, Rome, Latina and Frosinone, but he does not know the weather in Potenza and Matera. At time $t_2 > t_1$, he finds the weather report on Internet, and he knows that it is cloudy with a chance of rain in Viterbo and Rieti, it is cloudy without rain in Latina and Frosinone, and it is sunny in Matera and Potenza. Since he decides that the restaurant will be close in Rome, he does not look for any information about the weather there. This situation is synthesized in Table 13, where C, C + R, C−R and S denote respectively cloudy, cloudy with rain, cloudy without rain and sunny. Moreover, the symbol × means that the restaurant owner excludes Rome from all cities he is interested in knowing the weather, and ? means that he has not information about the respective cities.

Table 13. Information about the weather

	Viterbo	Rieti	Rome	Latina	Frosinone	Potenza	Matera
t_1	C	C	C	C	C	?	?
t_2	C + R	C + R	×	C − R	C − R	S	S

Table 13 corresponds to a refinement sequence made of the partial partitions C_1 and C_2, where

- $C_1 = \{\{\text{Viterbo, Rieti, Rome, Latina, Frosinone}\}, \{\text{Potenza, Matera}\}\}$ and
- $C_2 = \{\{\text{Viterbo, Rieti}\}, \{\text{Latina, Frosinone}\}, \{\text{Potenza, Matera}\}\}$.

Then, each block of C_1 is the set of the cities that, at time t_1, have the same weather with respect to the knowledge of the restaurant owner, and C_2 is made of the cities that, at time t_2, have the same weather with respect to the knowledge of the restaurant owner. We underline that the owner has more information about the weather in cities of Table 13 at time t_2 than at time t_1 (for example, at time t_1, he knows that it is cloudy in Viterbo, and at time t_2, he knows that it is cloudy with rain there); however, he is interested in knowing the weather in less cities at time t_2 than at time t_1 (precisely, at time t_2, he excludes Rome).

The finite sequences $(\Box_1, \ldots, \Box_n)$ and $(\bigcirc_1, \ldots, \bigcirc_n)$ of SO_n correspond to a sequence $(t_1, \ldots, t_n)$ made of consecutive instants of time, or of consecutive time intervals. In addition, let $i \in \{1, \ldots, n\}$, the interpretation of the modality $\Box_i$ with respect to an orthopaired Kripke model allows us to represent the knowledge of an agent at time t_i. Furthermore, the semantic interpretation of the modality $\bigcirc_i$ establishes whether the agent is interested in knowing the truth or falsity of a sentence at each initial possible world at time t_i. Thus, each Kripke frame $\mathcal{M} = (U, (R_1, \ldots, R_n))$ of SO_n is associated with a pair $(\mathcal{A}, (t_1, \ldots, t_n))$ such that $\mathcal{A}$ is an agent, and $(t_1, \ldots, t_n)$ is a sequence of successive instants of time. More precisely, let $u \in U, i \in \{1, \ldots, n\}$ and $\varphi \in \mathsf{Form}$, if $u \models \Box_i \varphi$, we can say that

"at time t_i, the agent $\mathcal{A}$ knows that φ is true at u".

Moreover, if $u \models \bigcirc_i \varphi$, then we can say that

"φ is true at u, but at time t_i, $\mathcal{A}$ is not interested in knowing it".

When $R_i(u) \neq \{u\}$ (i.e. $u \models \bigcirc_i \top$), at time t_i, the agent $\mathcal{A}$ is not able to distinguish the elements of $R_i(u)$ from one another; on the contrary, that is $R_i(u) = \{u\}$ (i.e. $u \models \neg \bigcirc_i \top$), at time t_i, the agent $\mathcal{A}$ ignores whether a formula is true or false at u. The epistemic interpretation that we give to modal logic SO_n is better explained through the following example.

Example 44. We consider a game where a player selects a card x in $\mathcal{D}$ that is a deck of French playing cards which are left face down, and he/she tries to guess the identity of x. He/she repeats these actions (i.e. select and try to guess a card) for up to three times, exactly at times t_1, t_2 and t_3, with $t_1 < t_2 < t_3$. If he/she guesses the identity of the choice card at least once, then he/she wins; otherwise, he/she loses. Trivially, let $i \in \{1, 2\}$, if he/she guesses the selected card at time t_i, then the game finishes without considering the time t_{i+1}. Furthermore, during the game, a referee, that knows the identity of all cards of $\mathcal{D}$, provides the player with information on several properties of the cards in $\mathcal{D}$ at each time of the sequence (t_1, t_2, t_3), as it will be shown.

We suppose that Alice and Bob are respectively the player and the referee of this game. Then, it occurs that

1. at time t_1, Bob divides the deck $\mathcal{D}$ into two stacks: red cards and black cards;
2. at time $t_2 > t_1$, he also brings together all cards that have the same suit in each group of cards that have the same colours;
3. at time $t_3 > t_2$, he divides each group of cards obtained at time t_2 into two stacks: the cards whose number is less than 7 and the cards whose number is greater or equal to 7.

The classification made by Bob to cards of $\mathcal{D}$ at times t_1, t_2 and t_3 is represented in the following figure, where $c(x)$ and $s(x)$ respectively denote the colour and the suit of card x.

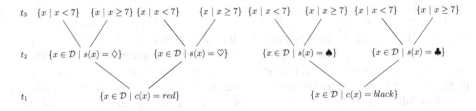

Fig. 34. Forest of Bob's classification at times t_1, t_2 and t_3

We set $B_1 = \{x \in \mathcal{D} \mid c(x) = red\}$, $B_2 = \{x \in \mathcal{D} \mid c(x) = black\}$, $B_3 = \{x \in \mathcal{D} \mid s(x) = \Diamond\}$, $B_4 = \{x \in \mathcal{D} \mid s(x) = \heartsuit\}$, $B_5 = \{x \in \mathcal{D} \mid s(x) = \spadesuit\}$, $B_6 = \{x \in \mathcal{D} \mid s(x) = \clubsuit\}$, $B_7 = \{x \in \mathcal{D} \mid s(x) = \Diamond \text{ and } x < 7\}$, $B_8 = \{x \in \mathcal{D} \mid s(x) = \Diamond \text{ and } x \geq 7\}$, $B_9 = \{x \in \mathcal{D} \mid s(x) = \heartsuit \text{ and } x < 7\}$, $B_{10} = \{x \in \mathcal{D} \mid s(x) = \heartsuit \text{ and } x \geq 7\}$, $B_{11} = \{x \in \mathcal{D} \mid s(x) = \spadesuit \text{ and } x < 7\}$, $B_{12} = \{x \in \mathcal{D} \mid s(x) = \spadesuit \text{ and } x \geq 7\}$, $B_{13} = \{x \in \mathcal{D} \mid s(x) = \clubsuit \text{ and } x < 7\}$, $B_{14} = \{x \in \mathcal{D} \mid s(x) = \clubsuit \text{ and } x \geq 7\}$.

We also assume that, let $i \in \{1, 2, 3\}$, at time t_i, Bob informs Alice about the properties that characterize each cards group corresponding to t_i. For example, at time t_2, he says to Alice that the cards of B_4 are all cards of $\mathcal{D}$ whose suit is $\heartsuit$ (then they are also red). Consequently, when Alice chooses a card x in B_i, despite she does not know the identity of x, she knows that x has the proprieties characterizing B_i. Thus, if she chooses a card x at time t_2 in B_4, then she knows that the suit of x is $\heartsuit$, and so that the colour of x is red.

In this framework, Alice represents the agent of the knowledge, and $\mathcal{D}$ is the universe of possible worlds of the Kripke frame assigned to Alice. We notice that each block of the forest in the previous figure is a set of cards which are indistinguishable for Alice at the respective time. For example, at time t_2, she still does not have enough information to distinguish $2\heartsuit$ from $8\heartsuit$. Moreover, it is easy to notice that the information that Bob gives to Alice defines three equivalence relations on $\mathcal{D}$, one for each time in (t_1, t_2, t_3), as follows: let $x, y \in \mathcal{D}$

- $xR_1y \Leftrightarrow c(x) = c(y)$,
- $xR_2y \Leftrightarrow s(x) = s(y)$,
- $xR_3y \Leftrightarrow xR_2y$ and $\{max(x, y) < 7 \text{ or } min(x, y) \geq 7\}$.

Now, we imagine that at time t_2, in order to further help Alice, Bob removes from $\mathcal{D}$ a group $\mathcal{D}^2$ of cards. Again, at time t_3, he removes from $\mathcal{D} \setminus \mathcal{D}^2$ the group $\mathcal{D}^3$ of cards. We suppose that he also informs Alice what cards belong to $\mathcal{D}^2$ (at time t_2) and $\mathcal{D}^3$ (at time t_3). These actions allow us to define three new equivalent relations, R_1', R_2' and R_3', as follows. Let $x, y \in \mathcal{D}$

- $xR_1'y \Leftrightarrow xR_1y$

- $xR_2'y \Leftrightarrow \begin{cases} xR_2y, & \text{if } x, y \notin \mathcal{D}_2 \\ x = y, & \text{otherwise} \end{cases}$

- $xR_3'y \Leftrightarrow \begin{cases} xR_3y, & \text{if } x, y \notin \mathcal{D}_2 \cup \mathcal{D}_3 \\ x = y, & \text{otherwise} \end{cases}$

We suppose that Bob chooses $\mathcal{D}^2$ and $\mathcal{D}^3$ so that each group B_i without the cards of $\mathcal{D}^2 \cup \mathcal{D}^3$ is not made of one card.

Then, we can observe that, let $i \in \{1, 2, 3\}$, a cards is removed from $\mathcal{D}$ at time t_i if and only if its equivalent class with respect to R_i' is a singleton.

From now on, we indicate the card with number or face i, and suit j with ij, and we write $[ij]_k$ to denote the equivalence class of ij with respect to R_k'. Therefore, let φ be the proposition "the card is black", trivially, we have that

$$i\diamondsuit, i\heartsuit \models \Box_1 \neg\varphi \text{ and } i\spadesuit, i\clubsuit \models \Box_1\varphi,$$

for each $i \in \{1, \ldots, 10\} \cup \{J, Q, K\}$. We respectively read the previous expressions as follows.

- "At time t_1, Alice knows that $i\diamondsuit$ is not black";
- "at time t_1, Alice knows that $i\heartsuit$ is not black";
- "at time t_1, Alice knows that $i\spadesuit$ is black";
- "at time t_1, Alice knows that $i\clubsuit$ is black".

On the other hand, if φ' is the proposition "the card is a two" and $j \in \{\diamondsuit, \heartsuit, \spadesuit, \clubsuit\}$, we have that

$$2j \models \neg\Box_1\varphi',$$

since $[2j]_1$ is equal to $\{ij \in \mathcal{D} \mid c(ij) = red\}$ or $\{ij \in \mathcal{D} \mid c(ij) = black\}$, and both are not contained in $||\varphi'|| = \{2j \mid j \in \{\diamondsuit, \heartsuit, \spadesuit, \clubsuit\}\}$. Then, $2j \models \neg\Box_1\varphi'$ means that

"at time t_1, Alice does not know that the number of $2j$ is a two".

We recall that all cards od $\mathcal{D}$ are left face down, and so Alice does not know the identity of $2j$. The previous sentences correspond to the fact that, at time t_1, Alice only knows the colour of all cards of $\mathcal{D}$, but she does not have more information about them; for example, she knows that $2\heartsuit$ is red, but no that it is a two. We suppose that $\mathcal{D}^2$ is made of all cards of $\mathcal{D}$ with face J, Q, K. Consequently, let ψ be the proposition "the suit of the card is a spade", the sentence

$$K\spadesuit \models \neg\Box_2\psi$$

that we read as follows,

"at time t_2, Alice does not know that the suit of card is a spade",

is true, since $[K\spadesuit]_2$ is a singleton.

Moreover, the sentence

$$K\spadesuit \models \neg \bigcirc_2 \psi$$

that we read as follows,

"the suit of card is a spade, but at time t_2, Alice is not interested in knowing it",

is also true.

The latter two propositions correspond to the fact that at time t_2 Alice has information on suit of cards of $\mathcal{D}$, but she ignores $K\spadesuit$, since it is removed from the deck.

Furthermore,

$$5\heartsuit \models \bigcirc_2 \neg \varphi$$

holds, and we read it as "the card is not black and at time t_2 Alice is interested to know it".

At this point, we assume that at time t_3 Bob removes $1\diamondsuit$, $2\diamondsuit$, $6\diamondsuit$, $8\diamondsuit$, $10\diamondsuit$, $2\heartsuit$, $4\heartsuit$, $5\heartsuit$, $6\heartsuit$, $7\heartsuit$, $1\spadesuit$, $2\spadesuit$, $3\spadesuit$, $7\spadesuit$, $10\spadesuit$, $3\clubsuit$, $5\clubsuit$, $6\clubsuit$, $7\spadesuit$ and $8\spadesuit$ from $\mathcal{D} \setminus D^2$. Then, let ψ' be the proposition "the number of the card is greater than or equal to 7", these sentences hold:

$$7\diamondsuit \models \square_3 \psi' \quad \text{and} \quad 9\spadesuit \models \bigcirc_3 \psi'.$$

On the other hand, we have that

$$9\spadesuit \models \neg \square_2 \psi' \quad \text{and} \quad 7\heartsuit \models \neg \bigcirc_3 \psi'.$$

They say that

- "at time t_3, Alice knows that the number of $7\diamondsuit$ is greater than or equal to 7",
- the number of $9\spadesuit$ is greater than or equal to 7, and at time t_3, Alice is interested in knowing it",
- "at time t_2, Alice does not know that the number of $9\spadesuit$ is greater than or equal to 7",
- "$7\heartsuit$ is greater than or equal to 7, but at time t_3, Alice is not interested in knowing it".

The pair $(\mathcal{D}, (R_1', R_2', R_3'))$ is a Kripke frame of SO_3 logic, and it is assigned to Alice and to the sequence (t_1, t_2, t_3). Furthermore, $(\mathcal{D}, (R_1', R_2', R_3'))$ corresponds to the refinement sequence whose forest is represented in the following figure.

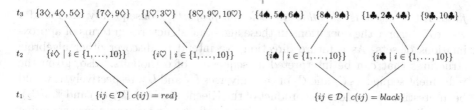

Fig. 35. Forest corresponding to $(\mathcal{D}, (R_1', R_2', R_3'))$

The next proposition states that at time t_i, Alice has the information acquired at time t_i, plus all information acquired at previous times.

Proposition 17. *Let φ be a formula, for each $i \geq j$, $\vdash \Box_i \Box_j \varphi \leftrightarrow \Box_j \varphi$.*

Finally, we can notice that by using theorems of SO_n, we can investigate on the properties of the knowledge of Alice during the sequence (t_1, t_2, t_3). For example, by Schema $\Box_i \varphi \rightarrow \bigcirc_i \varphi$, we can deduce that "at time t_i, if Alice knows φ, then she is also interested in knowing it".

6   Conclusions and Future Directions

I hope that we continue with exploration

Margaret H. Hamilton

In this thesis, we developed and studied a generalization of the rough set theory. In detail, we introduced the *sequences of orthopairs* generated by *refinement sequences*, that are special sequences of coverings representing situations where new information is gradually provided on smaller and smaller sets of objects. Refinement sequences can be viewed as formal contexts, so in the future, we propose to explore the connections between sequences of orthopairs and the *fuzzy concept lattices* [102]. Moreover, we want to consider *fuzzy sequences of orthopairs*, by generalizing the notion of *fuzzy rough sets* [43]. In particular, we would like to define novel sequences of orthopairs starting from the *Atanassov intuitionistic fuzzy sets* [5]. Another way to introduce novel sequences of orthopairs is to consider pairs of disjoint upsets such that intersection between their components has cardinality equal to an integer $k \geq 0$. In this case, the identity $\mathsf{K}_O(\mathcal{C}) = \mathsf{K}(\mathcal{C})$ could also hold for a refinement sequence $\mathcal{C}$ that is not complete and safe.

Also, we would like to deepen the relationships between sequences of orthopairs and decision trees by considering the so-called *three-way decision trees* [25, 72].

In Sect. 4, we investigated several operations between sequences of orthopairs, that allowed us to provide concrete representations of the following classes of many-valued structures: *finite centered Kleene algebras with interpolation property, finite centered Nelson algebras with the interpolation property, finite centered Nelson lattices with the interpolation property, finite IUML-algebras* and *finite KLI*-algebras with the interpolation property.* Consequently, we found a way to interpret the operations in these algebraic structures in terms of approximations of sets. As a future direction, we intend to discover other algebraic structures that can be interpreted as sequences of orthopairs. Also, given the refinement sequences $\mathcal{C}_1$ and $\mathcal{C}_2$ of the universes U_1 and U_2, respectively, it would be interesting to consider the product of the Kleene algebras $\mathbb{K}_O(\mathcal{C}_1)$ and $\mathbb{K}_O(\mathcal{C}_2)$, and to discover the universe and the class of refinement sequences corresponding it. Moreover, we can notice that rough sets can also be interpreted by a temporal semantics, as done for NM-algebras in [13]. Therefore, another topic of future works is to provide a pure logical temporal semantics in these structures and their related logics.

Furthermore, we will focus on the novel operations between orthopairs $\odot_4$ and $\hookrightarrow_4$, defined by Eqs. 36 and 37, in order to connect them with a three-valued propositional logic having a non-deterministic semantics [37].

In the previous chapter, we presented the original modal logic SO_n, with semantics based on sequences of orthopairs. The Kripke models of SO_n are characterized by a sequence $(R_1, \ldots, R_n)$ of equivalence relations corresponding to a refinement sequence of partitions. In the future, we intend to consider a new modal logic, that extends SO_n, since the sequences of the accessibility relations of its Kripke models are related to refinement sequences of coverings.

Sequences of orthopairs corresponds to decision trees with three outcomes, so we could investigate their relationship. Also, we could employ operations between sequences of orthopairs to combine several decision trees.

Eventually, we interpreted SO_n logic as an epistemic logic; namely, we used SO_n to represent the knowledge of an agent that increases over time, as new information is provided. Then, we also wish to compare SO_n with some other existing epistemic logics, especially the logics where time and multiple epistemic operators are involved [44,46], and to investigate the potential extensions of SO_n. As a future application, we also intend to study SO_n to predict the interest of users of a social network for a given piece of advertisement in a given time window. Indeed, in this case, each block of a partition can represent topics that received the same amount of interest by a user [18,41]. By refining the information about the user, it is possible to obtain a refinement sequence of partitions. Hence, the logic permits to express complex sentences about the user's interests and to tailor advertisements in a very effective way.

Acknowledgement. Firstly, I would like to express my sincere gratitude to my supervisor Professor Brunella Gerla for the constant support of my Ph.D. study and related research, for her patience, suggestions, motivation, and for sharing her knowledge. Her guidance helped me all the time of research and writing of this thesis.

I am very thankful to Professor Ricardo Oscar Rodriguez for his scientific advice, and many insightful discussions and suggestions about modal logic during my visit period at the University of Buenos Aires.

I would like to thank Professor Antonio Di Nola for encouraging me to start Ph.D. study, and for his insightful suggestions about research.

I am thankful to the co-authors of my first papers: Professor Stefano Aguzzoli, Professor Mimmo Parente, Professor Davide Ciucci, Doctor Carmen De Maio and Doctor Anna Rita Ferraioli, for sharing their knowledge with me.

I gratefully acknowledge the members of my Ph.D. committee for their time and valuable feedback on my thesis.

I am thankful to my Ph.D. colleagues, especially, Naeimeh, Zulfiker and Alberto; we supported each other by discussing over our research problems and results, and also by talking about things other than just our papers.

I would like to thank my all my friends, especially, Giovanna, Eva, Annalidia, Adelia, Iva, Rosa, Aurora, Marco, Angela and Ilaria, for standing by my side in every time.

I am very thankful to my boyfriend Matteo for his constant love and for always supporting me, even in the most difficult moments.

I am deeply thankful to my family: my parents, my sister Rosa, my brother Angelo and his girlfriend Roberta, for their unconditional love and spiritual support in all my life.

Stefania Boffa

References

1. Aguzzoli, S., Boffa, S., Ciucci, D., Gerla, B.: Refinements of orthopairs and IUML-algebras. In: Flores, V., et al. (eds.) IJCRS 2016. LNCS (LNAI), vol. 9920, pp. 87–96. Springer, Cham (2016). https://doi.org/10.1007/978-3-319-47160-0_8
2. Aguzzoli, S., Boffa, S., Ciucci, D., Gerla, B.: Finite iuml-algebras, finite forests and orthopairs. Fundamenta Informaticae 163, 139–163 (2018)
3. Aguzzoli, S., Flaminio, T., Marchioni, E.: Finite forests. Their algebras and logics, pp. 139–163 (2018)
4. Ahn, B., Cho, S., Kim, C.: The integrated methodology of rough set theory and artificial neural network for business failure prediction. Expert Syst. Appl. 18(2), 65–74 (2000)
5. Atanassov, K., Gargov, G.: Elements of intuitionistic fuzzy logic. Part i. Fuzzy Sets Syst. 95(1), 39–52 (1998)
6. Aumann, R.J.: Agreeing to disagree. Ann. Stat. 1236–1239 (1976)
7. Bacharach, M., Gerard-Varet, L.A., Mongin, P., Shin, H.S.: Epistemic Logic and the Theory of Games and Decisions, vol. 20. Springer, Heidelberg (2012). https://doi.org/10.1007/978-1-4613-1139-3
8. Banerjee, M., Chakraborty, M.K.: Rough consequence and rough algebra. In: Ziarko, W.P. (ed.) Rough Sets, Fuzzy Sets and Knowledge Discovery. Workshops in Computin, pp. 196–207. Springer, Heidelberg (1994). https://doi.org/10.1007/978-1-4471-3238-7_24
9. Banerjee, M., Chakraborty, M.K.: Algebras from rough sets (2004)
10. Belohlavek, R.: What is a fuzzy concept lattice? II. In: Kuznetsov, S.O., Ślęzak, D., Hepting, D.H., Mirkin, B.G. (eds.) RSFDGrC 2011. LNCS (LNAI), vol. 6743, pp. 19–26. Springer, Heidelberg (2011). https://doi.org/10.1007/978-3-642-21881-1_4
11. Bezhanishvili, N., Gehrke, M.: Free Heyting algebras: revisited. In: Kurz, A., Lenisa, M., Tarlecki, A. (eds.) CALCO 2009. LNCS, vol. 5728, pp. 251–266. Springer, Heidelberg (2009). https://doi.org/10.1007/978-3-642-03741-2_18
12. Bialynicki-Birula, A., Rasiowa, H.: On the representation of quasi-Boolean algebras (1957)

13. Bianchi, M.: A temporal semantics for nilpotent minimum logic. Int. J. Approx. Reason. **55**(1), 391–401 (2014)
14. Birkhoff, G.: Lattice Theory, vol. 25. American Mathematical Society (1940)
15. Birkhoff, G., et al.: Rings of sets. Duke Math. J. **3**(3), 443–454 (1937)
16. Blackburn, P., De Rijke, M., Venema, Y.: Modal Logic: Graph. Darst, vol. 53. Cambridge University Press, Cambridge (2002)
17. Boffa, S., De Maio, C., Di Nola, A., Fenza, G., Ferraioli, A.R., Loia, V.: Unifying fuzzy concept lattice construction methods. In: 2016 IEEE International Conference on Fuzzy Systems (FUZZ-IEEE), pp. 209–216. IEEE (2016)
18. Boffa, S., De Maio, C., Gerla, B., Parente, M.: Context-aware advertisment recommendation on twitter through rough sets. In: 2018 IEEE International Conference on Fuzzy Systems (FUZZ-IEEE), pp. 1–8. IEEE (2018)
19. Boffa, S., Gerla, B.: Sequences of orthopairs given by refinements of coverings. In: Petrosino, A., Loia, V., Pedrycz, W. (eds.) WILF 2016. LNCS (LNAI), vol. 10147, pp. 95–105. Springer, Cham (2017). https://doi.org/10.1007/978-3-319-52962-2_8
20. Boffa, S., Gerla, B.: Kleene algebras as sequences of orthopairs. In: Kacprzyk, J., Szmidt, E., Zadrożny, S., Atanassov, K.T., Krawczak, M. (eds.) IWIFSGN/EUSFLAT -2017. AISC, vol. 641, pp. 235–248. Springer, Cham (2018). https://doi.org/10.1007/978-3-319-66830-7_22
21. Boffa, S., Gerla, B.: How to merge opinions by using operations between sequences of orthopairs. In: 2019 IEEE International Conference on Fuzzy Systems (FUZZ-IEEE). IEEE (2019)
22. Brignole, D., Monteiro, A.: Caracterisation des algèbres de nelson par des egalités. I. Proce. Japan Acad. **43**(4), 279–283 (1967)
23. Busaniche, M., Cignoli, R.: Constructive logic with strong negation as a substructural logic. J. Logic Comput. **20**(4), 761–793 (2008)
24. Calegari, S., Ciucci, D.: Granular computing applied to ontologies. Int. J. Approx. Reason. **51**(4), 391–409 (2010)
25. Campagner, A., Ciucci, D.: Three-way and semi-supervised decision tree learning based on orthopartitions. In: Medina, J., et al. (eds.) IPMU 2018. CCIS, vol. 854, pp. 748–759. Springer, Cham (2018). https://doi.org/10.1007/978-3-319-91476-3_61
26. Carpineto, C., Romano, G.: Concept Data Analysis: Theory and Applications. Wiley, Hoboken (2004)
27. Castiglioni, J.L., Celani, S.A., San Martín, H.J.: Kleene algebras with implication. Algebra Universalis **77**(4), 375–393 (2017). https://doi.org/10.1007/s00012-017-0433-4
28. Celani, S.: Bounded distributive lattices with fusion and implication. Southeast Asian Bull. Math. **28**(6) (2004)
29. Chellas, B.F.: Modal Logic: An Introduction. Cambridge University Press, Cambridge (1980)
30. Cignoli, R.: Injective De Morgan and Kleene algebras. In: Proceedings of the American Mathematical Society, pp. 269–278 (1975)
31. Cignoli, R.: The class of Kleene algebras satisfying an interpolation property and nelson algebras. Algebra Universalis **23**(3), 262–292 (1986). https://doi.org/10.1007/BF01230621
32. Ciucci, D.: Orthopairs: a simple and widely used way to model uncertainty. Fundamenta Informaticae **108**(3–4), 287–304 (2011)
33. Ciucci, D.: Orthopairs and granular computing. Granul. Comput. **1**(3), 159–170 (2016). https://doi.org/10.1007/s41066-015-0013-y

34. Ciucci, D., Dubois, D.: Three-valued logics, uncertainty management and rough sets. In: Peters, J.F., Skowron, A. (eds.) Transactions on Rough Sets XVII. LNCS, vol. 8375, pp. 1–32. Springer, Heidelberg (2014). https://doi.org/10.1007/978-3-642-54756-0_1

35. Ciucci, D., Mihálydeák, T., Csajbók, Z.E.: On definability and approximations in partial approximation spaces. In: Miao, D., Pedrycz, W., Ślęzak, D., Peters, G., Hu, Q., Wang, R. (eds.) RSKT 2014. LNCS (LNAI), vol. 8818, pp. 15–26. Springer, Cham (2014). https://doi.org/10.1007/978-3-319-11740-9_2

36. Ciucci, D., Yao, Y.: Advances in rough set theory (2011)

37. Crawford, J.M., Etherington, D.W.: A non-deterministic semantics for tractable inference. In: AAAI/IAAI, pp. 286–291 (1998)

38. Cresswell, M.J., Hughes, G.E.: A New Introduction to Modal Logic. Routledge, Abingdon (2012)

39. Csajbók, Z.: Approximation of sets based on partial covering. Theoret. Comput. Sci. 412(42), 5820–5833 (2011)

40. Dai, J.H.: Rough 3-valued algebras. Inf. Sci. 178(8), 1986–1996 (2008)

41. De Maio, C., Boffa, S.: Discovery of interesting users in twitter by using rough sets. In: Park, J.J., Loia, V., Choo, K.-K.R., Yi, G. (eds.) MUE/FutureTech - 2018. LNEE, vol. 518, pp. 671–677. Springer, Singapore (2019). https://doi.org/10.1007/978-981-13-1328-8_86

42. Ditmarsch, H., Halpern, J.Y., van der Hoek, W., Kooi, B.P.: Handbook of Epistemic Logic. College Publications, London (2015)

43. Dubois, D., Prade, H.: Rough fuzzy sets and fuzzy rough sets. Int. J. Gener. Syst. 17(2–3), 191–209 (1990)

44. Dubois, D., Prade, H., Schockaert, S.: Generalized possibilistic logic: foundations and applications to qualitative reasoning about uncertainty. Artif. Intell. 252, 139–174 (2017)

45. Dwinger, P., Balbes, R.: Distributive lattices (1974)

46. Fagin, R., Halpern, J.Y., Moses, Y., Vardi, M.: Reasoning About Knowledge. MIT Press, Cambridge (2004)

47. Fidel, M.M.: An algebraic study of a propositional system of Nelson. Bull. Sect. Logic 7(2), 89 (1978)

48. Friedl, M.A., Brodley, C.E.: Decision tree classification of land cover from remotely sensed data. Remote Sens. Environ. 61(3), 399–409 (1997)

49. Fussner, W., Galatos, N.: Categories of models of r-mingle. arXiv preprint arXiv:1710.04256 (2017)

50. Ganter, B., Wille, R.: Formal Concept Analysis: Mathematical Foundations. Springer, Heidelberg (2012). https://doi.org/10.1007/978-3-642-59830-2

51. Garson, J.: Modal logic. In: Zalta, E.N. (ed.) The Stanford Encyclopedia of Philosophy. Metaphysics Research Lab, Stanford University, fall 2018 edn. (2018)

52. Golan, R.H., Ziarko, W.: A methodology for stock market analysis utilizing rough set theory. In: Proceedings of the IEEE/IAFE Computational Intelligence for Financial Engineering, pp. 32–40. IEEE (1995)

53. Grätzer, G., Wehrung, F.: Lattice Theory: Special Topics and Applications. Springer, Heidelberg (2016). https://doi.org/10.1007/978-3-319-44236-5

54. Grzymala-Busse, J.W.: LERS-A system for learning from examples based on rough sets. In: Słowiński, R. (ed.) Intelligent Decision Support. Series D: System Theory, Knowledge Engineering and Problem Solving, pp. 3–18. Springer, Heidelberg (1992). https://doi.org/10.1007/978-94-015-7975-9_1

55. Haack, S.: Philosophy of Logics. Cambridge University Press, Cambridge (1978)

56. Halpern, J.Y., Moses, Y.: Knowledge and common knowledge in a distributed environment. J. ACM (JACM) **37**, 549–587 (1990)
57. Hendricks, V.: New waves in epistemology (2008)
58. Hintikka, J.: Knowledge and Belief. An Introduction to the Logic of the Two Notions (1965)
59. Holliday, W.H.: Epistemic logic and epistemology. In: Hansson, S.O., Hendricks, V.F. (eds.) Introduction to Formal Philosophy. SUTP, pp. 351–369. Springer, Cham (2018). https://doi.org/10.1007/978-3-319-77434-3_17
60. Järvinen, J.: Knowledge representation and rough sets. Turku Centre for Computer Science (1999)
61. Järvinen, J., Radeleczki, S.: Rough sets determined by tolerances. Int. J. Approx. Reason. **55**(6), 1419–1438 (2014)
62. Järvinen, J., Radeleczki, S.: Representing regular pseudo complemented Kleene algebras by tolerance-based rough sets. J. Aust. Math. Soc. **105**(1), 57–78 (2018)
63. Kalman, J.A.: Lattices with involution. Trans. Am. Math. Soc. **87**(2), 485–491 (1958)
64. Krajicek, J., Krajíček, J., et al.: Bounded Arithmetic, Propositional Logic and Complexity Theory. Cambridge University Press, Cambridge (1995)
65. Krysiński, J.: Rough sets approach to the analysis of the structure-activity relationship of quaternary imidazolium compounds. Arzneimittelforschung **40**(7), 795–799 (1990)
66. Kryszkiewicz, M.: Rules in incomplete information systems. Inf. Sci. **113**(3–4), 271–292 (1999)
67. Ladner, R.E.: The computational complexity of provability in systems of modal propositional logic. SIAM J. Comput. **6**(3), 467–480 (1977)
68. Lewis, C.I., Langford, C.H., Lamprecht, P.: Symbolic Logic. Dover Publications, New York (1959)
69. Liau, C.-J.: Modal reasoning and rough set theory. In: Giunchiglia, F. (ed.) AIMSA 1998. LNCS, vol. 1480, pp. 317–330. Springer, Heidelberg (1998). https://doi.org/10.1007/BFb0057455
70. Lin, G., Liang, J., Qian, Y.: Multigranulation rough sets: from partition to covering. Inf. Sci. **241**, 101–118 (2013)
71. Lin, T.Y., Liu, Q.: Rough approximate operators: axiomatic rough set theory. In: Ziarko, W.P. (ed.) Rough Sets, Fuzzy Sets and Knowledge Discovery. Workshops in Computing, pp. 256–260. Springer, Heidelberg (1994). https://doi.org/10.1007/978-1-4471-3238-7_31
72. Liu, Y., Xu, J., Sun, L., Du, L.: Decisions tree learning method based on three-way decisions. In: Yao, Y., Hu, Q., Yu, H., Grzymala-Busse, J.W. (eds.) RSFDGrC 2015. LNCS (LNAI), vol. 9437, pp. 389–400. Springer, Cham (2015). https://doi.org/10.1007/978-3-319-25783-9_35
73. Metcalfe, G., Montagna, F.: Substructural fuzzy logics. J. Symb. Logic **72**(3), 834–864 (2007)
74. Moisil, G.C.: Recherches sur l'algèbre de la logique. Ann. Sci. Univ. Jassy **22**, 1–117 (1935)
75. Monteiro, A.: Matrices de morgan caractéristiques pour le calcul propositionnel classique (1963)
76. Monteiro, A.: Construction des algébres de Nelson finies. Univ. (1964)
77. Mrozek, A.: Rough sets and some aspects of expert systems realigation. The 7th International Workshop on Expert Systems and Application (1987)
78. Nation, J.B.: Notes on lattice theory. In: Cambridge Studies in Advanced Mathematics, vol. 60 (1998)

79. Orlowska, E.: A logic of indiscernibility relations. In: Skowron, A. (ed.) SCT 1984. LNCS, vol. 208, pp. 177–186. Springer, Heidelberg (1985). https://doi.org/10.1007/3-540-16066-3_17

80. Orlowska, E.: Logical aspects of learning concepts. Int. J. Approx. Reason. $2(4)$, 349–364 (1988)

81. Pagliani, P.: Rough set theory and logic-algebraic structures. In: Orlowska, E. (ed.) Incomplete information: Rough set analysis. Studies in Fuzziness and Soft Computing, vol. 13, pp. 109–190. Springer, Heidelberg (1998). https://doi.org/10.1007/978-3-7908-1888-8_6

82. Pal, S.K., Skowron, A.: Rough-Fuzzy Hybridization: A New Trend in Decision Making. Springer, New York (1999)

83. Pawlak, Z.: Rough sets. Int. J. Comput. Inf. Sci. $11(5)$, 341–356 (1982). https://doi.org/10.1007/BF01001956

84. Pawlak, Z.: Imprecise categories, approximations and rough sets. In: Pawlak, Z. (ed.) Rough sets. Theory and Decision Library. (Series D: System Theory, Knowledge Engineering and Problem Solving), pp. 9–32. Springer, Heidelberg (1991). https://doi.org/10.1007/978-94-011-3534-4_2

85. Polkowski, L.: Rough Sets in Knowledge Discovery 2: Applications, Case Studies and Software Systems, vol. 19. Physica (2013)

86. Priestley, H.: Ordered sets and duality for distributive lattices. In: North-Holland Mathematics Studies, vol. 99, pp. 39–60. Elsevier (1984)

87. Rasiowa, H.: N-lattices and constructive logic with strong negation (1969)

88. Rasiowa, H.: An algebraic approach to non-classical logics, vol. 78. North-Holland, Amsterdam (1974)

89. Rasiowa, H.: Rough concepts and multiple valued logic. In: Proceedings of the 16th ISMVL, vol. 86, pp. 282–288 (1986)

90. Reeves, S., Clarke, M.: Logic for Computer Science. International Computer Science Series (1990)

91. Shi, K.: S-rough sets and its applications in diagnosis-recognition for disease. In: 2002 International Conference on Machine Learning and Cybernetics, Proceedings, vol. 1, pp. 50–54. IEEE (2002)

92. Skowron, A., Stepaniuk, J.: Tolerance approximation spaces. Fundamenta Informaticae $27(2,3)$, 245–253 (1996)

93. Slowinski, R., Vanderpooten, D.: A generalized definition of rough approximations based on similarity. IEEE Trans. Knowl. Data Eng. $12(2)$, 331–336 (2000)

94. Slowinski, R., Zopounidis, C.: Application of the rough set approach to evaluation of bankruptcy risk. Intell. Syst. Account. Finan. Manag. $4(1)$, 27–41 (1995)

95. Snyder, D.P.: Modal logic and its applications (1971)

96. Sobociński, B.: Axiomatization of a partial system of three-value calculus of propositions. Institute of Applied Logic (1952)

97. Spinks, M., Veroff, R.: Constructive logic with strong negation is a substructural logic. I. Studia Logica $88(3)$, 325–348 (2008). https://doi.org/10.1007/s11225-008-9113-x

98. Tripathy, B., Acharjya, D., Cynthya, V.: A framework for intelligent medical diagnosis using rough set with formal concept analysis. arXiv preprint arXiv:1301.6011 (2013)

99. Vakarelov, D.: Notes on n-lattices and constructive logic with strong negation. Stud. Logica $36(1–2)$, 109–125 (1977). https://doi.org/10.1007/BF02121118

100. Ward, M., Dilworth, R.P.: Residuated lattices. Trans. Am. Math. Soc. $45(3)$, 335–354 (1939)

101. Whitesitt, J.E.: Boolean Algebra and its Applications. Courier Corporation (1995)
102. Yahia, S.B., Jaoua, A.: Discovering knowledge from fuzzy concept lattice. In: Kandel, A., Last, M., Bunke, H. (eds.) Data Mining and Computational Intelligence. Studies in Fuzziness and Soft Computing, pp. 167–190. Springer, Heidelberg (2001). https://doi.org/10.1007/978-3-7908-1825-3_7
103. Yao, Y.: Three-way decision: an interpretation of rules in rough set theory. In: Wen, P., Li, Y., Polkowski, L., Yao, Y., Tsumoto, S., Wang, G. (eds.) RSKT 2009. LNCS (LNAI), vol. 5589, pp. 642–649. Springer, Heidelberg (2009). https://doi.org/10.1007/978-3-642-02962-2_81
104. Yao, Y.: Rough sets and three-way decisions. In: Ciucci, D., Wang, G., Mitra, S., Wu, W.-Z. (eds.) RSKT 2015. LNCS (LNAI), vol. 9436, pp. 62–73. Springer, Cham (2015). https://doi.org/10.1007/978-3-319-25754-9_6
105. Yao, Y.: Two views of the theory of rough sets in finite universes. Int. J. Approx. Reason. 15(4), 291–317 (1996)
106. Yao, Y., Lin, T.Y.: Generalization of rough sets using modal logics. Intell. Autom. Soft Comput. 2(2), 103–119 (1996)
107. Zhu, W.: Generalized rough sets based on relations. Inf. Sci. 177(22), 4997–5011 (2007)
108. Zhu, W., Wang, F.Y.: Reduction and axiomization of covering generalized rough sets. Inf. Sci. 152, 217–230 (2003)
109. Ziarko, W.: Variable precision rough set model. J. Comput. Syst. Sci. 46(1), 39–59 (1993)

A Study of Algebras and Logics of Rough Sets Based on Classical and Generalized Approximation Spaces

Arun Kumar[✉]

Department of Mathematics, Institute of Science, Banaras Hindu University,
Varanasi 221005, India
arunk2956@gmail.com,arunkm@bhu.ac.in

Abstract. The seminal work of Z. Pawlak [60] on rough set theory has attracted the attention of researchers from various disciplines. Algebraists introduced some new algebraic structures and represented some old existing algebraic structures in terms of algebras formed by rough sets. In Logic, the rough set theory serves the models of several logics. This paper is an amalgamation of algebras and logics of rough set theory.

We prove a structural theorem for Kleene algebras, showing that an element of a Kleene algebra can be looked upon as a rough set in some appropriate approximation space. The proposed propositional logic $\mathcal{L}_K$ of Kleene algebras is sound and complete with respect to a 3-valued and a rough set semantics.

This article also investigates some negation operators in classical rough set theory, using Dunn's approach. We investigate the semantics of the Stone negation in perp frames, that of dual Stone negation in exhaustive frames, and that of Stone and dual Stone negations with the regularity property in K_- frames. The study leads to new semantics for the logics corresponding to the classes of Stone algebras, dual Stone algebras, and regular double Stone algebras. As the perp semantics provides a Kripke type semantics for logics with negations, exploiting this feature, we obtain duality results for several classes of algebras and corresponding frames.

In another part of this article, we propose a granule-based generalization of rough set theory. We obtain representations of distributive lattices (with operators) and Heyting algebras (with operators). Moreover, various negations appear from this generalized rough set theory and achieved new positions in Dunn's Kite of negations.

Keywords: Approximation spaces · Rough sets · Kleene algebras · Perp semantics · Discrete duality · Stone algebras · Regular double Stone algebras · Heyting algebras

A. Kumar—The paper is based on the work carried out by the author for his Ph.D. degree at the Indian Institute of Technology Kanpur, 208016, India.

© Springer-Verlag GmbH Germany, part of Springer Nature 2020
J. F. Peters and A. Skowron (Eds.): TRS XXII, LNCS 12485, pp. 123–251, 2020.
https://doi.org/10.1007/978-3-662-62798-3_4

1   Introduction

The work presented in this paper revolves around the theme of Rough Set Theory, in particular that of algebras and logics related to it. The theory was introduced by Pawlak in 1982 [60]. It drew the attention of algebraists and logicians soon after, and since then, a lot of research has been carried out in the areas related to algebras and logics. Some classical algebras such as 3-valued Łukasiewicz-Moisil (LM) algebras, regular double Stone algebras or semi-simple Nelson algebras have been given a new look with the help of the rough set theoretic framework, through representation theorems. On the other hand, some new algebraic structures surfaced during the algebraic studies of rough sets, examples being topological quasi Boolean algebras, pre-rough and rough algebras. It was shown then that 3-valued LM algebras, regular double Stone algebras, semi-simple Nelson algebras and pre-rough algebras are all equivalent, modulo some transformations. Study on logics of rough sets has taken its own course over the years, but a substantial intersection with the study on algebras evolved, wherein we find logics corresponding to the different (but possibly equivalent, as just mentioned) classes of algebraic structures formed by rough sets. Part of this paper involves a study of certain classes of algebras emanating from rough sets, proving their rough set representations. Corresponding logics with rough set semantics are thereby obtained.

Pawlak was motivated by practical applications. As he has mentioned at several places, rough set theory is a kind of implementation of Frege's 'boundary-line' approach to vagueness: a vague concept is characterized by a boundary region consisting of all elements of the domain of discourse which cannot be classified to the concept or its complement. An underlying indiscernibility is referred to, with respect to which a classification of the domain is obtained. The indiscernibility was assumed by Pawlak to be represented, mathematically, by an equivalence relation on the domain, so that the classification of the domain is given by the partition induced by the equivalence relation. The domain with a classification was termed an *approximation space*.

However, as studies in 'Pawlakian' rough set theory progressed, generalizations of the theory also emerged. These were proposed in various ways – by replacing the equivalence relation on the domain with any binary relation (e.g. [35–37, 40–42, 77, 82, 83]), or by replacing the partition with an arbitrary classification, say a covering (e.g. [13, 63, 64, 81, 85–87]). Yao [82] generalized rough set theory using modal logic. In fact, in the general scenario, questions such as (a) what are definable sets and rough sets, or (b) how to approximate a set in a granule based approximation space, still interest researchers. Significant topological connotations also arise. The second part of this paper deals with a particular generalization of rough set theory based on quasi-orders, the basic definitions being motivated by the 'granule-based' approach. We make a study of algebras of definable and rough sets, and corresponding logics, in this framework.

The other area of work that has engaged us is the study on negations. 'What is a Negation?', is still a point of discussion for philosophers, linguists, logicians and mathematicians. We have been particularly interested in the approach to the subject by Dunn [26,27,29]. Dunn's *perp semantics* provides a Kripke-type semantics for various logics with negations. Negations are looked upon as modal operators in this framework. Just as in modal logic, through duality results, this study also provides set representations for various classes of algebras. We noticed that there was no work on the interactions of rough sets, its logics or algebras, and perp semantics, and embarked on this line of work for the paper. We characterize various negations that appear in rough set theoretic algebraic structures (both Pawlakian and generalized), in compatibility, exhaustive and K_- frames. Some new negations have also come up during the investigations.

We have organized this introductory section as follows. In Sect. 1.1, we provide the basic notions of algebras and logics used in this work. In Sect. 1.2, preliminaries of classical rough set theory are presented. In Sect. 1.3, we present a summary of some representation results and some logics in the context of Pawlakian rough set theory. This is followed by a brief introduction to perp semantics, in Sect. 1.4. In Sect. 1.5, some generalizations of rough set theory are mentioned. Finally, in Sect. 1.6, we present the objectives of this paper, including a sectionwise summary.

1.1 Basic Notions from Algebras and Logics

There has always been a close relationship between logic and algebra. In this section, we review some fundamental notions and results on algebras, in particular lattice based algebras, and logic. For the detail study about the history of mentioned definitions and results, we refer to [2,21].

Definition 1. *Let A be a set. A binary relation $\leq$ is called an ordering in A if $\leq$ is reflexive, antisymmetric and transitive relation on A, i.e., for all $x, y, z \in A$,*

1. $x \leq x$,
2. $x \leq y$ and $y \leq x$ implies $x = y$,
3. $x \leq y$, $y \leq z$ implie $x \leq z$.

Definition 2. *A map ϕ from a poset $(A, \leq)$ to poset $(B, \leq)$ is called an order embedding if and only if*

$$a \leq b \Longleftrightarrow \phi(a) \leq \phi(b), \ a, b \in A.$$

Let $(A, \leq)$ be a poset. If for each pair $\{x, y\}$ of elements of A, $glb\{x, y\}$ and $lub\{x, y\}$ exists then $(A, \leq)$ is called a *lattice*. Usually, $glb\{x, y\}$ is denoted as $x \vee y$ and $lub\{x, y\}$ is denoted as $x \wedge y$. There is another equivalent way to define lattice.

Definition 3. *An abstract algebra $(A, \vee, \wedge)$ is called a lattice if for all $x, y, z \in A$*

1. $x \vee y = y \vee x$, $x \wedge y = y \wedge x$.
2. $x \vee (y \vee z) = (x \vee y) \vee z$, $x \wedge (y \wedge z) = (x \wedge y) \wedge z$.
3. $x \wedge (x \vee y) = x$, $(x \wedge y) \vee y = y$.

A *lattice L is called* bounded *if there exist largest and least elements. More-over, L is called* complete *if for any $A \subseteq L$, lub of A and glb of A exist. L is called* distributive *if for all $x, y, z \in L$, $x \wedge (y \vee z) = (x \wedge y) \vee (x \wedge z)$ and $x \vee (y \wedge z) = (x \vee y) \wedge (x \vee z)$. L is said to be* completely distributive *if L is complete and for any doubly indexed subset $\{a_{i,j} : i \in I, j \in J\}$ of L,*

$$\bigwedge_{i \in I} \bigvee_{j \in J} a_{i,j} = \bigvee_{f : I \to J} \bigwedge_{i \in I} a_{i,f(i)}.$$

Definition 4. *Let $\mathcal{L} := (L, \vee, \wedge, 0, 1)$ be a complete lattice.*

(i) An element $a \in L$ is said to be completely join irreducible, *if $a = \bigvee S$ implies that $a \in S$, for every subset S of L.*

Notation 1. *Let $\mathcal{J}_L$ denote the set of all completely join irreducible elements of L, and $J(x) := \{a \in \mathcal{J}_L : a \leq x\}$, for any $x \in L$.*

(ii) A set S is said to be join dense *in $\mathcal{L}$, provided for every element $a \in L$, there is a subset S' of S such that $a = \bigvee S'$.*

Completely distributive lattices in which the set of completely join irreducible elements is join dense, will be of special interest to us. In the literature, such lattices have been studied under various names, e.g. perfect, doubly algebraic, bi-algebraic or completely prime-algebraic distributive lattices (cf. [33]).

Lemma 1. [9] *Let L and K be two completely distributive lattices. Further, suppose $\mathcal{J}_L$ and $\mathcal{J}_K$ are join dense in L and K respectively. Let $\phi : \mathcal{J}_L \to \mathcal{J}_K$ be an order isomorphism. Then the extension map $\Phi : \mathcal{J}_L \to \mathcal{J}_K$,*

$$\Phi(x) = \bigvee(\phi(J(x))),$$

is a lattice isomorphism.

Definition 5. $\mathcal{K} := (K, \vee, \wedge, \sim, 0, 1)$ *is a* De Morgan algebra, *if*

(i) $(K, \vee, \wedge, 0, 1)$ is a bounded distributive lattice, and for all $a, b \in K$,
(ii) $\sim (a \wedge b) = \sim a \vee \sim b$ (De Morgan property),
(iii) $\sim\sim a = a$ (involution).

Definition 6. *An abstract algebra $(H, \vee, \wedge, \to, 0, 1)$ is called a* Heyting algebra *if it satisfies the following properties.*

(i) $(H, \vee, \wedge, 0, 1)$ is a bounded distributive lattice.
(ii) For each $a, b \in H$, $a \to b$ is defined as,

$$a \to b := max\{c \in H : a \wedge c \leq b\}.$$

In a Heyting algebra H, $a \to 0$ is called the *pseudo complement* of a, and is denoted as $\sim a$. Let us write the expression for $\sim a$ explicitly, as we will be using it in this paper heavily.

$$\sim a := a \to 0 = max\{c \in H : a \wedge c = 0\}.$$

A bounded *pseudo complemented lattice* L is a lattice in which $\sim a$ is defined for every element a of L, as above, i.e. $\sim a := max\{c \in L : a \wedge c = 0\}$.

Definition 7. *An algebra* $\mathcal{A} := (A, \vee, \wedge, \sim, 0, 1)$ *is a* Stone algebra *if*

(i) $(A, \vee, \wedge, \sim, 0, 1)$ *is a bounded distributive pseudo complemented lattice,*
(ii) $\sim a \vee \sim\sim a = 1$, *for all* $a \in A$ *(Stone property).*

A bounded *dual pseudo complemented lattice* L is a lattice in which a unary operator $\neg$ is defined such that for every element a of L, $\neg a := min\{c \in L : a \vee c = 1\}$ exists.

Definition 8. *An algebra* $\mathcal{A} := (A, \vee, \wedge, \neg, 0, 1)$ *is a* dual Stone algebra, *if*

(i) $(A, \vee, \wedge, \sim, 0, 1)$ *is a bounded distributive dual pseudo complemented lattice,*
(ii) $\neg a \wedge \neg\neg a = 0$, *for all* $a \in A$ *(Dual Stone property).*

Definition 9. *An algebra* $\mathcal{A} := (A, \vee, \wedge, \sim, \neg, 0, 1)$ *is a* regular double Stone algebra *if*

(i) $(A, \vee, \wedge, \sim, 0, 1)$ *is a bounded distributive lattice,*
(ii) $(A, \vee, \wedge, \sim, 0, 1)$ *is a Stone algebra,*
(iii) $(A, \vee, \wedge, \neg, 0, 1)$ *is double Stone algebra,*
(iv) $a \wedge \neg a \leq b \vee \sim b$, *for all* $a, b \in A$ *(Regularity).*

Note that regularity in a regular double Stone algebra $\mathcal{A}$ can also be characterized as [75]:

$$\sim a =\sim b \text{ and } \neg a = \neg b \text{ imply } a = b, \ a, b \in A.$$

Distributive lattices are algebraic models of the logic *DLL* introduced by Dunn [25]. The study of logics in this paper is almost completely based on *DLL*. Let us present the logic. The language consists of

– Propositional variables: $p, q, r, \ldots$.
– Logical connectives: $\vee, \wedge$.

The well-formed formulas of the logic are then given by the scheme:

$$p \mid \alpha \vee \beta \mid \alpha \wedge \beta.$$

Notation 2. *Denote the set of propositional variables by* $\mathcal{P}$, *and that of well-formed formulas by* $\mathcal{F}$.

Let α and β be two formulas. The pair (α, β) is called a *consequence pair*. The rules and postulates of the logic *DLL* are presented in terms of consequence pairs. Intuitively, the consequence pair (α, β) reflects that β is a consequence of α. In the representation of a logic, a consequence pair (α, β) is denoted by $\alpha \vdash \beta$ (called a *consequent*). The logic is now given through the following postulates and rules, taken from [27] and [29]. These define reflexivity and transitivity of $\vdash$, introduction, elimination principles and the distributive law for the connectives $\wedge$ and $\vee$.

Definition 10. (*DLL*- postulates)

1. $\alpha \vdash \alpha$ (Reflexivity).

2. $\dfrac{\alpha \vdash \beta \quad \beta \vdash \gamma}{\alpha \vdash \gamma}$ (Transitivity).

3. $\alpha \wedge \beta \vdash \alpha, \, \alpha \wedge \beta \vdash \beta$ (Conjunction Elimination)

4. $\dfrac{\alpha \vdash \beta \quad \alpha \vdash \gamma}{\alpha \vdash \beta \wedge \gamma}$ (Conjunction Introduction)

5. $\alpha \vdash \alpha \vee \beta, \, \beta \vdash \alpha \vee \beta$ (Disjunction Introduction)

6. $\dfrac{\alpha \vdash \gamma \quad \beta \vdash \gamma}{\alpha \vee \beta \vdash \gamma}$ (Disjunction Elimination)

7. $\alpha \wedge (\beta \vee \gamma) \vdash (\alpha \wedge \beta) \vee (\alpha \wedge \gamma)$ (Distributivity)

Further Dunn in [25] extended the language of *DLL* by adding,

– Propositional constants: $\top, \bot$.

Then, he added the following postulate to extend *DLL* to give a logic *BDLL*, whose algebraic models are bounded distributive lattices.

– $\alpha \vdash \top$ (Top); $\bot \vdash \alpha$ (Bottom).

In this paper, we study several extensions of the logic *BDLL*.

Let $\mathcal{L}$ be a logic which is an extension of *BDLL*. By $\alpha \vdash_{\mathcal{L}} \beta$, we shall mean that the consequent $\alpha \vdash \beta$ is derivable in the logical system $\mathcal{L}$.

1.2   Classical Rough Set Theory

A practical source of an approximation space in rough set theory, is an *information system*. It is a tuple $<U, Att, Val, f>$, where

1. U is a non empty set finite set of objects,
2. Att is a finite set of attributes,
3. $Val = \cup_{A \in Att} Val_A$, where Val_A is a non empty finite set of values for attribute A,
4. $f : U \times Att \rightarrow Val$ is an *information function*, where $f(a, A) \in Val_A$.

Given an information system $<U, Att, Val, f>$, let $F \subseteq Att$. Define a relation R on U as: aRb if and only if $f(a, A) = f(b, A)$, for all $A \in F$. Then R is an equivalence relation on U. The pair (U, R) is then an approximation space.

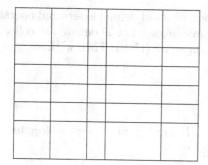

Outer rectangle represents U.

Each block represents an
equivalence class.

Fig. 1. (U, R)

Although, in an information system $<U, Att, Val, f>$, U is a finite non empty set, mathematically U can be taken as an infinite set.

As our approach in this paper is abstract, we work on approximation spaces (U, R) only. An approximation space can be visualized as in Fig. 1.   Given an approximation space (U, R), each equivalence class in U due to R is called an *elementary set,* and a union of elementary sets is called a *definable set.* Let $X \subseteq U$, Pawlak in [60, 61], proposed a pair of approximation operators, called *lower* and *upper approximation* operators, denoted L, U, respectively, and defined on the power set $\mathcal{P}(U)$ as follows.

$$\mathsf{L}X := \bigcup\{[x] : [x] \subseteq X\}.$$
$$\mathsf{U}X := \bigcup\{[x] : [x] \cap X \neq \emptyset\}.$$

Elementary sets in the approximation space are also regarded as 'granules', as their unions define definable sets, and in particular, the lower and upper approximations of sets.

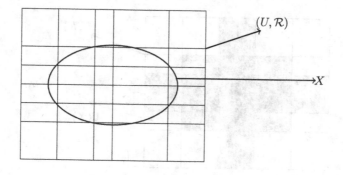

Fig. 2. $(U, R), X$

Before Pawlak used the approximation operators L, U in the context of rough set theory, similar operators were studied in the context of modal logic ($\mathbf{S}_5$), topology and algebras. Let us mention properties of definable sets and operators L and U, from the topological and algebraic angles. Let $\mathcal{D}$ denote the collection of definable sets for a given approximation space (U, R). Then we have

Proposition 1.

1. *Topologically,*
   (a) $\mathcal{D}$ *forms a clopen topology on U.*
   (b) L *is an interior operator on $\mathcal{P}(U)$. Hence L generates a topology on U in which definable sets are open sets.*
2. *Algebraically,*
   (a) $\mathcal{D}$ *forms a complete atomic Boolean algebra.*
   (b) $(\mathcal{P}(U), \cup, \cap, {}^c, \mathsf{L})$ *is a Boolean algebra with a modal operator.*

Note that L and U are dual operators, i.e., $\mathsf{L}X = (\mathsf{U}X^c)^c$. Hence L and U generate the same topology, in which U is the closure operator. The pictorial representation of $X, \mathsf{L}X$ and $\mathsf{U}X$ is given in the Figs. 2, 3 and 4 respectively.

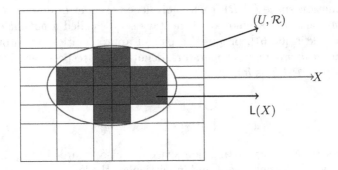

Fig. 3. $(U, R), X, \mathsf{L}(X)$

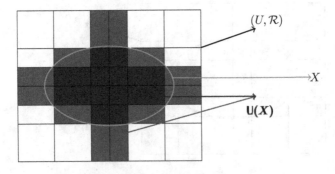

Fig. 4. $(U, R), X, \mathsf{U}(X)$

Let (U, R) be an approximation space. For each $X \subseteq U$, the ordered pair $(\mathsf{L}X, \mathsf{U}X)$ is called a *rough set*. One must mention here that there are alternative definitions of rough sets as well. A foundational study on this may be found in [14]. There are some works, for instance by Banerjee and Chakraborty [4], where a generalized version of rough sets in a Pawlakian approximation space has been studied. These are ordered pairs (D_1, D_2), where D_1, D_2 are definable sets, and $D_1 \subseteq D_2$. Let us denote by $\mathcal{RS}$, the collection of all rough sets in (U, R), and by $\mathcal{R}$, that of all the generalized rough sets, both as mentioned above. In other words, $\mathcal{RS} := \{(\mathsf{L}X, \mathsf{U}X) : X \subseteq U\}$, and $\mathcal{R} := \{(D_1, D_2) \in \mathcal{D} \times \mathcal{D} : D_1 \subseteq D_2\}$. In this work, we adopt the above definitions of rough sets only.

1.3   Algebras and Logics from Classical Rough Set Theory

Let (U, R) be an approximation space. As mentioned above, $\mathcal{D}$ is a complete atomic Boolean algebra. Then a generalized rough set is just a monotone ordered pair of sets. Now let B be any Boolean algebra, and let us consider the more general picture of generalized rough sets. Consider the set $B^{[2]} := \{(a, b) : a \leq b, a, b \in B\}$. Hence, the set $\mathcal{R}$ is the same as the set $\mathcal{D}^{[2]}$. A lot of work has been done on algebraic structures based on the set $B^{[2]}$. In fact, Moisil was the first one, who considered the set $B^{[2]}$, while constructing an example of a 3-valued LM algebra. In the context of our work on classical rough set theory, we shall refer to 3-valued LM algebras, so let us give the definition. We follow the notations of [11].

Definition 11. ([11]) *An algebra* $(A, \vee, \wedge, N, 0, 1, \phi_2)$ *is a 3-valued Łukasiewicz-Moisil (LM) algebra if*

1. $(A, \vee, \wedge, N, 0, 1)$ *is a De Morgan algebra,*
2. $\phi_2(x \vee y) = \phi_2 x \vee \phi_2 y$,
3. $\phi_2(x \wedge y) = \phi_2 x \wedge \phi_2 y$,
4. $\phi_2 x \wedge N\phi_2 x = 0$,
5. $\phi_2\phi_2 x = \phi_2 x$,
6. $\phi_2 N\phi_2 x = N\phi_2 x$,
7. $N\phi_2 Nx \leq \phi_2 x$,
8. $\phi_2 x = \phi_2 y$ *and* $\phi_2 Nx = \phi_2 Ny$ *imply* $x = y$,

for any $x, y \in A$.

Theorem 1. (*cf.* [17])

1. *Let* $\mathcal{B} := (B, \vee, \wedge, ^c, 0, 1)$ *be a Boolean algebra.*
   The structure $(B^{[2]}, \vee, \wedge, N, (0, 0), (1, 1), \phi_2)$ *is a 3-valued LM algebra, where*

$$(a, b) \vee (c, d) := (a \vee c, b \vee d),$$
$$(a, b) \wedge (c, d) := (a \wedge c, b \wedge d),$$
$$N(a, b) := (b^c, a^c),$$
$$\phi_2(a, b) := (b, b).$$

2. *Let* $(B_M, \vee, \wedge, N, 0, 1, \phi_2)$ *be a 3-valued LM algebra. Then there exists a Boolean algebra* B *such that* B_M *is embeddable into* $B^{[2]}$.

The important aspect of the above theorem is the fact that any arbitrary 3-valued LM algebra can also be looked upon as a 3-valued LM algebra in which elements are pairs of sets.

As mentioned earlier, 3-valued LM algebras, semi simple Nelson algebras and regular double Stone algebras are equivalent, in the sense that either of the algebras can be obtained from the other through some transformations. To put it more explicitly, let us define the operators $\sim_1, \sim_2, \sim_3, \mathsf{I}, \mathsf{C} : B^{[2]} \rightarrow B^{[2]}$ as,

$$\sim_1 (a, b) := (b^c, a^c),$$
$$\sim_2 (a, b) := (b^c, b^c),$$
$$\sim_3 (a, b) := (a^c, a^c),$$

$a, b \in B, \ a \leq b.$
Then, we have the following.

- The structure $(B^{[2]}, \vee, \wedge, \sim_2, \sim_3, (0, 0), (1, 1))$ is a Stone, double Stone and regular double Stone algebra.
- Moreover, the structure $(B^{[2]}, \vee, \wedge, \sim_1, \sim_2, (0, 0), (1, 1))$ is a semi simple Nelson algebra.

Hence, one can deduce that for a given approximation space (U, R), $\mathcal{R}$ forms a 3-valued LM algebra, regular double Stone algebra as well as a semi simple Nelson algebra, and using Stone's representation theorem, Theorem 1 and versions of Theorem 1 in the context of regular double Stone algebra and semi simple Nelson algebra, can be re-phrased in terms of generalized rough sets.

$\mathcal{R}$ in the context of rough set theory has been extensively studied by many authors. In particular, the investigations of Banerjee and Chakrborty have led to the introduction of topological quasi Boolean algebras, pre-rough and rough algebras [4]. Representation of rough algebra in terms of generalized rough sets has been proved in [4]. Let us give the definitions.

Definition 12. ([5]) *An algebra* $(A, \vee, \wedge, \neg, L, 0, 1)$ *is a* pre-rough algebra *if*

1. $(A, \vee, \wedge)$ *is a distributive lattice,*
2. $\neg\neg a = a,$
3. $\neg(a \vee b) = \neg a \wedge \neg b,$
4. $La \leq a,$
5. $L(a \wedge b) = La \wedge Lb,$
6. $LLa = La,$
7. $L1 = 1,$
8. $MLa = La,$ *where* $Ma = \neg L \neg a,$
9. $\neg La \vee La = 1,$
10. $L(a \vee b) = La \vee Lb,$
11. $La \leq Lb, Ma \leq Mb$ *imply* $a \leq b,$

$a, b \in A.$

Theorem 2. [4] *Any pre-rough algebra* $\mathcal{P} := (A, \leq, \vee, \wedge, \neg, L, 0, 1)$ *is isomorphic to the subalgebra of* $\mathcal{L}(A)$ *formed by the the set* $\{(La, Ma) : a \in A\}$, *where* $\mathcal{L}(A) = \{La : a \in A\}.$

An abstract algebra $(A, \vee, \wedge, \neg, L, 0, 1)$ is called a *topological quasi Boolean algebra* (tqBa) if it satisfies axioms 1-8 of a pre-rough algebra. An algebra $(A, \leq, \vee, \wedge, \neg, L, 0, 1)$ is a *rough algebra* if

1. $(A, \leq, \vee, \wedge, \neg, L, 0, 1)$ is a pre-rough algebra,
2. $L(A)$ is complete and completely distributive.

Saha, Sen and Chakraborty have subsequently done extensive work on tqBa, pre-rough and other 'intermediate' algebras [67,68].

Theorem 3. [4]

1. *Let (U, R) be an approximation space. Then the structure $(\mathcal{R}, \vee, \wedge, \sim_1, L, M, 0, 1)$ forms a topological quasi Boolean algebra, pre-rough algebra and rough algebra, where L, M are defined as follows:*

$$L(D_1, D_2) := (D_1, D_1) \text{ and } M(D_1, D_2) := (D_2, D_2).$$

2. *Any rough algebra is isomorphic to a subalgebra of the approximation space algebra $\mathcal{R}$ corresponding to some approximation space (U, R).*

Later, it was shown in [3] that pre-rough algebras and 3-valued LM algebras are equivalent. It is still an open problem to obtain representations of topological quasi Boolean algebras in terms of $\mathcal{R}$.

We already have $\mathcal{RS} \subseteq \mathcal{R}$. So, it is natural to ask, what kind of algebraic structure can be inherited in $\mathcal{RS}$ from $\mathcal{R}$. The first paper in this regard appeared by Pomykała and Pomykała [62] in which he showed that $\mathcal{RS}$ forms a Stone algebra. However, representation could not be obtained. Later, Comer in [19] proved that $\mathcal{RS}$ forms a regular double Stone algebra and obtained a representation result in terms of rough sets. Pagliani in [58,59] showed that the set $\mathcal{RS}' := \{(\mathsf{L}A, (\mathsf{U}A)^c) : A \subseteq U\}$ can be turned into a semi simple Nelson algebra and obtained a representation result for the class of finite semi simple Nelson algebras. Note that, there is a close connection between $\mathcal{RS}'$ and $\mathcal{RS}$: both are order isomorphic. Hence the structure of $\mathcal{RS}'$ can be transferred to $\mathcal{RS}$ and vice-versa.

Further, in [59], Pagliani showed that for any approximation space (U, R), $\mathcal{RS}$ may be turned into a 3-valued Post algebra, but $\mathcal{R}$ may not. Moreover, $\mathcal{RS}$ may form a Boolean algebra, but $\mathcal{R}$ may not form a Boolean algebra. In all the above representation results, negations have been defined using set theoretic complement. So, we can say that negations in these algebras are describable by set theoretic complement, or that the negations arise from Boolean negation.

The algebraic study of rough set theory leads to emergence of several logic. For instance, Banerjee and Chakraborty in [3, 4] obtained the propositional system of *pre-rough logic* for the class of pre-rough algebras. Pre-rough logic, by virtue of algebraic representation, can also be given a Kripke-type rough set semantics. There are several logics that emerge from rough set theory with semantics based on information systems (cf. [6, 45, 46]).

Let us recall the approximation operators in Pawlakian rough set theory. Let (U, R) be an approximation space. $\mathsf{L}, \mathsf{U} : \mathcal{P}U \to \mathcal{P}(U)$ are such that for $A \subseteq U$,

$$\mathsf{L}A := \bigcup\{[x] : [x] \subseteq A\},$$
$$\mathsf{U}A := \bigcup\{[x] : [x] \cap A \neq \emptyset\}. \qquad (*)$$

In literature L and U have been interpreted in the following way.

1. x *certainly* belongs to A, if $x \in \mathsf{L}A$, i.e. all objects which are indiscernible to x are in A.
2. x *certainly does not* belong to A, if $x \notin \mathsf{U}A$, i.e. all objects which are indiscernible to x are not in A.
3. Belongingness of x to A is *not certain, but possible*, if $x \in \mathsf{U}A$ but $x \notin \mathsf{L}A$. In rough set terminology, this is the case when x is in the *boundary* of A: some objects indiscernible to x are in A, while some others, also indiscernible to x, are in A^c.

These interpretations have led to much work in the study of connections between 3-valued algebras or logics and rough sets, see for instance [1,3,18,30,38,59]. It was shown by Banerjee in [3] that 3-valued Łukasiewicz logic and pre-rough logic are equivalent, thereby imparting a rough set semantics to the former. In fact, Obtułowicz was the first who made the connection of rough sets to multi-valued aspects. In [1], Avron and Konikowska have studied 3-valued logic with respect to the above interpretations, obtained a non-deterministic logical matrix and studied the 3-valued logic generated by this matrix. A simple predicate language is used, with no quantifiers or connectives, to express membership in rough sets. Connections, in special cases, with 3-valued Kleene, Łukasiewicz and two paraconsistent logics are established.

As we mentioned above, rough set theory provides a framework to look at elements of certain algebras as pairs of sets, through representation results. Such representations have a logical importance. Dunn's 4-valued semantics [27] is also motivated by his representation of De Morgan algebras. In fact, he considers pairs of the form (A^+, A^-), with no restrictions on completeness or consistency. He proved the following (we present his result in our language).

Theorem 4 [23]. *Given a De Morgan algebra* $\mathcal{K} := (K, \vee, \wedge, \sim, 0, 1)$, *there exists a set* U *such that* $\mathcal{K}$ *can be embedded into the De Morgan whose elements are of the form* (A^+, A^-), *where* $A^+, A^- \subseteq U$.

Further, he established the soundness and completeness results for De Morgan logic and a 4-valued semantics. In [30], Düntsch explicitly explains the 3-valued aspects of rough set theory. The logic he considers is that for regular double Stone algebras. He assigns formulas a generalized rough set, i.e., if α is a formula in the given language and v is the assignment, then $v(\alpha) = (A, B)$. These assignments induce the following 3-values:

$$v_\alpha(x) = \begin{cases} 1 & \text{if } x \in A \\ 1/2 & \text{if } x \in B, x \notin A \\ 0 & \text{if } x \notin B. \end{cases}$$

This v_α is a valuation in 3 element algebra. But Düntsch does not use this valuation in any exploration of logic. A sequent calculus of logic of regular double Stone algebra and its rough set semantics is discussed by Dai and Banerjee and Khan in [6, 20].

In this paper, we have discussed a logic for the class of Kleene algebras, and are able to give it a rough set as well as a 3-valued semantics.

1.4 Negations as Modal Operators

The perp semantics introduced by John M. Dunn in [24, 26] provides a framework of studying various negations as modal operators. Negations are looked upon as *impossibility* operators in 'compatibility' frames in this semantics. The notion of perp ($\perp$) has also been used by Goldblatt in 1974 in his seminal paper 'Semantic analysis of orthologic'. As mentioned by Dunn, the motivation (cf. [27]) of perp negation lies in Birkhoff-von Neumann's work of quantum logic. Dunn considered negations in the perp frame, and arrived at his well known 'Dunn's (Lopsided) Kite of Negations' (Fig. 5). He has also studied various logics with negations in the *dual* of compatibility (perp) frames, where negations can be interpreted as *unnecessity*. Later, he combined the perp and and its dual semantics [29] to study Galois negations.

As the perp semantics provides a Kripke type semantics for logics with negations, one can establish dualities between classes of algebras and frames. We exploit this feature in our work, to get duality results for several classes of algebras and corresponding frames.

The basic notions of perp, dual perp and combined semantics are presented in Sect. 3.

1.5 Generalized Rough Sets: Definitions and Properties

As mentioned earlier, generalization of rough set theory has been done in different ways in literature. In the relation based approach of generalizing rough sets, one uses an arbitrary binary relation in the definition of classical lower and upper approximations [35–37, 40–42, 77, 82, 83]. In covering based generalizations of rough sets, one uses a covering in place of a partition in the definition of Pawlakian lower and upper approximation operators [52, 63, 64, 77–79, 81, 83, 85–87]. For a good summary on generalizations of rough sets, one can refer to the work of Samanta and Chakraborty [69]. Pawlakian lower and upper approximation operators can be represented in different ways, and each representation has its own interpretation [76]. For example, the definition we provide in (*) (cf. Sect. 1.2) is a *granule based* definition of the approximation operators.

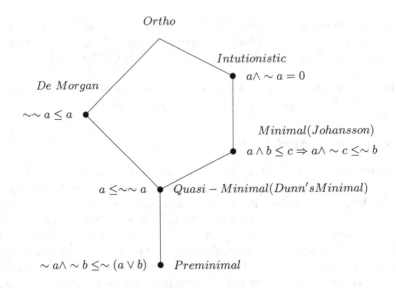

Fig. 5. Dunn's Lopsided Kite of Negations

The essence of classical rough set theory is to approximate a set with the help of granules, where a granule is an equivalence class in the domain. A granule is just a collection of objects having some properties in common. So, granules may be overlapping. In fact, one can generate granules from any given binary relation or a given covering. For example, in case of a binary relation R, $R(x) := \{y : xRy\}$ is a granule. In case of a covering, granules can be generated in many ways. For instance, let U be a set and $\mathcal{C}$ a given covering on U.

$$N_x^{\mathcal{C}} := \cup\{C_i \in \mathcal{C} : x \in C_i\} = \text{Friends(x) [63]}.$$
$$N(x) := \cap\{C_i \in \mathcal{C} : x \in C_i\} = \text{Neighbour(x) [85]}.$$

$N_x^{\mathcal{C}}$ and $N(x)$ are examples of granules. Now, lower and upper approximation operators can be defined in many ways too. The definition of rough sets is taken to be the same as in the classical case, i.e. $\mathcal{RS} := \{(\mathsf{L}X, \mathsf{U}X) : X \subseteq U\}$.

In covering based generalizations of rough sets, we could find only one work by Bonikowski [12] in which algebraic structures of $\mathcal{RS}$ have been dealt with. In this work, he provides a condition on the approximation space under which $\mathcal{RS}$ becomes a lattice.

There is notable work by Järvinen and Radeleczki, on algebraic studies in relation based rough sets. He considers the approximation space (U, R), where R is a quasi order (reflexive and transitive relation) on U, and defines the following lower and upper approximation operators. $\mathsf{L}, \mathsf{U} : \mathcal{P}(U) \to \mathcal{P}(U)$ such that for $A \subseteq U$,

$$\mathsf{L}A := \{x : R(x) \subseteq A\},$$
$$\mathsf{U}A := \{x : R(x) \cap A \neq \emptyset\}.$$

In [42] they prove that $\mathcal{RS}$ is a complete and completely distributive lattice. Moreover, in the same paper, they characterize the completely join irreducible elements and proves the join density of the set. Later, they have extended their results in [40], and proved the following.

Theorem 5 [40]. *Let (U, R) be an approximation space. The structure $(\mathcal{RS}, \vee, \wedge, c, (\emptyset, \emptyset), (U, U))$ is a Nelson algebra, where $\vee, \wedge$ are component wise join and meet respectively. c is the Kleene negation defined as $c(\mathsf{L}A, \mathsf{U}A) := ((\mathsf{U}A)^c, (\mathsf{L}A)^c)$.*

He also obtains a rough set representation of Nelson algebras over algebraic lattices.

Theorem 6 [40]. *Let $\mathbb{A} := (A, \vee, \wedge, c, 0, 1)$ be a Nelson algebra defined over an algebraic lattice. Then there exists an approximation space (U, R) such that $\mathbb{A} \cong \mathcal{RS}$.*

Further, Järvinen and Pagliani in [39] provide a rough set semantics for constructive logic with strong negation through this representation result.

Nagarajan and Umadevi [53] also consider the same definition of lower and upper approximation operators as Järvinen, and investigate algebraic structures of rough sets in the framework.

1.6 Objectives of This Paper

Representation results of 3-valued LM algebra, regular double Stone algebra, semi-simple Nelson algebra and rough algebras have been obtained in terms of either $\mathcal{R}$ or $\mathcal{RS}$ for some approximation space (U, R). Through the rough set representation, negations of the concerned algebras are described or defined by set theoretic complement. We observe that all the previously mentioned algebras are based on *Kleene algebra*. In Sect. 2 of this paper, we prove a structural theorem for the class of Kleene algebras, which was an unaddressed question so far. More precisely, we provide a solution to the following problem.

Question 1. *Given a Kleene algebra (Definition 13) $\mathcal{K} := (K, \vee, \wedge, \sim, 0, 1)$, does there exist an approximation space (U, R) such that K can be embedded into the Kleene algebra $\mathcal{R}$ ($\mathcal{RS}$) of some approximation space?*

The affirmative answer to this question provides a representation of Kleene algebras in terms of $\mathcal{R}$, and hence in terms of $\mathcal{RS}$ for some approximation space (U, R). This also reflects that each element of a Kleene algebra can be looked upon as a monotone ordered pair of sets. As mentioned earlier, such representations have a logical importance. Motivated by Dunn's representations of De Morgan algebras leading to his 4-valued semantics of De Morgan logic, in Sect. 2, we deal with the following issues.

Question 2.

1. Study the logical aspect of Question 1.

2. *Find a multi valued propositional logic which represents a logic for rough sets.*

We consider a logical system $\mathcal{L}_K$, and provide a rough set semantics for it through which we obtain a 3-valued semantics. This gives us an affirmative answer to Question 2.

For a given approximation space (U, R), the unary operators $\sim, \neg_s, \neg_{ds}$: $\mathcal{R}(\mathcal{RS}) \longrightarrow \mathcal{R}(\mathcal{RS})$ defined as,

(i) $\sim (D_1, D_2) := (D_2^c, D_1^c)$ $(\sim (LX, UX) = ((UX)^c, (LX)^c))$.

(ii) $\neg_s(D_1, D_2) := (D_2^c, D_2^c)$ $(\neg_s(LX, UX) = ((UX)^c, (UX)^c))$.

(iii) $\neg_{ds}(D_1, D_2) := (D_1^c, D_2^c)$ $(\neg_{ds}(LX, UX) = ((LX)^c, (LX)^c))$.

satisfy Kleene, Stone and double Stone property. Moreover, $\neg_s$ and $\neg_{ds}$ are regular in the sense that

$$\neg_s(D_1, D_2) \leq \neg_s(D_3, D_4) \text{ and } \neg_{ds}(D_1, D_2) \leq \neg_{ds}(D_3, D_4) \Rightarrow (D_1, D_2) \leq (D_3, D_4).$$

Dunn's model of negations provides a framework to study various logics with negations. In Sect. 3, we give a semantic analysis of negations with Kleene, Stone, double Stone and regularity properties. More precisely, we address the following issues.

Question 3.

1. *Can Kleene and Stone properties be looked upon as perp?*
2. *Semantic analysis of Kleene and double Stone properties in exhaustive frames (dual perp).*
3. *Semantic analysis of regular double Stone property of negation in K_- frame.*

Due to Kripke type behavior of perp semantics, one can establish discrete dualities between various lattice based algebras with negations and classes of compatibility, exhaustive or K_- frames. In Sect. 4, we study the following.

Question 4. *Establish discrete duality results for Kleene, Stone, double Stone and regular double Stone algebras.*

We next pass on to generalized rough set theory, focussing on generalizations based on quasi orders. However, we first address the issue of defining approximation operators maintaining the essence of classical rough set theory, namely by adopting a granule based approach. In Sect. 5, we deal with the following.

Question 5. *What would be the minimum requirement from a pair of approximation operators for them to qualify as 'approximating by granules'? More fundamentally, what do we mean by 'approximation by granules'?*

We put forth such criteria, and discuss a pair of approximation operators in quasi order-generated covering-based approximation spaces ($QOCAS$). Then, we ask the natural algebraic questions.

Question 6.

1. *Study the algebraic structures of definable sets in QOCAS.*
2. *Study the algebraic structures of $\mathcal{R}$ and $\mathcal{RS}$ for any QOCAS.*

In fact, we prove that $\mathcal{R}$ will always form a Heyting algebra, completely distributive lattice in which the set of join irreducible elements is dense. Moreover, we obtain rough set representation results for such classes of algebras. Further, we provide a condition on a QOCAS (U, R) under which $\mathcal{RS}$ forms a distributive lattice, which also gives the representation of $\mathcal{R}$ in terms $\mathcal{RS}$.

While constructing the algebraic models of 3-valued Łukasiewicz logic, Moisil defined a unary operator Δ on the set $B^{[2]}$, as,

$$\Delta(a, b) := (a, a), \ a, b \in B, a \le b,$$

which yields a representative class of 3-valued LM algebras. In the context of rough set theory, Banerjee and Chakraborty defined operators L and M on the set $\mathcal{R}$ and $\mathcal{RS}$ for a given approximation space, as:

$$L(D_1, D_2) := (D_1, D_1), L(\mathsf{L}A, \mathsf{U}A) = (\mathsf{L}A, \mathsf{L}A),$$
$$M(D_1, D_2) := (D_2, D_2), L(\mathsf{L}A, \mathsf{U}A) = (\mathsf{U}A, \mathsf{U}A).$$

Taking a cue from the work of Moisil and Banerjee and Chakraborty, we investigate the following in Sect. 6.

Question 7. *Define unary operators I and C on the set $L^2 := \{(a, b) : a \le b\}$, where L may be a distributive lattice, a Heyting algebra or a completely distributive lattice in which the set of join irreducible elements is join dense.*

1. *What kind of enhanced algebraic structures based on the set $L^{[2]}$ do we get?*
2. *Provide rough set representation of the algebraic structures obtained in 1.*

For a given Boolean algebra $(B, \vee, \wedge, {}^c, 0, 1)$, the Boolean negation ^{c} induces many negations on the set $B^{[2]}$, e.g., De Morgan, Kleene, or Stone negations. In the context of rough set theory, for a given approximation space (U, R), the set theoretic complement on the definable sets of $\mathcal{D}$ induces many negations on the collection $\mathcal{R}$ ($\mathcal{RS}$) of rough sets, such as negations of the Kleene or of the regular double Stone algebra.

Following similar ideas, some natural questions arise.

Question 8.

1. *What kind of unary operators are induced on $L^{[2]}$ by pseudo complementation or dual pseudo complemention defined on L?*
2. *Characterize the unary operators obtained in 1, in compatibility or exhaustive frames.*

We address these questions in Sect. 6, and find that the unary operators are indeed characterizable, and yield new positions in Dunn's Lopsided Kite and dual Kite of negations.

1.7   Convention

In this paper, while working with the examples, we will denote a set $A = \{a, b, c, d, e....\}$ as $abcde....$

2   Kleene Algebras: Boolean and Rough Set Representations

Algebraists, since the beginning of work on lattice theory, have been keenly interested in representing lattice-based algebras as algebras based on *set* lattices. Some such well-known representations are the Birkhoff representation for finite lattices, Stone representation for Boolean algebras, or Priestley representation for distributive lattices. It is also well-known that such representation theorems for classes of lattice-based algebras play a key role in studying set-based semantics of logics 'corresponding' to the classes. In this paper, we pursue this line of investigation, and focus on *Kleene algebras* and their representations. We then move to the corresponding propositional logic, denoted $\mathcal{L}_K$, and define a 3-valued and rough set semantics for it.

Kleene algebras were introduced by Kalman [43] and have been studied under different names such as normal i-lattices, Kleene lattices and normal quasi-Boolean algebras, e.g. cf. [15,16]. The algebras are defined as follows.

Definition 13. *An algebra* $\mathcal{K} := (K, \vee, \wedge, \sim, 0, 1)$ *is called a* Kleene algebra *if the following hold.*

1. $(K, \vee, \wedge, \sim, 0, 1)$ *is a De Morgan algebra, i.e.,*
   (i) $(K, \vee, \wedge, 0, 1)$ *is a bounded distributive lattice, and for all* $a, b \in K$,
   (ii) $\sim (a \wedge b) = \sim a \vee \sim b$ *(De Morgan property),*
   (iii) $\sim\sim a = a$ *(involution).*
2. $a \wedge \sim a \leq b \vee \sim b$, *for all* $a, b \in K$(Kleene property).

Note that in literature the structure 'idempotent semirings with a closure operation' have been also termed as Kleene algebras [47]. In this paper, by Kleene algebras we always mean the above structure.

In order to investigate a representation result for Kleene algebras, it would be natural to first turn to the known representation results for De Morgan algebras, as Kleene algebras are based on them. One finds the following, in terms of sets.

- Rasiowa [65] represented De Morgan algebras as set-based De Morgan algebras, where De Morgan negation of a set is defined by taking the set-theoretic complement of its image under an involution.
- In Dunn's [23,27] representation, each element of a De Morgan algebra can be identified with an ordered pair of sets, where De Morgan negation is defined as reversing the order in the pair.

On the other hand, we also find that there are algebras *based on* Kleene algebras which can be represented by ordered pairs of sets, and where negations are described by set-theoretic complements. Consider the set $B^{[2]} := \{(a, b) : a \leq b, a, b \in B\}$, for any partially ordered set $(B, \leq)$. We have already mentioned the following in Sect. 1.

- (Moisil (cf. [17])) For each 3-valued Łukasiewicz-Moisil (LM) algebra $\mathcal{A}$, there exists a Boolean algebra B such that $\mathcal{A}$ can be embedded into $B^{[2]}$.
- (Katriňák [44], cf. [11]) Every regular double Stone algebra can be embedded into $B^{[2]}$ for some Boolean algebra B.

Rough set theory also provides a way to represent algebras as pairs of sets. In rough set terminology (that will be elaborated on in Sect. 2.3), we have the following results for algebraic structures based on Kleene algebras.

- (Comer [19]) Every regular double Stone algebra is isomorphic to an algebra of rough sets in a Pawlak approximation space.
- (Järvinen [40]) Every Nelson algebra defined over an algebraic lattice is isomorphic to an algebra of rough sets in an approximation space based on a quasi order.

It must be mentioned that there are similar representation results in rough set theory for other structures as well, e.g. for the class of rough algebras [4], or finite semi-simple Nelson algebras [59].

In this section, the following representation results are established for Kleene algebras.

Theorem 7.

(i) Given a Kleene algebra $\mathcal{K}$, there exists a Boolean algebra $\mathcal{B}_{\mathcal{K}}$ such that $\mathcal{K}$ can be embedded into $\mathcal{B}_{\mathcal{K}}^{[2]}$.

(ii) Every Kleene algebra is isomorphic to an algebra of rough sets in a Pawlak approximation space.

This section is organized as follows. In Sect. 2.1, we prove (i) of Theorem 7. In Sect. 2.2, we establish a rough set representation of Kleene algebras, that is, (ii) of Theorem 7. The logic $\mathcal{L}_K$ and its 3-valued semantics are introduced in Sect. 2.3, and soundness and completeness results are proved. Rough set semantics of the logic $\mathcal{L}_K$ is presented in Sect. 2.4.

The content of this section is based on the article [50].

2.1   Boolean Representation of Kleene Algebras

Construction of new types of algebras from a given algebra has been of prime interest for algebraists, especially in the context of algebraic logic. Some well known examples of such construction are:

- Nelson algebra from a given Heyting algebra (Vakarelov [74], Fidel [32]).
- Kleene algebras from distributive lattices (Kalman [43]).
- 3-valued Łukasiewicz-Moisil (LM) algebra from a given Boolean algebra (Moisil, cf. [17]).
- Regular double Stone algebra from a Boolean algebra (Katriňák [44], cf. [11]).

Our work is based on Moisil's construction of a 3-valued LM algebra (which is, in particular, a Kleene algebra). Let us present this construction. Let $\mathcal{B} :=$ $(B, \vee, \wedge, ^c, 0, 1)$ be a Boolean algebra. Consider again, the set

$$B^{[2]} := \{(a, b) : a \leq b, a, b \in B\}.$$

Proposition 2 [11]. $\mathcal{B}^{[2]} := (B^{[2]}, \vee, \wedge, \sim, (0, 0), (1, 1))$ *is a Kleene algebra, where, for* $(a, b), (c, d) \in B^{[2]}$,

$$(a, b) \vee (c, d) := (a \vee c, b \vee d),$$
$$(a, b) \wedge (c, d) := (a \wedge c, b \wedge d),$$
$$\sim (a, b) := (b^c, a^c).$$

Proof. Let us only demonstrate the Kleene property for $\sim$.
$(a, b) \wedge \sim (a, b) = (a, b) \wedge (b^c, a^c) = (a \wedge b^c, b \wedge a^c) = (0, b \wedge a^c).$
$(c, d) \vee \sim (c, d) = (c, d) \vee (d^c, c^c) = (c \vee d^c, d \vee c^c) = (c \vee d^c, 1).$
Hence $(a, b) \wedge \sim (a, b) \leq (c, d) \vee \sim (c, d).$

In this section, we prove the representation Theorem 7(i).

Using Stone's representation, each Boolean algebra is embeddable in a power set algebra, so that $B^{[2]}$, for any Boolean algebra B, is embeddable in the Kleene algebra formed by $\mathcal{P}(U)^{[2]}$ for some set U. Thus, because of Theorem 7(i), one can say that each element of a Kleene algebra can also be looked upon as a pair of sets.

Now observe that we already have the following well-known representation theorem, due to the fact that **1, 2** and **3** (Fig. 6) are the only subdirectly irreducible (Kleene) algebras in the variety of Kleene algebras.

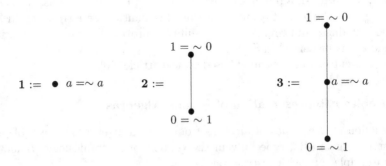

Fig. 6. Subdirectly irreducible Kleene algebras

Theorem 8 [2]. *Let $\mathcal{K}$ be a Kleene algebra. There exists a (index) set I such that $\mathcal{K}$ can be embedded into $\mathbf{3}^I$.*

So, to prove Theorem 7(i), we establish the following.

Theorem 9. *For the Kleene algebra $\mathbf{3}^I$ corresponding to any index set I, there exists a Boolean algebra $\mathcal{B}_{\mathbf{3}^I}$ such that $\mathbf{3}^I \cong (\mathcal{B}_{\mathbf{3}^I})^{[2]}$.*

We use the following steps:

1. Characterize the completely join irreducible elements of the Kleene algebras $\mathbf{3}^I$ and $(\mathbf{2}^I)^{[2]}$ and prove their join density in the respective lattices.
2. Establish an order isomorphism between $\mathcal{J}_{\mathbf{3}^I}$ and $\mathcal{J}_{(\mathbf{2}^I)^{[2]}}$.
3. Extend the order isomorphism to Kleene isomorphism to establish the result.

Completely join irreducible elements play a fundamental role in establishing isomorphisms between some well-known lattice based algebras.

Firstly, let us characterize the completely join irreducible elements of the Kleene algebras $\mathbf{3}^I$ and $(\mathbf{2}^I)^{[2]}$, for any given index set I, and prove their join density in the respective lattices.

Let $i \in I$ and let $i, k \in I$. Denote by f_i^x, $x \in \{a, 1\}$, the following element in $\mathbf{3}^I$.

$$f_i^x(k) := \begin{cases} x & \text{if } k = i \\ 0 & otherwise \end{cases}$$

Then we have,

Proposition 3. *The set of completely join irreducible elements of $\mathbf{3}^I$ is given by:*

$$\mathcal{J}_{\mathbf{3}^I} = \{f_i^a, f_i^1 : i \in I\}.$$

Moreover, $\mathcal{J}_{\mathbf{3}^I}$ is join dense in $\mathbf{3}^I$.

Proof. Let $f_i^a = \vee_{k \in K} f_k$, $K \subseteq I$. This implies that $f_i^a(j) = \vee_{k \in K} f_k(j)$, for each $j \in I$. If $j \neq i$, by the definition of f_i^a, $f_i^a(j) = 0$. So $\vee_{k \in K} f_k(j) = 0$, whence $f_k(j) = 0$, for each $k \in K$. If $j = i$, then $f_i^a(j) = a$, which means $\vee_{k \in K} f_k(j) = a$. But as a is join irreducible in $\mathbf{3}$, there exists a $k' \in K$ such that $f_{k'}(j) = a$. Hence $f_i^a = f_{k'}$. A similar argument works for f_i^1.

Now let $f \in \mathbf{3}^I$. Take $K := \{j \in I : f(j) \neq 0\}$, and for each $j \in K$, define the element f_j of $\mathbf{3}^I$ as

$$f_j(k) := \begin{cases} f(j) & \text{if } k = j \\ 0 & otherwise \end{cases}$$

Clearly, we have $f = \vee_{j \in K} f_j$, where $f_j \in \mathcal{J}_{\mathbf{3}^I}$.

Let us note that for each $i, j \in I$, $f_i^a \leq f_i^1$, and if $i \neq j$, neither $f_i^x \leq f_j^y$ nor $f_j^x \leq f_i^y$ holds for $x, y \in \{a, 1\}$. The order structure of $\mathcal{J}_{\mathbf{3}^I}$ can be visualized by Fig. 7.

Example 1. Let us consider the Kleene algebra $\mathbf{3}^3$. The set $\mathcal{J}_{\mathbf{3}^3}$ of completely join irreducible elements of $\mathbf{3}^3$ is then given by
$\mathcal{J}_{\mathbf{3}^3} = \{f_1^a := (a, 0, 0), f_1^1 := (1, 0, 0), f_2^a := (0, a, 0), f_2^1 := (0, 1, 0), f_3^a := (0, 0, a), f_3^1 := (0, 0, 1)\}$.
Let $f := (0, a, 1) \in \mathbf{3}^3$. Then $f = f_2 \vee f_3$, where $f_2 = (0, a, 0)$ and $f_3 = (0, 0, 1)$.

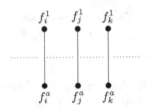

Fig. 7. Hasse diagram of $\mathcal{J}_{3^I}$

As any complete atomic Boolean algebra is isomorphic to $\mathbf{2}^I$ for some index set I, henceforth, we shall identify any complete atomic Boolean algebra B with $\mathbf{2}^I$. Now, for any such algebra, $B^{[2]}$ is a Kleene algebra (cf. Proposition 2); in fact, it is a completely distributive Kleene algebra.

Proposition 4. *Let B be a complete atomic Boolean algebra. The set of completely join irreducible elements of $B^{[2]}$ is given by*

$$\mathcal{J}_{B^{[2]}} = \{(0,a),(a,a) : a \in \mathcal{J}_B\}.$$

Moreover, $\mathcal{J}_{B^{[2]}}$ is join dense in $B^{[2]}$.

Proof. Let $a \in \mathcal{J}_B$ and let $(a,a) = \vee_{k \in K}(x_k, y_k)$, $K \subseteq I$, where $(x_k, y_k) \in B^{[2]}$ for each $k \in K$. $(a,a) = \vee_{k \in K}(x_k, y_k)$ implies $a = \vee_k x_k$. As $a \in \mathcal{J}_B$, $a = x_{k'}$ for some $k' \in K$. We already have $x_{k'} \leq y_{k'} \leq a$, hence combining with $a = x_{k'}$, we get $(a,a) = (x_{k'}, y_{k'})$. With similar arguments one can show that for each $a \in \mathcal{J}_B$, $(0,a)$ is completely join irreducible.

Now, let $(x,y) \in B^{[2]}$. Consider the sets $J(x)$, $J(y)$ (cf. Notation 1, Definition 4). Then $(x,y) = \vee_{a \in J(x)}(a,a) \vee \vee_{b \in J(y)}(0,b)$. Hence $\mathcal{J}_{B^{[2]}}$ is join dense in $B^{[2]}$.

For $a, b \in \mathcal{J}_B$, $(0,a) \leq (a,a)$, and if $a \neq b$, $x, y \in \{a, b\}$ with $x \neq y$, neither $(0,x) \leq (0,y),(y,y)$ nor $(x,x) \leq (0,y),(y,y)$ holds. Then, similar to the case of 3^I, the completely join irreducible elements of $B^{[2]}$ can be visualized by Fig. 8.

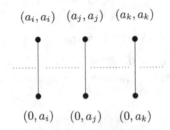

Fig. 8. Hasse diagram of $\mathcal{J}_{B^{[2]}}$

Example 2. Consider the Boolean algebra **4** of four elements with atoms a and b. The set of completely join irreducible elements of $\mathbf{4}^{[2]}$ is given by
$\mathcal{J}_{\mathbf{4}^{[2]}} = \{(0, a), (a, a), (0, b), (b, b)\}$.
Let $(a, 1) \in \mathbf{4}^{[2]}$. Then $J(a) = \{a\}$ and $J(1) = \{a, b\}$. Hence
$(a, 1) = (a, a) \vee (0, a) \vee (0, b)$.

Now, let us first present the basic lattice-theoretic definitions and results that will be required to arrive at the proof of Theorem 9.

Definition 14.

1. *A* complete lattice of sets *is a family $\mathcal{F}$ such that $\bigcup \mathcal{H}$ and $\bigcap \mathcal{H}$ belong to $\mathcal{F}$ for any $\mathcal{H} \subseteq \mathcal{F}$.*
2. *Let L be a complete lattice.*
   (a) *$a \in L$, is said to be* compact *if for every subset S of L,*
   $$a \leq \bigvee S \Rightarrow a \leq T \text{ for some finite subset } T \text{ of } S.$$
   (b) *L is said to be* algebraic *if any element $x \in L$ is the join of a set of compact elements of L.*
   (c) *L is said to satisfy the* Join-Infinite Distributive Law, *if for any subset $\{y_j\}_{j \in J}$ of L and any $x \in L$,*
   $$(JID) \quad x \wedge \bigvee_{j \in J} y_j = \bigvee_{j \in J} x \wedge y_j.$$

Theorem 10 [21]. *Let L be a lattice. The following are equivalent.*

(i) *L is complete, satisfies (JID) and the set of completely join irreducible elements is join dense in L.*
(ii) *L is completely distributive and L is algebraic.*

It can be easily seen that both the lattices $\mathbf{3}^I$ and $(\mathbf{2}^I)^{[2]}$ are complete and satisfy (JID). We have already observed from above that the sets of completely join irreducible elements of $\mathbf{3}^I$ and $(\mathbf{2}^I)^{[2]}$ are join dense in the respective lattices. So Theorem 10(i) holds for $\mathbf{3}^I$ and $(\mathbf{2}^I)^{[2]}$, and therefore, $\mathbf{3}^I$ and $(\mathbf{2}^I)^{[2]}$ are completely distributive and algebraic lattices.
For the remaining study of this Section, let us fix an index set I. Now we can write $\mathcal{J}_{\mathbf{3}^I} = \{f_i^a, f_i^1 : i \in I\}$ and $\mathcal{J}_{(\mathbf{2}^I)^{[2]}} = \{(0, g_i^1), (g_i^1, g_i^1) : i \in I\}$, where g_i^1's are the atoms or completely join irreducible elements of the Boolean algebra $\mathbf{2}^I$, defined as f_i^1 with domain restricted to **2**. In other words,

$$g_i^1(k) := \begin{cases} 1 \text{ if } k = i \\ 0 \text{ } otherwise \end{cases}$$

Theorem 11. *The sets of completely join irreducible elements of $\mathbf{3}^I$ and $(\mathbf{2}^I)^{[2]}$ are order isomorphic.*

Proof. We define the map $\phi : \mathcal{J}_{3^I} \to \mathcal{J}_{(2^I)^{[2]}}$ as follows. For $i \in I$,

$$\phi(f_i^a) := (0, g_i^1),$$
$$\phi(f_i^1) := (g_i^1, g_i^1).$$

One can show that ϕ is an order isomorphism due to the following.

- $f_i^x \leq f_j^y$ if and only if $i = j$ and $x, y = a$ or $x, y = 1$, or $x = a, y = 1$. In any case, by definition of ϕ, $\phi(f_i^x) \leq \phi(f_j^y)$.
- Let $\phi(f_i^x) \leq \phi(f_j^y)$ and assume $\phi(f_i^x) = (g_k^1, g_l^1)$ and $\phi(f_j^y) = (g_m^1, g_n^1)$. But then again: $k = l = m = n$ or $g_k^1 = g_m^1 = 0, l = n$ or $g_k^1 = 0, l = m = n$. Again, following the definition of ϕ, we have $f_i^x \leq f_j^y$.
- If $(0, g_i^1) \in \mathcal{J}_{(2^I)^{[2]}}$, then $\phi(f_i^a) = (0, g_i^1)$. Similarly for (g_i^1, g_i^1). Hence ϕ is onto.

Using Theorem 11 and Lemma 1, we have the following.

Theorem 12. *The algebras 3^I and $(2^I)^{[2]}$ are lattice isomorphic.*

In order to obtain Theorem 9, one would like to extend the above lattice isomorphism to a Kleene isomorphism. We use the technique of Järvinen in [40]. Let us present the preliminaries.

Let $\mathbb{K} := (K, \vee, \wedge, \sim, 0, 1)$ be a completely distributive De Morgan algebra. Define for any $j \in \mathcal{J}_K$,

$$j^* := \bigwedge \{x \in K : x \not\leq \sim j\}.$$

Then $j^* \in \mathcal{J}_K$. For complete details on j^*, one may refer to [40]. Further, it is shown that Lemma 1 can be extended to De Morgan algebras defined over algebraic lattices.

Theorem 13. *Let $\mathbb{L} := (L, \vee, \wedge, \sim, 0, 1)$ and $\mathbb{K} := (K, \vee, \wedge, \sim, 0, 1)$ be two De Morgan algebras defined on algebraic lattices. If $\phi : \mathcal{J}_L \to \mathcal{J}_K$ is an order isomorphism such that*

$$\phi(j^*) = \phi(j)^*, \text{ for all } j \in \mathcal{J}_L,$$

then Φ is an isomorphism between the algebras $\mathbb{L}$ and $\mathbb{K}$.

Now, let $f_i^a \in \mathcal{J}_{3^I}$. By definition, $(f_i^a)^* = \bigwedge \{f \in 3^I : f \not\leq \sim (f_i^a)\}$, where for each $i \in I$,

$$\sim (f_i^a)(k) = \begin{cases} a \text{ if } k = i \\ 1 \text{ otherwise} \end{cases}$$

Clearly, we have $f_i^1 \not\leq \sim (f_i^a)$. Now let $f \not\leq \sim (f_i^a)$. Then what does f look like? If $k \neq i$, $f(k) \leq \sim (f_i^a)(k) = 1$. So, for $f \not\leq \sim (f_i^a)$, $f(i)$ has to be 1 (otherwise $f(i) = 0$ or a will lead to $f \leq \sim (f_i^a)$). Hence, $f_i^1 \leq f$ and $(f_i^a)^* = f_i^1$.

Similarly, one can easily show that $(f_i^1)^* = f_i^a$.

On the other hand, let us consider $(0, g_i^1) \in \mathcal{J}_{(2^I)^{[2]}}$. Then, $(0, g_i^1)^* = \bigwedge \{(g, g') \in (2^I)^{[2]} : (g, g') \not\leq \sim (0, g_i^1)\}$. By definition of $\sim$, we have $\sim (0, g_i^1) =$

$((g_i^1)^c, 0^c) = ((g_i^1)^c, 1)$. Observe that $(g_i^1, g_i^1) \not\leq ((g_i^1)^c, 1)$, as, $g_i^i \not\leq (g_i^1)^c$ is true in a Boolean algebra. Now, let $(g, g') \in \mathcal{J}_{(2^I)^{[2]}}$ be such that $(g, g') \not\leq_\sim (0, g_i^1) = ((g_i^1)^c, 1)$. But we have $g' \leq 1$, so for $(g, g') \not\leq_\sim (0, g_i^1)$ to hold, we must have $g \not\leq (g_i^1)^c$. g_i^1 is an atom of 2^I and $g \not\leq (g_i^1)^c$ imply $g_i^1 \leq g$. Hence $(g_i^1, g_i^1) \leq (g, g')$, and we get $(0, g_i^1)^* = (g_i^1, g_i^1)$. Similarly, we have $(g_i^1, g_i^1)^* = (0, g_i^1)$. Let us summarize these observations in the following lemma.

Lemma 2. *The completely distributive De Morgan algebra 3^I has the following properties. For each $i \in I$,*

1. $(f_i^a)^* = f_i^1$, $(0, g_i^1)^* = (g_i^1, g_i^1)$.
2. $(f_i^1)^* = f_i^a$, $(g_i^1, g_i^1)^* = (0, g_i^1)$.

We return to Theorem 9.

Proof of Theorem 9:

Let the Kleene algebra 3^I be given. Consider 2^I as a Boolean subalgebra of 3^I. Using the definition of ϕ (cf. Theorem 11) and its extension (cf. Lemma 1), and Lemma 2 we have, for each $i \in I$,

$$\phi((f_i^a)^*) = \phi(f_i^1) = (g_i^1, g_i^1) = \phi(f_i^a)^*, \phi((f_i^1)^*) = \phi(f_i^a) = (0, g_i^1) = \phi(f_i^1)^*.$$

By Theorem 12, ϕ is an order isomorphism between $\mathcal{J}_{3^I}$ and $\mathcal{J}_{(2^I)^{[2]}}$. Hence using Theorem 13, Φ is an isomorphism between the De Morgan algebras 3^I and $(2^I)^{[2]}$. As both the algebras are Kleene algebras which are also equational algebras defined over De Morgan algebras, the De Morgan isomorphism Φ extends to Kleene isomorphism.   □

Let us illustrate the above theorem through examples:

Example 3. Consider the Kleene algebra $3 := \{0, a, 1\}$. Then $\mathcal{J}_3 = \{a, 1\}$. For $2 := \{0, 1\}$, $2^{[2]} = \{(0, 0), (0, 1), (1, 1)\}$ and $\mathcal{J}_{2^{[2]}} = \{(0, 1), (1, 1)\}$. Further, $a^* = 1$, $1^* = a$ and $(0, 1)^* = (1, 1)$ and $(1, 1)^* = (0, 1)$.
Then $\phi : \mathcal{J}_3 \to \mathcal{J}_{2^{[2]}}$ is defined as

$$\phi(a) := (0, 1),$$

$$\phi(1) := (1, 1).$$

Hence the extension map $\Phi : 3 \to 2^{[2]}$ is given as

$$\Phi(a) := (0, 1),$$

$$\Phi(1) := (1, 1),$$

$$\Phi(0) := (0, 0).$$

The diagrammatic illustration of this example is given in Fig. 9.

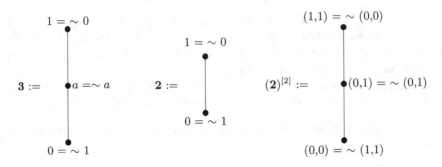

Fig. 9. $\mathbf{3} \cong (\mathbf{2})^{[2]}$

Example 4. Let us consider the Kleene algebra $\mathbf{3} \times \mathbf{3}$.
$\mathbf{3} \times \mathbf{3} := \{(0,0), (0,a), (0,1), (a,1), (1,1), (a,0), (1,0), (1,a), (a,a)\}$.
$\mathcal{J}_{3\times 3} = \{(0,a), (0,1), (a,0), (1,0)\}$ and
$(0,a)^* = (0,1), (0,1)^* = (0,a), (a,0)^* = (1,0), (1,0)^* = (a,0)$.
Take the Boolean subalgebra $\mathbf{2} \times \mathbf{2} := \{(0,0), (0,1), (1,0), (1,1)\}$ of $\mathbf{3} \times \mathbf{3}$. For
convenience, let us change the notations. We represent the set $\mathbf{2} \times \mathbf{2}$ and its
elements as $\mathbf{2}^2 = \{0, x, y, 1\}$, where $(0,0)$ is replaced by 0, $(0,1)$ is replaced by
x, $(1,0)$ is replaced by y, and $(1,1)$ is replaced by 1. Then
$(\mathbf{2}^2)^{[2]} = \{(0,0), (0,x), (0,1), (0,y), (x,x), (x,1), (y,1), (y,y), (1,1)\}$, and
$\mathcal{J}_{(\mathbf{2}^2)^{[2]}} = \{(0,x), (0,y), (x,x), (y,y)\}$.
Further, $(0,x)^* = (x,x), (x,x)^* = (0,x)$ and $(0,y)^* = (y,y), (y,y)^* = (0,y)$.
The diagrammatic illustration of the isomorphism between $\mathbf{3} \times \mathbf{3}$ and $(\mathbf{2}^2)^{[2]}$ is
given in Fig. 10.

Let us end this section with the following note. Gehrke and Walker in [34]
proved that $\mathcal{R}$ when considered as Stone algebra, is isomorphic to $\mathbf{2}^I \times \mathbf{3}^J$ for
appropriate index sets I and J. It is also observed that $\mathbf{3}$ is isomorphic to $\mathbf{2}^{[2]}$,
by $0, a, 1$ mapping to $(0,0), (0,1), (1,1)$ respectively, whence $\mathbf{3}^I$ is isomorphic to
$(\mathbf{2}^{[2]})^I$ for index set I. Then it can easily be proved that $(\mathbf{2}^{[2]})^I$ is isomorphic to
$(\mathbf{2}^I)^{[2]}$. This provides an alternative proof of Theorem 9.

2.2   Rough Set Representation of Kleene Algebras

Let us briefly recall from Sect. 1.2, some basic notions of rough set theory. For
an approximation space (U, R), one defines for each $A \subseteq U$, the lower and upper
approximations $\mathsf{L}A, \mathsf{U}A$ respectively as:

$$\mathsf{L}A := \bigcup\{[x] : [x] \subseteq X\},$$
$$\mathsf{U}A := \bigcup\{[x] : [x] \cap X \neq \emptyset\}. \qquad (*)$$

The ordered pair $(\mathsf{L}A, \mathsf{U}A)$ is called a rough set in (U, R). Let us also recall the
notations:

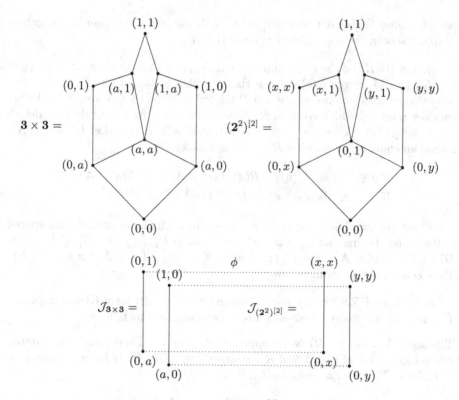

Fig. 10. $3^2 \cong (2^2)^{[2]}$

$$\mathcal{R} := \{(D_1, D_2) \in \mathcal{D} \times \mathcal{D} : D_1 \subseteq D_2\}, \mathcal{RS} := \{(\mathsf{L}A, \mathsf{U}A) : A \subseteq U\}.$$

The collection $\mathcal{D}$ of definable sets forms a complete atomic Boolean algebra in which atoms are the equivalence classes. The collection $\mathcal{RS}$ forms a distributive lattice – in fact, it forms a Kleene algebra. On the other hand, $\mathcal{R}$ is the set $\mathcal{D}^{[2]}$ and hence forms a Kleene algebra (cf. Proposition 2) as well.

In this section, we proceed to establish part (ii) of Theorem 7. Moreover, we formalize the connection of rough sets with the 3-valued semantics being considered in this work. We end this section with a rough set semantics for $\mathcal{L}_K$ (cf. Theorem 22), obtained as a consequence of the rough set representation result of Kleene algebras.

It has already been noted by many authors (for e.g., [4,5,59]) that, for an approximation space (U, R), sets $\mathcal{R}$ and $\mathcal{RS}$ may not be the same. So, it is natural to ask how $\mathcal{R}$ and $\mathcal{RS}$ differ as algebraic structures. The following result mentioned in [4] gives a connection between the two. The proof is not given in [4]; we sketch it here, as it is used in the sequel.

Theorem 14. *For any approximation space (U, R), there exists an approximation space (U', R') such that $\mathcal{R}$ corresponding to (U, R) is order isomorphic to $\mathcal{R}'$*

corresponding to (U', R'). Further, $\mathcal{R}' = \mathcal{RS}'$, the latter denoting the collection of rough sets in the approximation space (U', R') .

Proof. Let (U, R) be the given approximation space. Consider the set $\mathbf{A} := \{a \in U : |R(a)| = 1\}$, where $R(a)$ denotes the equivalence class of a in U. So $\mathbf{A}$ is the collection of all elements which are R-related only to themselves. Now, let us consider a set $\mathbf{A}'$ which consists of 'dummy' elements from outside U, indexed by the set $\mathbf{A}$, i.e. $\mathbf{A}' := \{a' : a \in \mathbf{A}\}$ such that $\mathbf{A}' \cap U = \emptyset$. Let $U' = U \cup \mathbf{A}'$. Define an equivalence relation R' on U' as follows.

$$\text{If } a \in U \text{ then } R'(a) := R(a) \cup \{x' \in \mathbf{A}' : x \in R(a) \cap \mathbf{A}\}.$$
$$\text{If } a' \in \mathbf{A}' \text{ then } R'(a') := R(a) \cup \{a'\} \ (= \{a, a'\}).$$

Note that the number of equivalence classes in both the approximation spaces is the same. Define the map $\phi : \mathcal{R} \to \mathcal{R}'$ as $\phi(D_1, D_2) := (D_1', D_2')$, where $D_1' := D_1 \cup \{x' \in \mathbf{A}' : x \in D_1 \cap \mathbf{A}\}$ and $D_2' := D_2 \cup \{x' \in \mathbf{A}' : x \in D_2 \cap \mathbf{A}\}$. Then ϕ is an order isomorphism.

Since $\mathcal{R}$ and $\mathcal{RS}$ for any approximation space (U, R) form Kleene algebras, Theorem 14 can easily be extended to Kleene algebras as follows.

Theorem 15. *Let (U, R) be an approximation space. There exists an approximation space (U', R') such that $\mathcal{R}$ corresponding to (U, R) is Kleene isomorphic to $\mathcal{RS}'$ $(= \mathcal{R}')$ corresponding to (U', R').*

Proof. Consider (U', R') and ϕ as in Theorem 14. ϕ is a lattice isomorphism, as the restriction of ϕ to the completely join irreducible elements of the lattices $\mathcal{D}^{[2]}$ and $\mathcal{D}'^{[2]}$ is an order isomorphism (using Proposition 4 and Lemma 1). Let us now show that $\phi(\sim (D_1, D_2)) = \sim (\phi(D_1, D_2))$. To avoid confusion, we follow these notations: for $X \subseteq U$ we use X^{c_1} for the complement in U and X^{c_2} for the complement in U'.

Now, $\phi(\sim (D_1, D_2)) = \phi(D_2^{c_1}, D_1^{c_1}) = ((D_2^{c_1})', (D_1^{c_1})')$. By definition of ϕ, we have:

$$(D_2^{c_1})' = D_2^{c_1} \cup \{x' \in \mathbf{A}' : x \in D_2^{c_1} \cap \mathbf{A}\}.$$
$$(D_1^{c_1})' = D_1^{c_1} \cup \{x' \in \mathbf{A}' : x \in D_1^{c_1} \cap \mathbf{A}\}.$$

Claim:
$(D_2^{c_1})' = (D_2')^{c_2}$, and $(D_1^{c_1})' = (D_1')^{c_2}$.
Proof of Claim: Let us first prove that $(D_2^{c_1})' = (D_2')^{c_2}$. Note that
$(D_2^{c_1})' = D_2^{c_1} \cup \{x' \in \mathbf{A}' : x \in D_2^{c_1} \cap \mathbf{A}\}$, and
$(D_2')^{c_2} = (D_2 \cup \{x' : x \in D_2 \cap \mathbf{A}\})^{c_2} = (D_2)^{c_2} \cap (\{x' \in \mathbf{A}' : x \in D_2 \cap \mathbf{A}\})^{c_2}$.
Let $X := \{x' \in \mathbf{A}' : x \in D_2^{c_1} \cap \mathbf{A}\}$ and $Y := \{x' \in \mathbf{A}' : x \in D_2 \cap \mathbf{A}\}$.
Consider $a \in (D_2^{c_1})' = D_2^{c_1} \cup X$.
Case 1 $a \in D_2^{c_1}$:
As $D_2 \subseteq U$, $D_2^{c_1} \subseteq D_2^{c_2}$. Hence $a \in D_2^{c_2}$. As $D_2^{c_1} \subseteq U$, $a \notin \mathbf{A}'$, whence $a \in Y^{c_2}$.
So $a \in (D_2')^{c_2}$.

Case 2 $a \in X$:

$a = x'$, where $x \in D_2^{c_1} \cap \mathbf{A}$. As, $x' \in R'(x)$ and $D_2^{c_2}$ is the union of equivalence classes, in particular it contains $R'(x)$. So $a = x' \in D_2^{c_2}$.

$x \in D_2^{c_1}$ implies $x \notin D_2$. Hence $a = x' \in Y^{c_2}$. So $a \in (D_2')^{c_2}$.

Conversely, let $a \in (D_2')^{c_2} = (D_2)^{c_2} \cap Y^{c_2}$.

Case 1 $a \in U$:

$a \in D_2^{c_2}$ implies that $a \in D_2^{c_1}$. Hence $a \in (D_2^{c_1})'$.

Case 2 $a \in \mathbf{A}'$:

$a \in Y^{c_2}$ implies $a \in \{x' \in \mathbf{A}' : x \in D_2^{c_1} \cap \mathbf{A}\}$. Hence $a \in (D_2^{c_1})'$.

Similar arguments as above show that $(D_1^{c_1})' = (D_1')^{c_2}$.

Proof of Theorem 15:

$\phi(\sim (D_1, D_2)) = \phi(D_2^{c_1}, D_1^{c_1}) = ((D_2^{c_1})', (D_1^{c_1})') = ((D_2')^{c_2}, (D_1')^{c_2}) = \sim \phi(D_1, D_2)$.

Hence ϕ is a Kleene isomorphism.

It is now not hard to see the correspondence between a complete atomic Boolean algebra and rough sets in an approximation space.

Theorem 16. *Let B be a complete atomic Boolean algebra.*

(i) *There exists an approximation space (U, R) such that*
   (a) $B \cong \mathcal{D}$.
   (b) $B^{[2]}$ *is Kleene isomorphic to $\mathcal{R}$.*
(ii) *There exists an approximation space (U', R') such that $B^{[2]}$ is Kleene isomorphic to $\mathcal{RS}'$.*

Proof. Let U denote the collection of all atoms of B, and R the identity relation on U. (U, R) is the required approximation space.

Thus we get **Theorem 7(ii)**: given a Kleene algebra $\mathcal{K}$, there exists an approximation space (U, R) such that $\mathcal{K}$ can be embedded into $\mathcal{RS}$. In other words, every Kleene algebra is isomorphic to an algebra of rough sets in a Pawlak approximation space.

2.3   Kleene Logic: 3-Valued Semantics

The De Morgan negation operator with the Kleene property (cf. Definition 13), is referred to as the *Kleene negation*. In literature, one finds various generalizations of the classical (Boolean) negation, including the De Morgan and Kleene negations. It is natural to ask the following question: do these generalized negations arise from (or can be described by) the Boolean negation? The representation result above (Theorem 7) for Kleene algebras shows that Kleene algebras always arise from Boolean algebras, thus answering the above question in the affirmative for the Kleene negation.

Representation of lattice-based algebras as algebras in which objects are pairs of sets, has proved to be of significance in the study of semantics for the logic corresponding to the class of algebras. For instance, such a representation of De

Morgan algebras leads to Dunn's well-known 4-valued semantics of De Morgan logic. In a similar way, the above representation results for Kleene algebras help us in the study of semantics for the logic $\mathcal{L}_K$ corresponding to the class of Kleene algebras. $\mathcal{L}_K$ is the De Morgan consequence system [27] with the negation operator satisfying the *Kleene axiom*: $\alpha \wedge \sim \alpha \vdash \beta \vee \sim \beta$. We show that $\mathcal{L}_K$ is sound and complete with respect to a 3-valued as well as a rough set semantics, making use of the representation results.

As mentioned earlier, Moisil in 1941 (cf. [17]) proved that $B^{[2]}$ forms a 3-valued LM algebra. So, while discussing the logic corresponding to the structures $B^{[2]}$, one is naturally led to 3-valued Łukasiewicz logic. Varlet (cf. [11]) noted the equivalence between regular double Stone algebras and 3-valued LM algebras, whence $B^{[2]}$ can be given the structure of a regular double Stone algebra as well. Here, due to Proposition 2 and Theorem 7(i), we focus on $B^{[2]}$ as a *Kleene algebra*, and study the (propositional) logic corresponding to the class of Kleene algebras and the structures $B^{[2]}$. We denote this system as $\mathcal{L}_K$, and present it in this section.

Our approach to the study is motivated by Dunn's 4-valued semantics of the De Morgan consequence system [27]. The 4-valued semantics arises from the fact that each element of a De Morgan algebra can be looked upon as a pair of sets. In our case, we have observed in Sect. 2.1 as a consequence of Theorem 7(i), that each element of a Kleene algebra can also be looked upon as a pair of sets. As demonstrated in Example 3 above, the Kleene algebra $\mathbf{3} \cong \mathbf{2}^{[2]}$. We exploit the fact that $\mathbf{3}$, in particular, can be represented as a Kleene algebra of pairs of sets, to get completeness of the logic $\mathcal{L}_K$ with respect to a 3-valued semantics.

The *Kleene* axiom $\alpha \wedge \sim \alpha \vdash \beta \vee \sim \beta$, given by Kalman [43], was studied by Dunn [27,28] in the context of examining 3-valued semantics for the first degree fragment (no nested implications) of a variety of logics including the semi-relevant logic RM. He showed that the De Morgan consequence system coupled with the Kleene axiom (the resulting consequence relation being denoted as $\vdash_{Kalman}$), is sound and complete with respect to a semantic consequence relation (denoted $\models_{0,1}^{3_R}$) defined on $\mathbf{3}_R$, the *right hand chain* of the De Morgan lattice $\mathbf{4}$ given in Fig. 11. $\mathbf{3}_R$ is the side of $\mathbf{4}$ in which the elements are interpreted as t(rue), f(alse) and b(oth), and $\models_{0,1}^{3_R}$ essentially incorporates truth *and* falsity preservation by valuations in its definition. He called this consequence system, the *Kalman consequence system*. The completeness result for the Kalman consequence system is obtained considering all 4-valued valuations restricted to $\mathbf{3}_R$: the proof makes explicit reference to valuations on $\mathbf{4}$.

The logic $\mathcal{L}_K$ (K for Kalman and Kleene) that we are considering in our work, is the Kalman consequence system with slight modifications. $\mathcal{L}_K$ is shown to be sound and complete with respect to a 3-valued semantics that is based on the same idea underlying the consequence relation $\models_{0,1}^{3_R}$, viz. that of truth as well as falsity preservation. However, the definitions and proofs in this case, *do not refer to* $\mathbf{4}$.

Let us present $\mathcal{L}_K$. The logic $\mathcal{L}_K$ is build upon the logic $BDLL$ by adding following rules and postulates.

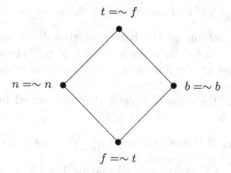

Fig. 11. De Morgan lattice 4

Definition 15. ($\mathcal{L}_K$- postulates)

1. $\dfrac{\alpha \vdash \beta}{\sim \beta \vdash \sim \alpha}$ (Contraposition).
2. $\sim \alpha \wedge \sim \beta \vdash \sim (\alpha \vee \beta)$ ($\vee$-linearity).
3. $\alpha \vdash \top$ (Top).
4. $\bot \vdash \alpha$ (Bottom).
5. $\top \vdash \sim \bot$ (Nor).
6. $\alpha \vdash \sim\sim \alpha$.
7. $\sim\sim \alpha \vdash \alpha$.
8. $\alpha \wedge \sim \alpha \vdash \beta \vee \sim \beta$ (Kalman/Kleene).

Let us now consider any Kleene algebra $(K, \vee, \wedge, \sim, 0, 1)$. We first define valuations on K.

Definition 16. *A map $v : \mathcal{F} \to K$ is called a* valuation *on K, if it satisfies the following properties for any $\alpha, \beta \in \mathcal{F}$.*

1. $v(\alpha \vee \beta) = v(\alpha) \vee v(\beta)$.
2. $v(\alpha \wedge \beta) = v(\alpha) \wedge v(\beta)$.
3. $v(\sim \alpha) = \sim v(\alpha)$.
4. $v(\bot) = 0$.
5. $v(\top) = 1$.

A consequent $\alpha \vdash \beta$ is *valid in K under the valuation v*, if $v(\alpha) \leq v(\beta)$. If the consequent is valid under all valuations on K, then it is *valid in K*. Let $\mathcal{A}$ be a class of Kleene algebras. If the consequent $\alpha \vdash \beta$ is valid in each algebra of $\mathcal{A}$, then we say $\alpha \vdash \beta$ *is valid in $\mathcal{A}$*, and denote it as $\alpha \vDash_{\mathcal{A}} \beta$.

Let $\mathcal{A}_K$ denote the class of *all* Kleene algebras. We have, in the classical manner,

Theorem 17. $\alpha \vdash_{\mathcal{L}_K} \beta$ *if and only if* $\alpha \vDash_{\mathcal{A}_K} \beta$, *for any* $\alpha, \beta \in \mathcal{F}$.

We now focus on valuations on the Kleene algebra $B^{[2]}$. For $\alpha \in \mathcal{F}$, $v(\alpha)$ is a pair of the form (a, b), $a, b \in B$. Suppose for $\beta \in \mathcal{F}$, $v(\beta) := (c, d)$, $c, d \in B$. By definition, the consequent $\alpha \vdash \beta$ is valid in $B^{[2]}$ under v, when $v(\alpha) \leq v(\beta)$, i.e., $(a, b) \leq (c, d)$, or $a \leq c$ and $b \leq d$.

Let $\mathcal{A}_{KB^{[2]}}$ denote the class of Kleene algebras formed by the sets $B^{[2]}$, for *all* Boolean algebras B.

Theorem 18. $\alpha \vDash_{\mathcal{A}_K} \beta$ *if and only if* $\alpha \vDash_{\mathcal{A}_{KB^{[2]}}} \beta$, *for any* $\alpha, \beta \in \mathcal{F}$.

Proof. Let $\alpha \vDash_{\mathcal{A}_{KB^{[2]}}} \beta$. Consider any Kleene algebra $(K, \vee, \wedge, \sim, 0, 1)$, and let v be a valuation on K. By Theorem 7(i), there exists a Boolean algebra B such that K is embedded in $B^{[2]}$. Let ϕ denote the embedding. It is a routine verification that $\phi \circ v$ is a valuation on $B^{[2]}$. The other direction is trivial, as $\mathcal{A}_{KB^{[2]}}$ is a subclass of $\mathcal{A}_K$.

On the other hand, as observed earlier, the structure $B^{[2]}$ is embeddable in the Kleene algebra formed by $\mathcal{P}(U)^{[2]}$ for some set U, utilizing Stone's representation. Hence if v is a valuation on $B^{[2]}$, it can be be extended to a valuation on $\mathcal{P}(U)^{[2]}$. Let $\mathcal{A}_{K\mathcal{P}(U)^{[2]}}$ denote the class of Kleene algebras of the form $\mathcal{P}(U)^{[2]}$, for *all* sets U. So, we get from Theorem 18 the following.

Corollary 1. $\alpha \vDash_{\mathcal{A}_K} \beta$ *if and only if* $\alpha \vDash_{\mathcal{A}_{K\mathcal{P}(U)^{[2]}}} \beta$, *for any* $\alpha, \beta \in \mathcal{F}$.

Following [27], we now consider semantic consequence relations defined by valuations $v : \mathcal{F} \to \mathbf{3}$ on the Kleene algebra $\mathbf{3}$. Let us re-label the elements of $\mathbf{3}$ as f, u, t, having the standard truth value connotations.

Definition 17. *Let* $\alpha, \beta \in \mathcal{F}$.

> $\alpha \vDash_t \beta$ *if and only if, if* $v(\alpha) = t$ *then* $v(\beta) = t$ (Truthpreservation).
> $\alpha \vDash_f \beta$ *if and only if, if* $v(\beta) = f$ *then* $v(\alpha) = f$ (Falsitypreservation).
> $\alpha \vDash_{t,f} \beta$ *if and only if,* $\alpha \vDash_t \beta$ *and* $\alpha \vDash_f \beta$.

We adopt $\vDash_{t,f}$ as the semantic consequence relation for the logic $\mathcal{L}_K$. Note that the consequence relation $\vDash_t$ is the consequence relation used in [73] to interpret the strong Kleene logic. In case of Dunn's 4-valued semantics, the consequence relations $\vDash_t$, $\vDash_f$ and $\vDash_{t,f}$ are defined using valuations on $\mathbf{4}$. As shown in [27], all the three turn out to be equivalent. Observe that for valuations on $\mathbf{3}$ that are being considered here, the consequence relations $\vDash_t$, $\vDash_f$ and $\vDash_{t,f}$ are not equivalent: $\alpha \wedge \sim \alpha \vDash_t \beta$, but $\alpha \wedge \sim \alpha \nvDash_f \beta$; $\beta \vDash_f \alpha \vee \sim \alpha$, but $\beta \nvDash_t \alpha \vee \sim \alpha$.

Theorem 19. $\alpha \vDash_{\mathcal{A}_{K\mathcal{P}(U)^{[2]}}} \beta$ *if and only if* $\alpha \vDash_{t,f} \beta$, *for any* $\alpha, \beta \in \mathcal{F}$.

Proof. Let $\alpha \vDash_{\mathcal{A}_{K\mathcal{P}(U)^{[2]}}} \beta$, and $v : \mathcal{F} \to \mathbf{3}$ be a valuation. By Example 3 and comments above, $\mathbf{3}$ is embeddable in (in fact, isomorphic to) the Kleene algebra of $\mathcal{P}(U)^{[2]}$ for some set U. If the embedding is denoted by ϕ, $\phi \circ v$ is a valuation on $\mathcal{P}(U)^{[2]}$. Then $(\phi \circ v)(\alpha) \leq (\phi \circ v)(\beta)$ implies $v(\alpha) \leq v(\beta)$. Thus if $v(\alpha) = t$, we have $v(\beta) = t$, and if $v(\beta) = f$, then also $v(\alpha) = f$.

Now, let $\alpha \models_{t,f} \beta$. Let U be a set, and $\mathcal{P}(U)^{[2]}$ be the corresponding Kleene algebra. Let v be a valuation on $\mathcal{P}(U)^{[2]}$ – we need to show $v(\alpha) \leq v(\beta)$. For any $\gamma \in \mathcal{F}$ with $v(\gamma) := (A, B)$, $A, B \subseteq U$, and for each $x \in U$, define a map $v_x : \mathcal{F} \to \mathbf{3}$ as

$$v_x(\gamma) := \begin{cases} t \text{ if } x \in A \\ u \text{ if } x \in B \setminus A \\ f \text{ if } x \notin B. \end{cases}$$

We show that v_x is a valuation.

Consider any $\gamma, \delta \in \mathcal{F}$, with $v(\gamma) := (A, B)$ and $v(\delta) := (C, D)$, $A, B, C, D \subseteq U$.

1. $v_x(\gamma \wedge \delta) = v_x(\gamma) \wedge v_x(\delta)$.
   Note that $v(\gamma \wedge \delta) = (A \cap C, B \cap D)$.
   <u>Case 1</u> $v_x(\gamma) = t$ and $v_x(\delta) = t$: Then $x \in A \cap C$, and we have $v_x(\gamma \wedge \delta) = t = v_x(\gamma) \wedge v_x(\delta)$.
   <u>Case 2</u> $v_x(\gamma) = t$ and $v_x(\delta) = u$: $x \in A$, $x \in D$ and $x \notin C$, which imply $x \notin A \cap C$ but $x \in B \cap D$. Hence $v_x(\gamma \wedge \delta) = u = v_x(\gamma) \wedge v_x(\delta)$.
   <u>Case 3</u> $v_x(\gamma) = t$ and $v_x(\delta) = f$: $x \in A$, $x \notin D$, which imply $x \notin B \cap D$. Hence $v_x(\gamma \wedge \delta) = f = v_x(\gamma) \wedge v_x(\delta)$.
   <u>Case 4</u> $v_x(\gamma) = u$ and $v_x(\delta) = f$: $x \notin A$ but $x \in B$ and $x \notin D$, which imply $x \notin B \cap D$. Hence $v_x(\gamma \wedge \delta) = f = v_x(\gamma) \wedge v_x(\delta)$.
   <u>Case 5</u> $v_x(\gamma) = u$, $v_x(\delta) = u$: $x \in B$ but $x \notin A$ and $x \in D$ but $x \notin C$. So, $x \in B \cap D$ and $x \notin A \cap C$. Hence $v_x(\gamma \wedge \delta) = u = v_x(\gamma) \wedge v_x(\delta)$.
   <u>Case 6</u> $v_x(\gamma) = f$, $v_x(\delta) = f$: $x \notin B$ and $x \notin D$. So, $x \notin B \cap D$. Hence $v_x(\gamma \wedge \delta) = f = v_x(\gamma) \wedge v_x(\delta)$.

2. $v_x(\gamma \vee \delta) = v_x(\gamma) \vee v_x(\delta)$.
   Observe that $v(\gamma \vee \delta) = (A \cup C, B \cup D)$.
   <u>Case 1</u> $v_x(\gamma) = t$ and $v_x(\delta) = t$: Then $x \in A$, $x \in C$, which imply $x \in A \cup C$. Hence $v_x(\gamma \vee \delta) = t = v_x(\gamma) \vee v_x(\delta)$.
   <u>Case 2</u> $v_x(\gamma) = t$ and $v_x(\delta) = u$: $x \in A$, $x \in D$ and $x \notin C$, in any way $x \in A \cup C$. Hence $v_x(\gamma \vee \delta) = t = v_x(\gamma) \vee v_x(\delta)$.
   <u>Case 3</u> $v_x(\gamma) = t$ and $v_x(\delta) = f$: $x \in A$, $x \notin D$, which imply $x \in A \cup C$. Hence $v_x(\gamma \vee \delta) = t = v_x(\gamma) \vee v_x(\delta)$.
   <u>Case 4</u> $v_x(\gamma) = u$ and $v_x(\delta) = f$: $x \notin A$ but $x \in B$ and $x \notin D$, which imply $x \notin A \cup C$ but $x \in B \cup D$. Hence $v_x(\gamma \vee \delta) = u = v_x(\gamma) \vee v_x(\delta)$.
   <u>Case 5</u> $v_x(\gamma) = u$, $v_x(\delta) = u$: $x \in B$ but $x \notin A$ and $x \in D$ but $x \notin C$. So, $x \in B \cup D$ and $x \notin A \cup C$. Hence $v_x(\gamma \vee \delta) = u = v_x(\gamma) \vee v_x(\delta)$.
   <u>Case 6</u> $v_x(\gamma) = f$, $v_x(\delta) = f$: $x \notin B$ and $x \notin D$. So, $x \notin B \cup D$. Hence $v_x(\gamma \wedge \delta) = f = v_x(\gamma) \wedge v_x(\delta)$.

3. $v_x(\sim \gamma) = \sim v_x(\gamma)$.
   Note that $v(\sim \gamma) = (B^c, A^c)$.
   <u>Case 1</u> $v_x(\gamma) = t$: Then $x \in A$, i.e. $x \notin A^c$. Hence $v_x(\sim \gamma) = f = \sim v_x(\gamma)$.
   <u>Case 2</u> $v_x(\gamma) = u$: $x \notin A$ but $x \in B$. So $x \in A^c$ and $x \notin B^c$. Hence $v_x(\sim \gamma) =$

$u = \sim v_x(\gamma)$.

Case 3 $v_x(\gamma) = f$: $x \notin B$, i.e. $x \in B^c$. So $v_x(\sim \gamma) = t = \sim v_x(\gamma)$.

Hence v_x is a valuation in 3. Now let us show that $v(\alpha) \leq v(\beta)$. Let $v(\alpha) := (A', B')$, $v(\beta) := (C', D')$, and $x \in A'$. Then $v_x(\alpha) = t$, and as $\alpha \vDash_{t,f} \beta$, by definition, $v_x(\beta) = t$. This implies $x \in C'$, whence $A' \subseteq C'$.
On the other hand, if $x \notin D'$, $v_x(\beta) = f$. Hence $v_x(\alpha) = f$, so that $x \notin B'$, giving $B' \subseteq D'$.

Note that the above proof cannot be applied on the Kleene algebra $B^{[2]}$ instead of $\mathcal{P}(U)^{[2]}$, as we have used set representations explicitly.

An immediate consequence of Theorem 17, Corollary 1 and Theorem 19 is

Theorem 20. $\alpha \vdash_{\mathcal{L}_K} \beta$ *if and only if* $\alpha \vDash_{t,f} \beta$, *for any* $\alpha, \beta \in \mathcal{F}$.

2.4   Rough Set Semantics for $\mathcal{L}_K$ and the Kleene Algebra 3

Now, one can apply the Theorem 16 to get a rough set semantics for the logic $\mathcal{L}_K$. Let $\mathcal{A}_{KRS}$ denote the class containing the collections $\mathcal{RS}$ of rough sets over *all* possible approximation spaces (U, R). Then using Theorems 16 and 17, we have:

Theorem 21. *(Rough Set Semantics) For any* $\alpha, \beta \in \mathcal{F}$,

(i) $\alpha \vDash_{\mathcal{A}_K} \beta$ *if and only if* $\alpha \vDash_{\mathcal{A}_{KRS}} \beta$,
(ii) $\alpha \vdash_{\mathcal{L}_K} \beta$ *if and only if* $\alpha \vDash_{\mathcal{A}_{KRS}} \beta$.

Thus, using Theorem 20, we can formally link the 3-valued semantics being considered here, and rough sets.

Theorem 22. *For any* $\alpha, \beta \in \mathcal{F}$, $\alpha \vDash_{\mathcal{A}_{KRS}} \beta$ *if and only if* $\alpha \vDash_{t,f} \beta$.

We would now like to explicate the relationship between rough sets and the 3-valued semantics of the logic $\mathcal{L}_K$ indicated in the above theorem. For this, let us again mention the interpretations yielded by the lower and upper approximations of a set A in an approximation space (U, R) (cf. Sect. 1).

1. x *certainly* belongs to A, if $x \in \mathsf{L}A$, i.e. all objects which are indiscernible to x are in A.
2. x *certainly does not* belong to A, if $x \notin \mathsf{U}A$, i.e. all objects which are indiscernible to x are not in A.
3. Belongingness of x to A is *not certain, but possible*, if $x \in \mathsf{U}A$ but $x \notin \mathsf{L}A$. In rough set terminology, this is the case when x is in the *boundary* of A: some objects indiscernible to x are in A, while some others, also indiscernible to x, are in A^c.

Consider a formula α in $\mathcal{L}_K$, and let v be a valuation in $\mathcal{RS}$ for some approximation space (U, R) such that $v(\alpha) := (\mathsf{L}A, \mathsf{U}A), A \subseteq U$. Let $x \in U$. We define the following semantic consequence relations.

$$v, x \vDash_t^{\mathcal{RS}} \alpha \text{ if and only if } x \in \mathsf{L}A.$$

$$v, x \vDash_f^{\mathcal{RS}} \alpha \text{ if and only if } x \notin \mathsf{U}A.$$

$$v, x \vDash_u^{\mathcal{RS}} \alpha \text{ if and only if } x \notin \mathsf{L}A, x \in \mathsf{U}A.$$

Then $v, x \vDash_t^{\mathcal{RS}} \alpha$ captures the interpretation 1., $v, x \vDash_f^{\mathcal{RS}} \alpha$ captures the interpretation 2. and $v, x \vDash_u^{\mathcal{RS}} \alpha$ captures the interpretation 3..
Next, let us define the following relations.
$\alpha \vDash_t^{\mathcal{RS}} \beta$ if and only if $v, x \vDash_t^{\mathcal{RS}} \alpha$ implies $v, x \vDash_t^{\mathcal{RS}} \beta$, for all valuations v in $\mathcal{RS}$ and $x \in U$.
$\alpha \vDash_f^{\mathcal{RS}} \beta$ if and only if $v, x \vDash_f^{\mathcal{RS}} \beta$ implies $v, x \vDash_f^{\mathcal{RS}} \alpha$, for all valuations v in $\mathcal{RS}$ and $x \in U$.
$\alpha \vDash_{t,f}^{\mathcal{RS}} \beta$ if and only if $\alpha \vDash_t^{\mathcal{RS}} \beta$ and $\alpha \vDash_f^{\mathcal{RS}} \beta$.
Now we link the syntax and semantics.

Definition 18. *Let $\alpha \vdash \beta$ be a consequent.*

- *$\alpha \vdash \beta$ is valid in an approximation space (U, R), if and only if $\alpha \vDash_{t,f}^{\mathcal{RS}} \beta$.*
- *$\alpha \vdash \beta$ is valid in a class $\mathcal{F}$ of approximation spaces if and only if $\alpha \vdash \beta$ is valid in all approximation spaces $(U, R) \in \mathcal{F}$.*

Theorem 23. *$\alpha \vDash_{A_{KRS}} \beta$ if and only if $\alpha \vdash \beta$ is valid in the class of all approximation spaces.*

Proof. Let $\alpha \vDash_{A_{KRS}} \beta$. Let (U, R) be an approximation space, and v be a valuation in $\mathcal{RS}$ with $v(\alpha) := (\mathsf{L}A, \mathsf{U}A)$ and $v(\beta) := (\mathsf{L}B, \mathsf{U}B)$, $A, B \subseteq U$.
By the assumption, $\mathsf{L}A \subseteq \mathsf{L}B$ and $\mathsf{U}A \subseteq \mathsf{U}B$. Let us first show that $\alpha \vDash_t^{\mathcal{RS}} \beta$. So, let $x \in U$ and $v, x \vDash_t^{\mathcal{RS}} \alpha$, i.e., $x \in \mathsf{L}A$. But we have $\mathsf{L}A \subseteq \mathsf{L}B$, hence $v, x \vDash_t^{\mathcal{RS}} \beta$.
Now let us show that $\alpha \vDash_f^{\mathcal{RS}} \beta$. So let $y \in U$ such that $v, y \vDash_f^{\mathcal{RS}} \beta$, i.e., $y \notin \mathsf{U}B$.
But we have $\mathsf{U}A \subseteq \mathsf{U}B$, hence $y \notin \mathsf{U}A$, i.e., $v, y \vDash_f^{\mathcal{RS}} \alpha$.
Now, suppose $\alpha \vdash \beta$ is valid in the class of all approximation spaces. We want to show that $\alpha \vDash_{A_{KRS}} \beta$. Let v be a valuation in $\mathcal{RS}$ as taken above. We have to show that $\mathsf{L}A \subseteq \mathsf{L}B$ and $\mathsf{U}A \subseteq \mathsf{U}B$. Let $x \in \mathsf{L}A$, i.e., $v, x \vDash_t^{\mathcal{RS}} \alpha$. Hence by our assumption, $v, x \vDash_t^{\mathcal{RS}} \beta$, i.e., $x \in \mathsf{L}B$. So $\mathsf{L}A \subseteq \mathsf{L}B$. Now, let $y \notin \mathsf{U}B$, i.e., $v, y \vDash_f^{\mathcal{RS}} \beta$. By our assumption, $v, y \vDash_f^{\mathcal{RS}} \alpha$, i.e., $y \notin \mathsf{U}A$.

Furthermore, let us spell out the natural connections of the Kleene algebra **3** with rough sets. Observe that **3**, being isomorphic to $\mathbf{2}^{[2]}$ (as noted earlier), can also be viewed as a collection of rough sets in an approximation space, due to Theorem 16(ii).

Proposition 5. *There exists an approximation space (U, R) such that $\mathbf{3} \cong \mathcal{RS}$.*

Proof. Let $U := \{x, y\}$ and consider the equivalence relation $R := U \times U$ on U. The correspondence is depicted in Fig. 12.
Note that we also have $(\emptyset, U) = (\mathsf{L}y, \mathsf{U}y) = \sim (\mathsf{L}y, \mathsf{U}y)$.

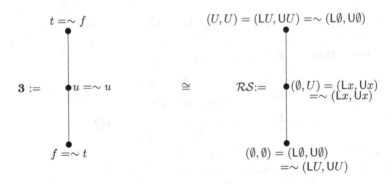

$$t = \sim f$$

$$(U,U) = (LU, UU) = \sim (L\emptyset, U\emptyset)$$

$$\mathbf{3} := \quad u = \sim u$$

$$\cong$$

$$\mathcal{RS} := \quad (\emptyset, U) = (Lx, Ux) \\ = \sim (Lx, Ux)$$

$$f = \sim t$$

$$(\emptyset, \emptyset) = (L\emptyset, U\emptyset) \\ = \sim (LU, UU)$$

Fig. 12. $\mathbf{3} \cong \mathcal{RS}$

3 Perp Semantics of Some Negations in Classical Rough Set-Theoretic Structures

Several classical negations have been described in terms of rough sets through algebraic representation results, e.g. Kleene negation (cf. Sect. 2.1) or double Stone negation with regularity (cf. [19]). In particular, for a given approximation space (U, R), the collections $\mathcal{R}$ of generalized rough sets and $\mathcal{RS}$ of rough sets on an approximation space (U, R) form Kleene, Stone, dual Stone and regular double Stone algebras (cf. Sect. 1). More precisely, recall that if we define the unary operators $\sim, \neg_1, \neg_2$ on $\mathcal{RS}$ as:

$$\sim (LA, UA) := ((UA)^c, (LA)^c),$$
$$\neg_1(LA, UA) := ((UA)^c, (UA)^c),$$
$$\neg_2(LA, UA) := ((LA)^c, (LA)^c), A \subseteq U,$$

then $\sim$ defines a Kleene negation, $\neg_1, \neg_2$ define Stone and dual Stone negations respectively. Moreover, negations $\neg_1, \neg_2$ are regular in the sense that, if $\neg_1(LA, UA) = \neg_1(LB, UB)$ and $\neg_2(LA, UA) = \neg_2(LB, UB)$ then $(LA, UA) = (LB, UB)$, for any $A, B \subseteq U$. Our interest is in studying the logical behavior of these negations, i.e. in a semantic analysis of these negations.

We follow Dunn's approach for the study, as briefly outlined in Sect. 1.4. We prove correspondence and completeness results of the classical negations mentioned above, with respect to perp, dual of perp and the 'combined semantics' involving perp and the dual of perp. More precisely, the negations we consider are those with the Kleene, Stone, and dual Stone with regularity properties.

We have organized this Section as follows. In Sect. 3.1, we present the preliminaries of perp semantics, and establish correspondence and completeness results for logics with negations having the Kleene and Stone properties in certain classes of compatibility frames. The perp semantics obtained for the logic $\mathcal{L}_K$ for Kleene algebras discussed in Sect. 2, thus imparts a semantics to the logic other than (but equivalent to) the algebraic, rough set and 3-valued semantics.

Preliminaries of the dual semantics are presented in Sect. 3.2, followed by a study of the dual of Kleene and Stone properties in exhaustive frames. We further prove correspondence and completeness results. Section 3.3 begins with the preliminaries on the united semantics. We combine Kleene property with its dual, Stone property with its dual, and study their semantical behavior in K_- frames in this section. A perp semantics for the logic of regular double Stone algebras is obtained, thus imparting this logic also with a semantics other than the rough set semantics.

The preliminaries used in this Section are taken from [29].

The content of this section is partly based on the article [51].

3.1 Semantics in Kite of Negations

In modal logic, the system K is minimal amongst all normal modal systems. Similarly Dunn [29] identifies the logic K_i which is minimal with respect to perp semantics (defined below).

The logic K_i is built upon the logic $BDLL$ (cf. Sect. 1), by adding the following rules and postulates.

1. $\dfrac{\alpha \vdash \beta}{\sim \beta \vdash \sim \alpha}$ (Contraposition)
2. $\sim \alpha \wedge \sim \beta \vdash \sim (\alpha \vee \beta)$ ($\vee$-linearity).
3. $\top \vdash \sim \bot$ (Nor).

Definition 19. *A* compatibility frame *is a triple* $(W, C, \leq)$ *with the following properties.*

1. $(W, \leq)$ *is a partially ordered set.*
2. C *is a binary relation on* W *such that for* $x, y, x', y' \in W$, *if* $x' \leq x$, $y' \leq y$ *and* xCy *then* $x'Cy'$.

C *is called a* compatibility relation *on* W.

A perp frame *is a tuple* $(W, \perp, \leq)$, *where* $\perp$, *the* perp relation *on* W, *is the complement of the compatibility relation* C.

As in [29], we do not distinguish between compatibility and perp frames, and present the results in the section in terms of the compatibility relation.

Definition 20. *A relation* $\vDash$ *between points of* W *and propositional variables in* $\mathcal{P}$ *is called an* evaluation, *if it satisfies the* hereditary *condition:*

- *if* $x \vDash p$ *and* $x \leq y$ *then* $y \vDash p$, *for any* $x, y \in W$.

Recursively, an evaluation $\vDash$ can be extended to $\mathcal{F}$ as follows. Let $x \in W$.

1. $x \vDash \alpha \wedge \beta$ if and only if $x \vDash \alpha$ and $x \vDash \beta$.
2. $x \vDash \alpha \vee \beta$ if and only if $x \vDash \alpha$ or $x \vDash \beta$.
3. $x \vDash \top$.
4. $x \nvDash \bot$.

5. $x \vDash \sim \alpha$ if and only if for all $y \in W$, xCy implies that $y \nvDash \alpha$.

Then one can easily show that $\vDash$ satisfies the hereditary condition for all formulae in $\mathcal{F}$. Thus, for each formula α in $\mathcal{F}$, an evaluation $\vDash$ gives a subset of W that is *upward closed* in the partially ordered set $(W, \leq)$. ($X \subseteq W$ is upward closed or a *cone*, if $x \in X$ and $x \leq y$, $y \in W$, imply $y \in X$.)

For the compatibility frame $\mathbf{F} := (W, C, \leq)$, the pair $(\mathbf{F}, \vDash)$ for an evaluation $\vDash$ is called a *model*. The notion of validity is introduced next in the following (usual) manner.

- A consequent $\alpha \vdash \beta$ is *valid in a model* $(\mathbf{F}, \vDash)$, denoted as $\alpha \vDash_{(\mathbf{F}, \vDash)} \beta$, if and only if, if $x \vDash \alpha$ then $x \vDash \beta$, for each $x \in W$.
- $\alpha \vdash \beta$ is *valid in the compatibility frame* $\mathbf{F}$, denoted as $\alpha \vDash_{\mathbf{F}} \beta$, if and only if $\alpha \vDash_{(\mathbf{F}, \vDash)} \beta$ for every model $(\mathbf{F}, \vDash)$.
- Let $\mathbb{F}$ denote a class of compatibility frames. $\alpha \vdash \beta$ is *valid in* $\mathbb{F}$, denoted as $\alpha \vDash_{\mathbb{F}} \beta$, if and only if $\alpha \vDash_{\mathbf{F}} \beta$ for every frame $\mathbf{F}$ belonging to $\mathbb{F}$.

In [29] it has been proved that K_i is the minimal logic which is sound and complete with respect to the class of all compatibility frames. Further, Dunn [24, 26, 29] established various correspondence and completeness results with respect to perp semantics and arrived at a *kite of negations*, commonly known as 'Dunn's Kite of Negations' (cf. Fig. 13). Dunn then extended this kite to the 'Lopsided Kite of Negations' (cf. Fig. 14).

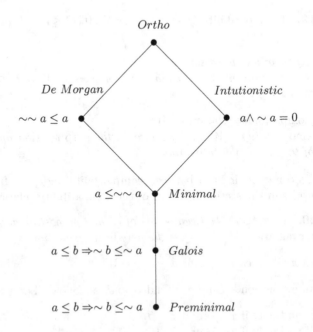

Fig. 13. Dunn's Kite of Negations

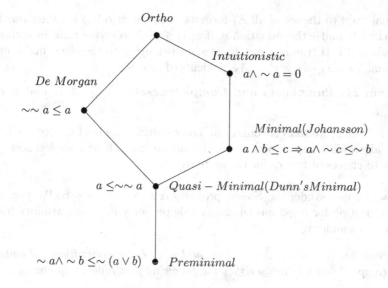

Fig. 14. Lopsided Kite of Negations

Frame completeness results for various logics with negation have been proved using the canonical frames for the logics. Let us give the definitions for the canonical frame. Let Λ denote any extension of the logic K_i.

Definition 21. *A set P of sentences in a logic Λ is called a* prime theory *if*

1. $\alpha \vdash \beta$ *holds and* $\alpha \in P$, *then* $\beta \in P$.
2. $\alpha, \beta \in P$ *then* $\alpha \wedge \beta \in P$.
3. $\top \in P$ *and* $\perp \notin P$.
4. $\alpha \vee \beta \in P$ *implies* $\alpha \in P$ *or* $\beta \in P$.

Let W_c be the collection of all prime theories of Λ. Define a relation C_c on W_c as $P_1 C_c P_2$ if and only if, for all sentences α of $\mathcal{F}$, $\sim \alpha \in P_1$ implies $\alpha \notin P_2$. The tuple $(W_c, C_c, \subseteq)$ is the *canonical frame* for Λ.
A logic Λ is called *canonical*, if its canonical frame is a frame for Λ.
Now, let us consider algebraic structures corresponding to the logic K_i.

Definition 22. *A K_i-algebra is a structure* $(K, \vee, \wedge, \sim, 0, 1)$ *such that for all* $a, b \in K$,

1. $(K, \vee, \wedge, 0, 1)$ *is a bounded distributive lattice.*
2. $a \leq b \Rightarrow \sim b \leq \sim a$.
3. $\sim a \wedge \sim b \leq \sim (a \vee b)$.
4. $\sim 0 = 1$.

Let us denote by $\mathcal{A}_{K_i}$ the class of all K_i-algebras. The algebraic semantics of the logic K_i is defined in the standard way with the help of valuations $v : \mathcal{P} \to K$ on any K_i-algebra K ($\mathcal{P}$ is the set of propositional variables of K_i) extended (in the usual way) to the set of all K_i-formulas. A sequent $\alpha \vdash \beta$ is called *true* in a K_i-algebra K under the valuation v, if $v(\alpha) \leq v(\beta)$. $\alpha \vdash \beta$ is *valid in a class of K_i-algebras* if it is true in every K_i-algebra belonging to the class, under every valuation. Validity of $\alpha \vdash \beta$ in $\mathcal{A}_{K_i}$ is denoted as $\alpha \vDash_{\mathcal{A}_{K_i}} \beta$.

Theorem 24. *(Soundness and Completeness)* $\alpha \vdash_{K_i} \beta$ *if and only if* $\alpha \vDash_{\mathcal{A}_{K_i}} \beta$.

In the following, we shall encounter different enhancements of the logic K_i, and obtain algebraic completeness as for K_i above, as well as completeness with respect to classes of frames for the systems.

Now, let us consider the Kleene property $\alpha \wedge \sim \alpha \vdash \beta \vee \sim \beta$. We provide a characterization for negations to satisfy this property in a compatibility frame, and prove canonicity.

Theorem 25. $\alpha \wedge \sim \alpha \vdash \beta \vee \sim \beta$ *is valid in a compatibility frame, if and only if the compatibility relation satisfies the following first order property:*

$$\forall x(xCx \vee \forall y(xCy \to y \leq x)). \tag{*}$$

Moreover, canonicity holds.

Proof. Consider any compatibility frame $(W, C, \leq)$, let (*) hold, and let $x \in W$. Suppose xCx, then $x \nvDash \alpha \wedge \sim \alpha$, and trivially, if $x \nvDash \alpha \wedge \sim \alpha$ then $x \vDash \beta \vee \sim \beta$.

Now suppose $\forall y(xCy \to y \leq x)$ is true. Let $x \nvDash \beta$ and xCz. Then $z \leq x$, whence $z \nvDash \beta$. So, by definition $x \vDash \sim \beta$. Hence $x \vDash \beta \vee \sim \beta$. Hence in any case if $x \vDash \alpha \wedge \sim \alpha$ then $x \vDash \beta \vee \sim \beta$.

Let (*) not hold. This implies that there exists x in W such that $not(xCx)$ and there exists y in W such that xCy and $y \nleq x$. Take such a pair x, y from W. Let us define, for any $z, w \in W$,

$$z \vDash p \text{ if and only if } x \leq z \text{ and } not(xCz),$$
$$w \vDash q \text{ if and only if } y \leq w.$$

We show that $\vDash$ is an evaluation (cf. Definition 20). Let $z \vDash p$ and $z \leq z'$. Then $x \leq z'$. If xCz', then by the frame condition on C we have

$$x \leq x, z \leq z' xCz' \text{ imply } xCz,$$

which is a contradiction to the fact that $z \vDash p$.
Furthermore, $x \vDash p$, as $x \leq x$ and $not(xCx)$. We also have $x \vDash \sim p$: if xCw for any $w \in W$ then by definition, $w \nvDash p$. Hence, $x \vDash \sim p$ and so $x \vDash p \wedge \sim p$.
On the other hand, $x \nvDash q$ as $y \nleq x$. By the assumption, xCy and $y \vDash q$, hence $x \nvDash \sim q$. So, we have $x \vDash p \wedge \sim p$ but $x \nvDash q \vee \sim q$.

Canonicity :
Let $not(PC_cP)$. Then, by definition of C_c, there exists an $\alpha \in \mathcal{F}$ such that α, $\sim \alpha \in P$. But this implies that $\alpha \wedge \sim \alpha \in P$. Hence for all $\beta \in \mathcal{F}$, $\beta \vee \sim \beta \in P$.

So, for all β, either $\beta \in P$ or $\sim \beta \in P$.
Now let PC_cQ and $\beta \in Q$. Then $\sim \beta \notin P$. But as from above, $not(PC_cP)$, we have $\beta \in P$. So, $Q \subseteq P$.

Hence a negation with the Kleene property can be treated as a modal operator. Let us recall the following from Sect. 2. The logic $\mathcal{L}_K$ contains K_i along with the postulates (1) $\alpha \vdash \sim\sim \alpha$, (2) $\sim\sim \alpha \vdash \alpha$ and (3) $\alpha \wedge \sim \alpha \vdash \beta \vee \sim \beta$ of Definition 15. It has been shown that the logic $\mathcal{L}_K$ is sound and complete with respect to class of Kleene algebras. Note that the class of all $K_i - algebras$ satisfying De Morgan and Kleene axioms coincides with the class of all Kleene algebras. We now present a perp semantics for the logic $\mathcal{L}_K$.

The consequents $\alpha \vdash \sim\sim \alpha$ and $\sim\sim \alpha \vdash \alpha$ have been characterized by Dunn (for e.g., [29]) and Restall [66] respectively as follows.

Theorem 26.

1. $\alpha \vdash \sim\sim \alpha$ is valid in the class of all compatibility frames satisfying the following frame condition:

$$\forall x \forall y (xCy \to yCx).$$

2. $\sim\sim \alpha \vdash \alpha$ is valid in the class of all compatibility frames satisfying the frame condition:

$$\forall x \exists y (xCy \wedge \forall z(yCz \to z \leq x)).$$

From Theorem 25, the correspondence and completeness for the logic K_i with Kleene property have been obtained. On combining the properties for the classes of frames corresponding to the De Morgan and Kleene property, we get the following definition.

Definition 23. *A compatibility frame* $(W, C, \leq)$ *is called a Kleene frame if it satisfies the frame conditions*

1. $\forall x \forall y (xCy \to yCx).$
2. $\forall x \exists y (xCy \wedge \forall z(yCz \to z \leq x)).$
3. $\forall x (xCx \vee \forall y(xCy \to y \leq x)).$

An example of a Kleene frame is given by the following.

Example 5. Let $W := \{a, b, c\}$. Let us define

(i) a partial order $\leq$ on U as:

$$a \leq a, c \leq a, c \leq c \text{ and } b \leq b.$$

(ii) a relation C on U as:

$$aCc, bCb, cCc, cCa.$$

By definition of compatibility relation: if $x \leq x', y \leq y'$ and $x'Cy'$ then xCy. We have:

$$c \leq c, c \leq a \text{ and } cCa \Rightarrow cCc.$$
$$c \leq c, a \leq a \text{ and } cCa \Rightarrow cCa.$$
$$a \leq a, c \leq c \text{ and } aCc \Rightarrow aCc.$$
$$c \leq a, c \leq c \text{ and } aCc \Rightarrow cCc.$$
$$c \leq c, c \leq c \text{ and } cCc \Rightarrow cCc.$$
$$b \leq b, b \leq b \text{ and } bCb \Rightarrow bCb.$$

Hence $(W, C, \leq)$ is a compatibility frame. Let us verify whether $(W, C, \leq)$ is a Kleene frame.

1. $\forall x \forall y (xCy \rightarrow yCx)$.

$$C \text{ is symmetric.}$$

2. $\forall x \exists y (xCy \wedge \forall z (yCz \rightarrow z \leq x))$.

$$aCc, cCc \text{ and } cCa \Rightarrow c, a \leq a.$$
$$bCb \text{ and } bCb \Rightarrow b \leq b.$$
$$cCa \text{ and } aCc \Rightarrow c \leq c.$$

3. $\forall x (xCx \vee \forall y (xCy \rightarrow y \leq x))$.

$$not(aCa) \text{ but } aCc \text{ and } c \leq a.$$

Also we have bCb and cCc.
Hence $(W, C, \leq)$ is a Kleene frame.

Denote by $\mathbb{F}_K$, the class of all Kleene frames. We have then arrived at

Theorem 27. *The following are all equivalent, for any $\alpha, \beta \in \mathcal{F}$.*

(a) $\alpha \vdash_{\mathcal{L}_K} \beta$.
(b) $\alpha \vDash_{\mathcal{A}_K} \beta$.
(c) $\alpha \vDash_{\mathcal{RS}_K} \beta$.
(d) $\alpha \vDash_{t,f} \beta$.
(e) $\alpha \vDash_{\mathbb{F}_K} \beta$.

Hence, the Kleene negation achieves a new position in Dunn's Lopsided Kite of Negations (Fig. 15).

In [29], an intuitionistic negation is defined as one having the Absurd ($a \wedge \sim a = 0$) along with Minimal (Johansson) property ($a \wedge b \leq c \Rightarrow a \wedge \sim c \leq \sim b$) properties. Now, the usual definition of pseudo complement in a bounded distributive lattice L is:

$$\sim a := max\{c \in L : a \wedge c = 0\}, \ a \in L. \tag{$*$}$$

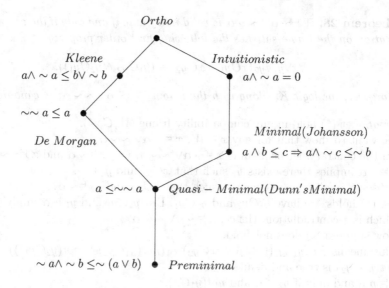

Fig. 15. Enhanced lopsided kite of negations

Before proceeding to the logic, we note through the results below that intuitionistic negation defined as Absurd with Minimal (Johansson) negation coincides with the pseudo complement defined in (*).

Proposition 6. *Let* $(K, \vee, \wedge, \sim, 0, 1)$ *be a* $K_i -$ *algebra with the following. For all* $a, b, c \in K$,

1. $a \wedge \sim a = 0$.
2. $a \wedge b \leq c \Rightarrow a \wedge \sim c \leq \sim b$.

Then $\sim a = max\{c \in K : a \wedge c = 0\}$.

Proof. Using 2: if $a \wedge z = 0$, then $z \leq \sim a$. By 1: we have $a \wedge \sim a = 0$ hence $\sim a = max\{c \in K : a \wedge c = 0\}$.

Proposition 7. *Let* $(K, \vee, \wedge, \sim, 0, 1)$ *be a bounded distributive pseudo complemented lattice, then for all* $a, b, c \in K$,

1. $a \wedge \sim a = 0$.
2. $a \wedge b \leq c \Rightarrow a \wedge \sim c \leq \sim b$.

Proof. Let us prove 2 only. Let $a \wedge b \leq c$, hence $a \wedge b \wedge \sim c \leq c \wedge \sim c = 0$. Using property of pseudo complement, we have $a \wedge \sim c \leq \sim b$.

Hence a Stone algebra can be defined as a $K_i -$ *algebra* with Absurd, Minimal (Johansson) and Stone property ($\sim a \vee \sim\sim a = 1$). Let us first prove a correspondence result for Stone property independently.

Theorem 28. $\top \vdash\sim \alpha\vee \sim\sim \alpha$ *is valid in a frame if and only if the compatibility relation on the frame satisfies the following first order property:*

$$\forall x \forall y_1 \forall y_2 (xCy_1 \wedge xCy_2 \rightarrow (y_1Cy_2 \wedge y_2Cy_1)). \qquad (*)$$

Moreover, the logic K_i *along with the axiom* $\top \vdash\sim \alpha\vee \sim\sim \alpha$, *is canonical.*

Proof. Let $(*)$ hold in any compatibility frame $(W, C, \leq)$.
We want to show that for any $x \in W$, $x \vDash\sim \alpha\vee \sim\sim \alpha$.
On the contrary, let us assume $x \nvDash\sim \alpha\vee \sim\sim \alpha$, i.e., $x \nvDash\sim \alpha$ and $x \nvDash\sim\sim \alpha$.
$x \nvDash\sim \alpha$ implies there exists y_1 such that xCy_1 and $y_1 \vDash \alpha$.
$x \nvDash\sim\sim \alpha$ implies there exists y_2 such that xCy_2 and $y_2 \vDash\sim \alpha$.
As, $(*)$ holds, we have y_1Cy_2 and y_2Cy_1. But y_2Cy_1 and $y_1 \vDash \alpha$ imply $y_2 \nvDash\sim \alpha$ which is a contradiction. Hence, $x \vDash\sim \alpha\vee \sim\sim \alpha$.
Now suppose $(*)$ does not hold.
This means $\exists x \exists y_1 \exists y_2 ((xCy_1 \wedge xCy_2) \wedge (not(y_1Cy_2) \vee not(y_2Cy_1)))$. Assume $not(y_1Cy_2)$ is true and define,
$z \vDash p$ if and only if $y_2 \leq z$ and $not(y_1Cz)$.
Let us first show that the relation $\vDash$ is well defined, i.e. hereditary. So let $z \vDash p$ and $z \leq z'$, hence, $y_2 \leq z'$. If y_1Cz' then using the frame condition we have y_1Cz, which is a contradiction.
We have $x \nvDash\sim p$ as xCy_2 and $y_2 \vDash p$ (as $not(y_1Cy_2)$). Also, $x \nvDash\sim\sim p$ as xCy_1 and y_1Cz imply $z \nvDash p$.
Hence we have $x \nvDash\sim p\vee \sim\sim p$.
Canonicity :
First observe that for any prime theory P and any formula α of this logic, $\sim \alpha\vee \sim\sim \alpha \in P$, which implies either $\sim \alpha \in P$ or $\sim\sim \alpha \in P$.
Now let PC_cQ_1 and PC_cQ_2, for any prime theories P, Q_1 and Q_2. We want to show that $Q_1C_cQ_2$ and $Q_2C_cQ_1$. Let us show $Q_1C_cQ_2$.
Let $\sim \alpha \in Q_1$ but we have PC_cQ_1 hence $\sim\sim \alpha \notin P$. This means $\sim \alpha \in P$. We also have PC_cQ_2 which will give us $\alpha \notin Q_2$. Hence $Q_1C_cQ_2$.
Similarly one can show $Q_2C_cQ_1$.

The characterizations of Absurd and Minimal (Johansson) properties are given in [24, 29].

Theorem 29. ([24, 29])

$$\frac{\alpha \wedge \beta \vdash \gamma}{\alpha \wedge \sim \gamma \vdash\sim \beta}$$

1. *The rule* $\dfrac{\alpha \wedge \beta \vdash \gamma}{\alpha \wedge \sim \gamma \vdash\sim \beta}$ *is valid in a compatibility frame if and only if the following frame condition holds:*

$$\forall x \forall y (xCy \rightarrow \exists z (x \leq z \wedge y \leq z \wedge xCz)).$$

2. $\alpha \wedge \sim \alpha \vdash \perp$ *is valid, precisely in the class of all compatibility frames satisfying the frame condition:*

$$\forall x (xCx).$$

Moreover, canonicity holds in respective cases.

Hence Stone negation can be visualized as 'impossibility' as well as modal operator. Let $\mathcal{L}_S$ denote the logic K_i along with following rules and postulates.

$$\frac{\alpha \wedge \beta \vdash \gamma}{}$$

1. $\alpha \wedge \sim \gamma \vdash \sim \beta$.
2. $\alpha \wedge \sim \alpha \vdash \bot$.
3. $\top \vdash \sim \alpha \vee \sim\sim \alpha$.

Definition 24. *Let us call a compatibility frame* $(W, C, \leq)$ *a Stone frame if it satisfies the frame conditions:*

1. $\forall x \forall y (xCy \rightarrow \exists z(x \leq z \wedge y \leq z \wedge xCz))$.
2. $\forall x(xCx)$.
3. $\forall x \forall y_1 \forall y_2 (xCy_1 \wedge xCy_2 \rightarrow (y_1 C y_2 \wedge y_2 C y_1))$.

Let us provide an example of such a frame.

Example 6. Let us consider the poset as in Example 5. Define the relation C on U as:

$$aCa, aCc, bCb, cCc, cCa.$$

As in Example 5, we can easily show that the structure $(W, C, \leq)$ is a compatibility frame. Let us show that it is a Stone frame.
1. $\forall x \forall y (xCy \rightarrow \exists z(x \leq z \wedge y \leq z \wedge xCz))$.

$$aCa \text{ and } a \leq a, aCa.$$
$$aCc \text{ and } c \leq a, a \leq a, aCa.$$
$$bCb \text{ and } b \leq b, bCb.$$
$$cCc \text{ and } c \leq c, cCc.$$
$$cCa \text{ and } a \leq a, c \leq a, cCa.$$

2. $\forall x(xCx)$.

$$C \text{ is reflexive relation.}$$

3. $\forall x \forall y_1 \forall y_2 (xCy_1 \wedge xCy_2 \rightarrow (y_1 C y_2 \wedge y_2 C y_1))$.

$$aCa, aCc \text{ and } cCa, aCc.$$
$$cCc, cCa \text{ and and } cCa, aCc.$$

Hence the frame $(W, C, \leq)$ is a Stone frame.

Let us denote by $\mathbb{F}_S$, the class of all Stone frames, and let $\mathcal{A}_S$ denote the class of all Stone algebras. Then we can conclude the following.

Theorem 30. *For any* $\alpha, \beta \in \mathcal{F}$. *The following are equivalent.*

(i) $\alpha \vdash_{\mathcal{L}_S} \beta$.
(ii) $\alpha \vDash_{\mathcal{A}_S} \beta$.
(iii) $\alpha \vDash_{\mathcal{F}_S} \beta$.

So, we can conclude that Stone negation can be positioned (Fig. 16) in Dunn's Lopsided Kite of Negations.

3.2   Semantics in Dual Kite of Negations

Perp semantics is defined using compatibility frames. Dunn shows [29] that the dual of these negations in the kite of negations can be studied via dual of compatibility frames, namely *exhaustive frames* (cf. Definition 26 below). Through the semantics in exhaustive frames, it has been shown that the negations in the dual Lopsided kite of negations (Fig. 17) can be treated as modal operators. But in this case, modalities are interpreted as 'unnecessity'. Let us first present the minimal logic in this framework.

The logic K_u contains all the postulates and rules of the logic $BDLL$ (cf. Sect. 1) along with:

$$\frac{\alpha \vdash \beta}{\neg\beta \vdash \neg\alpha}$$

1.   (Contraposition)
2. $\neg(\alpha \wedge \beta) \vdash \neg\alpha \vee \neg\beta$ ($\wedge$-linearity).
3. $\neg\top \vdash \bot$ (dual-Nor).

The algebraic semantics of the logic K_u is given by the class of all $K_u - algebras$ which can be defined as follows.

Definition 25. *A $K_u-algebra$ is a structure $(K, \vee, \wedge, \neg, 0, 1)$ with the following properties.*

1. $(K, \vee, \wedge, 0, 1)$ *is a bounded distributive lattice.*
2. $\neg(a \wedge b) = \neg a \vee \neg b.$
3. $a \leq b \Rightarrow \neg b \leq \neg a.$

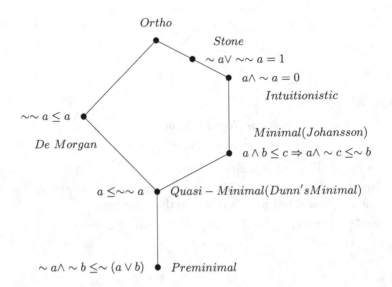

Fig. 16. Enhanced Lopsided Kite of Negations

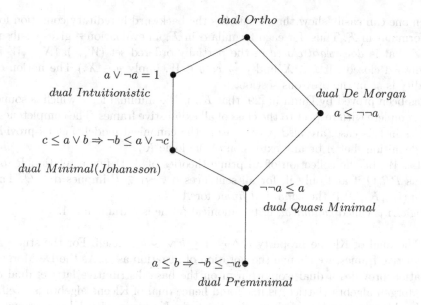

dual Ortho

$a \lor \neg a = 1$

dual Intuitionistic

dual De Morgan

$a \leq \neg\neg a$

$c \leq a \lor b \Rightarrow \neg b \leq a \lor \neg c$

dual Minimal(Johansson)

$\neg\neg a \leq a$

dual Quasi Minimal

$a \leq b \Rightarrow \neg b \leq \neg a$

dual Preminimal

Fig. 17. Dual Lopsided Kite of Negations

4. $\neg 1 = 0$.

It is observed in [29], that dual of compatibility frames are exhaustive frames. In fact, any compatibility frame is an exhaustive frame, but semantics when considering the interpretation of impossibility is studied in compatibility frames, while semantics when considering unnecessity is studied in exhaustive frames.

Definition 26. *An* exhaustive frame *is triple* $(W, R, \leq)$ *which satisfies the following conditions.*

1. $(W, \leq)$ *is a partially ordered set.*
2. $\leq \circ R \circ \leq^{-1} \subseteq R$.

Definition 27. *A relation* $\vDash$ *between points of* W *and propositional variables in* $\mathcal{P}$ *is called an* evaluation *, if it satisfies the* backward hereditary *condition:*

– *if* $x \vDash p$ *and* $y \leq x$ *then* $y \vDash p$, $x, y \in W$.

The evaluation relation $\vDash$ can be recursively extended to the set $\mathcal{F}$ as in the previous section. Let $x \in W$.

1. $x \vDash \alpha \land \beta$ if and only if $x \vDash \alpha$ and $x \vDash \beta$.
2. $x \vDash \alpha \lor \beta$ if and only if $x \vDash \alpha$ or $x \vDash \beta$.
3. $x \vDash \top$.
4. $x \nvDash \bot$.
5. $x \vDash \neg\alpha$ if and only if $\exists y (xRy \land y \nvDash \alpha)$.

Then one can easily show that $\vDash$ satisfies the backward hereditary condition for all formulas in $\mathcal{F}$. Thus, for each formula α in $\mathcal{F}$, an evaluation $\vDash$ gives a subset of W that is *downward closed* in the partially ordered set $(W, \leq)$. ($X \subseteq W$ is downward closed , if $x \in X$ and $y \leq x$, $y \in W$, imply $y \in X$.) The notion of validity is as in the previous section.

It has been proved by Dunn in [29] that K_u is the minimal logic which is sound and complete with respect to the class of all exhaustive frames. The completeness results in this case, are also proved using the canonical model. Let us provide the definition. Let Λ be any extension of the logic K_u.

Let W_c be the collection of all prime theories of Λ. Define a relation R_c on W_c as PR_cQ if and only if, for all sentences α , $\neg\alpha \notin P$ implies $\alpha \in Q$. The tuple $(W_c, R_c, \supseteq)$ is the *canonical frame* for Λ.

A logic Λ is called *canonical*, if its canonical frame is a frame for Λ.

The dual of Kleene property $a \wedge \sim a \leq b \vee \sim b$ is itself. For the study in exhaustive frames, we change the notation of negation as $\neg$. As the De Morgan negation provides a dual isomorphism on the base distributive lattice, dual of De Morgan algebra (lattice) is itself and hence dual of Kleene algebra is itself.

In this section we first show that the logic K_u along with Kleene property is sound and complete with respect to a class of exhaustive frames, i.e., Kleene negation can be visualized as a modal (unnecessity) operator. It turns out that the same first order property as in Theorem 25 is used for characterization.

Theorem 31. $\alpha \wedge \neg\alpha \vdash \beta \vee \neg\beta$ *is valid in an exhaustive frame, if and only if the frame satisfies the following first order property:*

$$\forall x(xRx \vee \forall y(xRy \rightarrow y \leq x)). \tag{*}$$

The canonical frame for the logic $K_u + \alpha \wedge \neg\alpha \vdash \beta \vee \neg\beta$ satisfies (*).

Proof. Consider any exhaustive frame $(W, R, \leq)$, let (*) hold, and let $x \in W$. Suppose xRx, then $x \vDash \beta \vee \neg\beta$. Hence, trivially if $x \vDash \alpha \wedge \sim \alpha$ then $x \vDash \beta \vee \sim \beta$.

Now suppose $\forall y(xRy \rightarrow y \leq x)$ is true. Let $x \vDash \alpha$ and xRy. Then $y \leq x$. This implies $y \vDash \alpha$ using backward hereditary property. Hence $x \nvDash \alpha \wedge \neg\alpha$. Trivially then, if $x \vDash \alpha \wedge \sim \alpha$ then $x \vDash \beta \vee \sim \beta$.

Let (*) not hold.

This implies $\exists x(not(xRx) \wedge \exists y(xRy \wedge y \nleq x))$. Let us define, for any $z, w \in W$,

$$z \vDash p \text{ if and only if } z \leq x.$$

$$w \vDash q \text{ if and only if } xRw.$$

We show that $\vDash$ is an evaluation. Let $z \vDash q$ and $z' \leq z$. Using frame condition: $x \leq x$, xRz and $z \leq^{-1} z'$ imply xRz'. So, $z' \vDash q$.

Furthermore, $x \vDash p$, as $x \leq x$. We also have $x \vDash \neg p$: $\exists y$ such that xRy and $y \nleq x$. So $x \vDash p \wedge \neg p$.

On the other hand, $x \nvDash q$ as $not(xRx)$. Let z such that xRz hence using definition of $\vDash$ we have $x \nvDash \neg q$. So, we have $x \vDash p \wedge \neg p$ but $x \nvDash q \vee \neg q$.

Canonicity :

Let $not(PR_cP)$. By definition of R_c, then there exists an $\alpha \in \mathcal{F}$ such that $\alpha, \neg\alpha \notin P$, but then $\alpha \vee \neg\alpha \notin P$. Hence for all $\beta \in \mathcal{F}$, $\beta \wedge \neg\beta \notin P$. So, for all β, either $\beta \notin P$ or $\neg\beta \notin P$.

Now let PR_cQ and let $\beta \notin Q$. Hence $\neg\beta \in P$ but then from above $\beta \notin P$. So, $P \subseteq Q$.

Now, consider the logic $\mathcal{L}_{K'}$ which contains all the rules and postulates of K_u along with the following postulates:

1. $\neg\neg\alpha \vdash \alpha$.
2. $\alpha \vdash \neg\neg\alpha$.
3. $\alpha \wedge \neg\alpha \vdash \beta \vee \neg\beta$.

The characterization of the postulates $\neg\neg\alpha \vdash \alpha$ and $\alpha \vdash \neg\neg\alpha$ in an exhaustive frame is given by Dunn in [29].

Theorem 32 ([29]).

1. $\alpha \vdash \neg\neg\alpha$ *is valid in an exhaustive frame* $(W, R, \leq)$ *if and only if the frame satisfies the following frame condition:*

$$\forall x \exists y (xRy \wedge \forall z(yRz \rightarrow z \leq x)).$$

2. $\neg\neg\alpha \vdash \alpha$ *is valid in an exhaustive frame* $(W, R, \leq)$ *if and only if the frame satisfies the following frame condition:*

$$\forall x \forall y (xRy \rightarrow yRx).$$

Moreover, canonicity holds in respective cases.

Definition 28. *Let us call an exhaustive frame* $(W, R, \leq)$ *a dual Kleene frame if it satisfies the same conditions as a Kleene frame, only in terms of the exhaustive relation* R, *i.e. we have,*

1. $\forall x \forall y (xRy \rightarrow yRx)$.
2. $\forall x \exists y (xRy \wedge \forall z(yRz \rightarrow z \leq x))$.
3. $\forall x (xRx \vee \forall y(xRy \rightarrow y \leq x))$.

Example 7. Consider the Kleene frame as in Example 5. It is easy to check that $(W, C, \leq)$ is also a dual Kleene frame. In other words, C is also an exhaustive relation on W.

Definition 29. *Let us call an algebra* $\mathcal{K} := (K, \vee, \wedge, \neg, 0, 1)$, *a dual Kleene algebra if it satisfies the following conditions. For all* $a, b \in K$,

1. $(K, \vee, \wedge, \neg, 0, 1)$ *is a* K_u-*algebra.*
2. $\neg\neg a \leq a$.
3. $a \leq \neg\neg a$.
4. $a \wedge \neg a \leq b \vee \neg b$.

It is easy to show that any dual Kleene algebra is a Kleene algebra (via the transformation $\neg \mapsto \sim$), and any Kleene algebra is a dual Kleene algebra (via the transformation $\sim \mapsto \neg$). Hence the class $\mathcal{A}_{K'}$, of all dual Kleene algebras coincides with the class $\mathcal{A}_K$ of all Kleene algebras (via, $\neg \leftrightarrow \sim$).

Denote by $\mathbb{F}_{DK}$, the class of all dual Kleene frames. Let α be a formula in the logic $\mathcal{L}_{K'}$. Let us denote by α^*, the formula in the logic $\mathcal{L}_K$ which is obtained from α on replacing each occurrence of $\neg$ by $\sim$.
We have then arrived at

Theorem 33. *The following are all equivalent, for any $\alpha, \beta \in \mathcal{F}$.*

(a) $\alpha \vdash_{\mathcal{L}_{K'}} \beta$.
(b) $\alpha^* \vdash_{\mathcal{L}_K} \beta^*$.
(c) $\alpha^* \vDash_{\mathcal{A}_K} \beta^*$.
(d) $\alpha^* \vDash_{\mathcal{RS}_K} \beta^*$.
(e) $\alpha^* \vDash_{t,f} \beta^*$.
(f) $\alpha^* \vDash_{\mathbb{F}_K} \beta^*$.
(g) $\alpha \vDash_{\mathbb{F}_{DK}} \beta$.
(h) $\alpha \vDash_{\mathcal{A}_{K'}} \beta$.

The dual of pseudo complement in a distributive lattice L is the dual pseudo complement, defined as follows:

$$\neg a := min\{c \in L : a \vee c = 1\}, \ a \in L.$$

Before proceeding, let us note the equivalence between dual Intuitionistic negation in the dual Kite of negations, and the dual pseudo complement. The proof of the following proposition is very similar to that of Propositions 6 and 7.

Proposition 8. *Let $(K, \vee, \wedge, \neg, 0, 1)$ be a distributive lattice with a unary operator $\neg$. The following are equivalent.*

1. $(K, \vee, \wedge, \neg, 0, 1)$ *is a K_u – algebra which satisfies the dual Intuitionistic and dual minimal (Johansson) properties.*
2. $(K, \vee, \wedge, \neg, 0, 1)$ *is a dual pseudo complemented lattice.*

Thus a dual Stone algebra is a K_u-algebra with negation satisfying dual Intuitionistic, dual minimal (Johansson) and double Stone ($\neg a \wedge \neg \neg a = 0$) properties.
The characterization for the dual Intuitionistic property is given in [29].

Theorem 34. (Dunn [29]).

1. $\top \vdash \alpha \vee \neg \alpha$ *is valid in an exhaustive frame if and only if the frame satisfies the following first order condition:*

$$\forall x (xRx).$$

2. *The rule* $\dfrac{\gamma \vdash \alpha \vee \beta}{\neg \beta \vdash \alpha \vee \neg \gamma}$ *is valid in an exhaustive frame if and only if the frame satisfies the following first order condition:*

$$\forall x \forall y (xRy \rightarrow \exists z (x \leq z \wedge y \leq z \wedge xRz)).$$

Now, let us characterize dual pseudo complement in an exhaustive frame.

Theorem 35. $\neg\alpha \wedge \neg\neg\alpha \vdash \bot$ *is valid in an exhaustive frame if and only if the frame satisfies the following first order property:*

$$\forall x \forall y_1 \forall y_2 (xRy_1 \wedge xRy_2 \rightarrow (y_1 Ry_2 \wedge y_2 Ry_1)). \qquad (*)$$

Moreover, it is canonical.

Proof. Let $(*)$ hold in an exhaustive frame $(W, R, \leq)$ and let $x \in W$.
We want to show that $x \nvDash \neg\alpha \wedge \neg\neg\alpha$. Suppose $x \vDash \neg\alpha \wedge \neg\neg\alpha$, this implies $x \vDash \neg\alpha$ and $x \vDash \neg\neg\alpha$. Hence there exist y_1, y_2 such that xRy_1, xRy_2 and $y_1 \nvDash \alpha$, $y_2 \nvDash \neg\alpha$. As $(*)$ holds, $y_2 Ry_1$. But then $y_2 \vDash \neg\alpha$, which is a contradiction. Hence, $x \nvDash \neg\alpha \wedge \neg\neg\alpha$.
Suppose $(*)$ does not hold.
This means $\exists x \exists y_1 \exists y_2 ((xRy_1 \wedge xRy_2) \wedge (not(y_1 Ry_2) \vee not(y_2 Ry_1)))$. Assume $not(y_1 Ry_2)$ and define,
$z \vDash p$ if and only if $y_1 Rz$.
Then $\vDash$ is well-defined: let $z \vDash p$ and $z' \leq z$, then using the property of exhaustive frames we have,

$$y_1 \leq y_1, y_1 Rz \text{ and } z \geq z' \text{ imply } y_1 Rz'.$$

Hence $z' \vDash p$.
Now, clearly we have

1. $x \vDash \neg p$, as, xRy_2 and $y_2 \nvDash p$.
2. $x \vDash \neg\neg p$ as, xRy_1 and $y_1 \nvDash \neg p$.

Hence $x \vDash \neg p \wedge \neg\neg p$. So, $\neg p \wedge \neg\neg p \vdash \bot$ is not valid in $(W, R, \leq)$.

Canonicity:

First observe that for any prime theory P and any formula α, $\neg\alpha \wedge \neg\neg\alpha \notin P$, as we have assumed our logic contains $\neg\alpha \wedge \neg\neg\alpha \vdash \bot$. This implies $\neg\alpha \notin P$ or $\neg\neg\alpha \notin P$.
Now let PR_cQ_1 and PR_cQ_2. We want to show that $Q_1 R_c Q_2$ and $Q_2 R_c Q_1$. Let us show $Q_1 R_c Q_2$:
So, let $\neg\alpha \notin Q_1$ but we have PR_cQ_1 hence $\neg\neg\alpha \in P$. This means $\neg\alpha \notin P$. We also have PR_cQ_2 which will give us $\alpha \in Q_2$. Hence $Q_1 R_c Q_2$.
Similarly one can show $Q_2 R_c Q_1$.

Let $\mathcal{L}_{DS}$ denote the logic K_u along with following rules and postulates.

$$\frac{\gamma \vdash \alpha \vee \beta}{}$$
1. $\neg\beta \vdash \alpha \vee \neg\gamma$.
2. $\top \vdash \alpha \vee \neg\alpha$.
3. $\neg\alpha \wedge \neg\neg\alpha \vdash \bot$.

Definition 30. *Let us call an exhaustive frame $(W, R, \leq)$ a dual Stone frame if it satisfies the following frame conditions,*

1. $\forall x \forall y(xRy \rightarrow \exists z(x \leq z \land y \leq z \land xRz))$.
2. $\forall x(xRx)$.
3. $\forall x \forall y_1 \forall y_2(xRy_1 \land xRy_2 \rightarrow (y_1Ry_2 \land y_2Ry_1))$.

Example 8. Let us consider the structure $(W, C, \leq)$ as in Example 5. It can be easily shown that the structure $(W, R, \leq)$ is a dual Stone frame, i.e., C is an exhaustive relation as well.

Let us denote by $\mathbb{F}_{DS}$, the class of all dual Stone frames, and let $\mathcal{A}_{DS}$ denote the class of dual Stone an abstract algebra. Then we can conclude the following.

Theorem 36. *For any* $\alpha, \beta \in \mathcal{F}$, *the following are equivalent.*

(a) $\alpha \vdash_{\mathcal{L}_{DS}} \beta$.
(b) $\alpha \vDash_{\mathcal{A}_{DS}} \beta$.
(c) $\alpha \vDash_{\mathcal{F}_{DS}} \beta$.

Let us end this section with the note that dual Kleene and dual Stone negations can be given positions in Dunn's dual (Lopsided) Kite of Negation (Fig. 18).

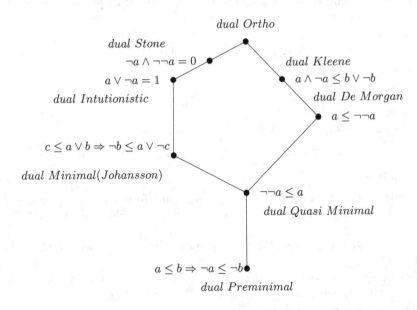

Fig. 18. Enhanced Dual Lopsided Kite of Negations

3.3   Semantics in United Kite of Negations

Dunn further provided a uniform semantics for combining both the kites to give the 'united kite' (Fig. 19). Let us first provide the rules and postulates of the minimal logic in this context. Let K_- be the logic which contains the logic $BDLL$ along with the following postulates and rules.

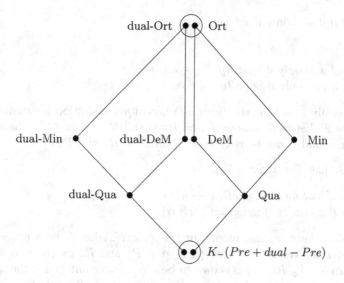

Fig. 19. Dunn's United Kite of Negations

1. $\sim \alpha \wedge \sim \beta \vdash \sim (\alpha \vee \beta)$.
2. $\top \vdash \sim \bot$.
3. $\neg(\alpha \wedge \beta) \vdash \neg \alpha \vee \neg \beta$.
4. $\neg \top \vdash \bot$.

$$\dfrac{\alpha \vdash \beta}{\sim \beta \vdash \sim \alpha}$$

5. $\sim \beta \vdash \sim \alpha$.

$$\dfrac{\alpha \vdash \beta}{\neg \beta \vdash \neg \alpha}$$

6. $\neg \beta \vdash \neg \alpha$.
7. $\sim \alpha \wedge \neg \beta \vdash \neg(\alpha \vee \beta)$.
8. $\sim (\alpha \wedge \beta) \vdash \sim \alpha \vee \neg \beta$.

The semantics of the logic K_- is defined in a K_- *frame*, which is a triple $(W, R, \leq)$ with the following properties.

1. $(W, \leq)$ is a partially ordered set.
2. $\leq^{-1} \circ R \subseteq R \circ \leq$.
3. $\leq \circ R \subseteq R \circ \leq^{-1}$.

The semantic clauses are as in the previous sections. Let us recall Definition 20.

Definition 31. *A relation $\vDash$ between points of W and propositional variables in $\mathcal{P}$ is called an* evaluation, *if it satisfies the* forward hereditary *condition:*

$-$ *if $x \vDash p$ and $x \leq y$ then $y \vDash p$, for all $x, y \in W$.*

Analogous to previous sections, an evaluation can be extended to $\mathcal{F}$. Let $x \in W$.

1. $x \vDash \alpha \wedge \beta$ if and only if $x \vDash \alpha$ and $x \vDash \beta$.

2. $x \vDash \alpha \vee \beta$ if and only if $x \vDash \alpha$ or $x \vDash \beta$.
3. $x \vDash \top$.
4. $x \nvDash \bot$.
5. $x \vDash \sim \alpha$ if and only if $\forall y(xRy \rightarrow y \nvDash \alpha)$.
6. $x \vDash \neg \alpha$ if and only if $\exists y(xRy \wedge y \nvDash \alpha)$.

It can be easily seen that the (forward) hereditary condition is satisfied by all formulas in $\mathcal{F}$. Let us denote by $R_\neg := R\circ \leq$ and $R_\sim := R\circ \leq^{-1}$. The semantic clauses for $\sim$ and $\neg$ can be re-defined in terms of $R_\neg$ and $R_\sim$:

Lemma 3. [29] *For any $x \in W$,*

1. *$x \vDash \sim \alpha$ if and only if $\forall y(xR_\sim y \rightarrow y \nvDash \alpha)$.*
2. *$x \vDash \neg \alpha$ if and only if $\exists y(xR_\neg y \wedge y \nvDash \alpha)$.*

Let us define the canonical model for the logic K_-. Let P be a prime theory, and consider the two sets: $P_\neg := \{\alpha : \neg \alpha \notin P\}$ and $P_\sim := \{\alpha : \sim \alpha \notin P\}$. In [29], the triple $(W_c, R_c, \subseteq_c)$ is shown to be a K_- frame and called the canonical model for the logic K_-, where W_c is the collection of all prime theories, $\subseteq_c$ is the inclusion and R_c is defined as follows: PR_cQ if and only if $P_\neg \subseteq Q \subseteq P_\sim$. The following lemma concerning prime theories is proved in [29].

Lemma 4.

1. *Assume that $P_\neg \subseteq Q$. Then there exists a prime theory S such that PR_cS and $S \subseteq Q$, i.e., $PR_{C\neg}Q$.*
2. *Assume that $Q \subseteq P_\sim$. Then there exists a prime theory S such that PR_cS and $Q \subseteq S$, i.e., $PR_{C\sim}Q$.*

The logic K_- is sound and complete with respect to the class of all $K_- - algebras$, which are defined as follows:

Definition 32. *A $K_- - algebra$ is a structure $(K, \vee, \wedge, \sim, \neg, 0, 1)$ with the following properties.*

1. *$(K, \vee, \wedge, 0, 1)$ is a bounded distributive lattice.*
2. *$\sim (a \vee b) = \sim a \wedge \sim b$.*
3. *$\sim 0 = 1$.*
4. *$a \leq b \Rightarrow \sim b \leq \sim a$.*
5. *$\neg(a \wedge b) = \neg a \vee \neg b$.*
6. *$a \leq b \Rightarrow \neg b \leq \neg a$.*
7. *$\neg 1 = 0$.*
8. *$(\sim a \wedge \neg b) \leq \neg(a \vee b)$.*
9. *$\sim (a \wedge b) \leq (\sim a \vee \neg b)$.*

In this section, we investigate the Kleene property $\alpha \wedge \sim \alpha \vdash \beta \vee \sim \beta$ and its dual $\alpha \wedge \neg \alpha \vdash \beta \vee \neg \beta$ in K_- frames.

Theorem 37. $\alpha \wedge \sim \alpha \vdash \beta \vee \sim \beta$ *is valid in a K_- frame $(W, R, \le)$ if and only if the K_- frame satisfies the following condition:*

$$\forall x(xR_\sim x \vee \forall y(xR_\sim y \rightarrow y \le x)). \qquad (*)$$

Moreover, canonicity holds.

Proof. The proof of the 'if' part of this theorem is very similar to the proof of Theorem 25. Let us present the proof of the 'only if' part.
Let $(*)$ not hold in a K_- frame $(W, R, \le)$.
This implies $\exists x(not(xR_\sim x) \wedge \exists y(xR_\sim y \wedge y \not\le x))$. Let us define, for any $z, w \in W$,

$$z \vDash p \text{ if and only if } x \le z \text{ and } not(xR_\sim z),$$

$$w \vDash q \text{ if and only if } y \le w.$$

Let us show that $\vDash$ is indeed an evaluation. Let $z \vDash p$ and $z \le z'$. Then $x \le z'$. If $xR_\sim z'$, then by the definition of $R_\sim$, there exists z'' such that xRz'' and $z' \le z''$. We have $z \le z' \le z''$ which implies $xR_\sim z$. which is a contradiction to the fact that $z \vDash p$.

Now, analogous to the proof of Theorem 25, it is easy to establish that $x \vDash p \wedge \sim p$ but $x \nvDash q \vee \sim q$.

Canonicity:

We have to show that in the canonical frame,

$$\forall P(PR_{c\sim}P \vee \forall Q(PR_{c\sim}Q \rightarrow Q \subseteq P)).$$

So, let $not(PR_{c\sim}P)$.

Assume $P \subseteq P_\sim$. Then using Lemma 4, $PR_{c\sim}P$, which is a contradiction. Hence $P \not\subseteq P_\sim$. There exists a formula $\alpha \in \mathcal{F}$ such that $\alpha \in P$ but $\alpha \notin P_\sim$. But $\alpha \notin P_\sim$ imply $\sim \alpha \in P$. We have $\alpha \wedge \sim \alpha \in P$. Hence for any $\beta \in \mathcal{F}$ we have either $\beta \in P$ or $\sim \beta \in P$.

Let $PR_{c\sim}Q$. There exists Q' such that PR_cQ' and $Q' \supseteq Q$.

Now, let $\gamma \in Q$ – this implies $\gamma \in Q'$. But as PR_cQ' we have $\gamma \in P_\sim$. Hence $\sim \gamma \notin P$. Hence $\gamma \in P$.

Theorem 38. $\alpha \wedge \neg\alpha \vdash \beta \vee \neg\beta$ *is valid in a K_- frame $(W, R, \le)$ if and only if the K_- frame satisfies the following first order condition:*

$$\forall x(xR_\neg x \vee \forall y(xR_\neg y \rightarrow x \le y)). \qquad (*)$$

Moreover, the canonical model satisfies this frame condition.

Proof. Assume $(*)$ holds in a K_- frame $(W, R, \le)$.
Let $x \in W$, $xR_\neg x$, and $x \nvDash \beta$. So $x \vDash \neg\beta$. Hence $x \vDash \beta \vee \sim \beta$. So, if $x \vDash \alpha \wedge \sim \alpha$ then $x \vDash \beta \vee \sim \beta$ is vacuously true.
Let $\forall y(xR_\neg y \rightarrow x \le y)$ be true. Let $x \vDash \alpha \wedge \neg\alpha$ i.e. $x \vDash \alpha$ and $x \vDash \neg\alpha$. But $x \vDash \neg\alpha$ implies there exists y such that $xR_\neg y$ and $y \nvDash \alpha$. Using hereditary property of

the semantic consequence relation, we have $x \nvDash \alpha$ which is a contradiction. Hence $x \nvDash \alpha \wedge \neg \alpha$. So, vacuously $x \vDash \beta \vee {\sim} \beta$.

Let us assume (*) does not hold.

This implies $\exists x (not(xR_\neg x) \wedge \exists y (xR_\neg y \wedge x \nleq y))$. Let us define, for any $z, w \in W$,

$$z \vDash p \text{ if and only if } x \leq z.$$

$$w \vDash q \text{ if and only if } xR_\neg w.$$

Similar to the proof of Theorem 31, one can easily show that $\vDash$ is indeed an evaluation and $x \vDash p \wedge \neg p$ but $x \nvDash q \vee \neg q$.

Canonicity:

Let $not(PR_{c\neg}P)$.

Assume $P_\neg \subseteq P$. Then using Lemma 4, $PR_{c\neg}P$, which is a contradiction. Hence $P_\neg \nsubseteq P$. There exists a formula $\alpha \in \mathcal{F}$ such that $\alpha \in P_\neg$ but $\alpha \notin P$. But $\alpha \in P_\neg$ imply $\neg \alpha \notin P$. We have $\alpha \vee \neg \alpha \notin P$. Hence for any $\beta \in \mathcal{F}$ we have either $\beta \notin P$ or $\neg \beta \notin P$.

Let $PR_{c\neg}Q$. There exists Q' such that PR_cQ' and $Q' \subseteq Q$.

Now, Let $\gamma \notin Q$ which implies $\gamma \notin Q'$. But as PR_cQ' we have $\gamma \notin P_\neg$. So $\neg \gamma \in P$, which means $\gamma \notin P$. Hence $P \subseteq Q$.

Hence we have the following.

Theorem 39.

1. The logic $K_- + \alpha \wedge {\sim} \alpha \vdash \beta \vee {\sim} \beta + \alpha \wedge \neg \alpha \vdash \beta \vee \neg \beta$ is sound and complete with respect to the class of all K_- frames which satisfy the first order conditions $\forall x (xR_{\sim} x \vee \forall y (xR_{\sim} y \to y \leq x))$ and $\forall x (xR_\neg x \vee \forall y (xR_\neg y \to x \leq y))$.
2. $K_- + \alpha \wedge {\sim} \alpha \vdash \beta \vee {\sim} \beta$ is sound and complete with respect to the class of all K_-−algebras satisfying the property $a \wedge {\sim} a \leq b \vee {\sim} b$ and $a \wedge \neg a \leq b \vee \neg b$.

Definition 33. A K_- − algebra $(K, \vee, \wedge, {\sim}, \neg, 0, 1)$ is called a double Kleene algebra if it satisfies the following properties.

(i) $a \leq {\sim}{\sim} a$, $\neg\neg a \leq a$.
(ii) ${\sim}{\sim} a \leq a$, $a \leq \neg\neg a$.
(iii) $a \wedge {\sim} a \leq b \vee {\sim} b$, $a \wedge \neg a \leq b \vee \neg b$.

Definition 34. A K_- frame $(W, R, \leq)$ is called a double Kleene frame if it satisfies the following properties.

(i) $\forall x \forall y (xR_{\sim} y \to yR_{\sim} x)$, $\forall x \forall y (xR_\neg y \to yR_\neg x)$.
(ii) $\forall x \exists y (xR_{\sim} y \wedge \forall z (yR_{\sim} z \to z \leq x))$, $\forall x \exists y (xR_\neg y \wedge \forall z (yR_\neg z \to x \leq z))$.
(iii) $\forall x (xR_{\sim} x \vee \forall y (xR_{\sim} y \to y \leq x))$, $\forall x (xR_\neg x \vee \forall y (xR_\neg y \to x \leq y))$

Let $\mathcal{F}_{doubleK}$ and $\mathcal{A}_{doubleK}$ denote the classes of all double Kleene frames and double Kleene algebras respectively.

Now consider the logic $\mathcal{L}_{doubleK}$, which is K_u along with the following rules and postulates.

(i) $\alpha \vdash \sim\sim \alpha$, $\neg\neg\alpha \vdash \alpha$.

(ii) $\sim\sim \alpha \leq \alpha$, $\alpha \leq \neg\neg\alpha$.

(iii) $\alpha \wedge \sim \alpha \leq \beta \vee \sim \beta$, $\alpha \wedge \neg\alpha \leq \beta \vee \neg\beta$.

Then we have

Theorem 40. *For any $\alpha, \beta \in \mathcal{F}$, the following are equivalent.*

1. $\alpha \vdash_{\mathcal{L}_{doubleK}} \beta$.
2. $\alpha \vDash_{\mathcal{F}_{doubleK}} \beta$.
3. $\alpha \vDash_{\mathcal{A}_{doubleK}} \beta$.

The positions of Kleene negation and its dual in the united kite can be seen in Fig. 20.

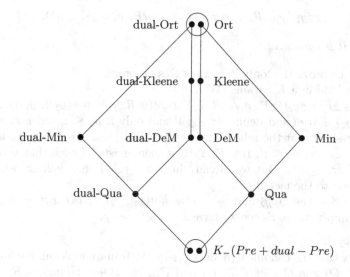

Fig. 20. Dunn's United Kite of Negations

The rough set representation of the class of regular double Stone algebras was obtained by Comer [19]. A sequent calculus for the logic of regular double Stone algebra and its rough set semantics was provided by Dai ([20], cf. Banerjee and Khan [6]). In this section we present another semantics for the logic of regular double Stone algebras. In Sect. 1, we defined these algebras. They can be re-defined using the $K_- - algebras$.

Theorem 41. *Let $(K, \vee, \wedge, \sim, \neg, 0, 1)$ be a bounded distributive lattice with unary operators $\sim, \neg$. The following are equivalent.*

1. $(K, \vee, \wedge, \sim, \neg, 0, 1)$ *is a regular double Stone algebra.*
2. $(K, \vee, \wedge, \sim, \neg, 0, 1)$ *is a $K_- - algebra$ with the following properties: $\forall a, b \in K$,*
   (i) $a \wedge \sim a = 0$.

(ii) $a \wedge b \leq c \Rightarrow a \wedge \sim c \leq \sim b$.
(iii) $\sim a \vee \sim\sim a = 1$.
(iv) $a \vee \neg a = 1$.
(v) $c \leq a \vee b \Rightarrow \neg b \leq a \vee \neg c$.
(vi) $\neg a \wedge \neg\neg a = 0$.
(vii) $a \wedge \neg a \leq b \vee \sim b$ *(regularity).*

Proof. Combining the proofs of Propositions 6, 7 and 8, we get the results.

Now let us characterize Stone, dual Stone and regularity properties in a K_- frame.

Theorem 42. $\top \vdash \sim \alpha \vee \sim\sim \alpha$ *is valid in a K_- frame $(W, R, \leq)$ if and only if the K_- frame satisfies the following first order property:*

$$\forall x \forall y_1 \forall y_2 (x R_\sim y_1 \wedge x R_\sim y_2 \rightarrow (y_1 R_\sim y_2 \wedge y_2 R_\sim y_1)). \tag{*}$$

Moreover, it is canonical.

Proof. Let us prove the 'only if' part of this theorem.
Let $(*)$ not hold in a K_- frame $(W, R, \leq)$.
This means $\exists x \exists y_1 \exists y_2 ((x R_\sim y_1 \wedge x R_\sim y_2) \wedge (not(y_1 R_\sim y_2) \vee not(y_2 R_\sim y_1)))$. Assume $not(y_1 R_\sim y_2)$ is true and define, '$z \vDash p$ if and only if $y_2 \leq z$ and $not(y_1 R_\sim z)$'.
Let us first show that the relation $\vDash$ is well defined, i.e., hereditary. So let $z \vDash p$ and $z \leq z'$, hence, $y_2 \leq z'$. If $y_1 R_\sim z'$ then there exists z'' such that $y_1 R z''$ and $z'' \leq^{-1} z'$ ($z' \leq z''$). But we already have $z \leq z'$. Hence $y_1 R_\sim z$, which is a contradiction to the fact that $z \vDash p$.
We have: $x \nvDash \sim p$ as $x C y_2$ and $y_2 \vDash p$ (as $not(y_1 C y_2)$). Also, $x \nvDash \sim\sim p$ as $x C y_1$ and $y_1 C z$ imply $z \nvDash p$. Hence we have $x \nvDash \sim p \vee \sim\sim p$.

Canonicity:
Let us show that the canonical model satisfies the frame condition. For any prime theories P, Q_1 and Q_2, let $P R_{c\sim} Q_1$ and $P R_{c\sim} Q_2$. Our claim is $Q_1 R_{c\sim} Q_2$ and $Q_2 R_{c\sim} Q_1$. Let us show $Q_1 R_{c\sim} Q_2$, the other will follow similarly.

1. $P R_{c\sim} Q_1$, i.e., $P R_c \circ \subseteq^{-1} Q_1$ implies there exists P_1 such that $P R_c P_1$ and $P_1 \supseteq Q_1$. By definition of R_c, we have $P R_c P_1$ implies $P_\neg \subseteq P_1 \subseteq P_\sim$.
2. $P R_{c\sim} Q_2$ implies that there exists P_2 such that $P R_c P_2$ and $P_2 \supseteq Q_2$. By definition of R_c, $P R_c P_2$ implies $P_\neg \subseteq P_2 \subseteq P_\sim$.

Let us show that $Q_2 \subseteq Q_{1\sim}$. Let $\alpha \notin Q_{1\sim}$.

$$\Rightarrow \sim \alpha \in Q_1, \Rightarrow \sim \alpha \in P_1$$
$$\Rightarrow \sim \alpha \in P_\sim, \Rightarrow \sim\sim \alpha \notin P$$
$$\Rightarrow \sim \alpha \in P, \Rightarrow \alpha \notin P_\sim$$
$$\Rightarrow \alpha \notin P_2, \Rightarrow \alpha \notin Q_2.$$

Hence $Q_2 \subseteq Q_{1\sim}$. Using Lemma 4, we have $Q_1 R_{c\sim} Q_2$.

Theorem 43. $\neg\alpha \wedge \neg\neg\alpha \vdash \bot$ *is characterized by the condition*
$\forall x \forall y_1 \forall y_2 (xR_\neg y_1 \wedge xR_\neg y_2 \rightarrow (y_1 R_\neg y_2 \wedge y_2 R_\neg y_1))$
in a K_- frame. This frame condition is also canonical.

Proof. Let the frame condition not hold in a K_- frame $(W, R, \leq)$.
This means $\exists x \exists y_1 \exists y_2 ((xR_\neg y_1 \wedge xR_\neg y_2) \wedge (not(y_1 R_\neg y_2) \vee not(y_2 R_\neg y_1)))$. Assume
$not(y_1 R_\neg y_2)$ and define
$z \vDash p$ if and only if $y_1 R_\neg z$.
Then $x \vDash \neg p \wedge \neg\neg p$. Hence $\neg p \wedge \neg\neg p \vdash \bot$ is not valid.
The other direction of this theorem is very similar to the proof of Theorem 35.

Canonicity:
Let us show that the canonical model satisfies the frame condition:

$$\forall P \forall Q_1 \forall Q_2 (PR_{c\neg}Q_1 \wedge PR_{c\neg}Q_2 \rightarrow (Q_1 R_{c\neg}Q_2 \wedge Q_2 R_{c\neg}Q_1)).$$

Let $P, Q_1, Q_2 \in W_c$ such that $PR_{c\neg}Q_1$ and $PR_{c\neg}Q_2$. $PR_{c\neg}Q_1$ implies that
there exists a prime theory P_1 such that $PR_c P_1$ and $P_1 \subseteq Q_1$. Let us show that
$Q_1 R_{c\neg}Q_2$. In other words, in view of Lemma 4, we have to show that $Q_{1\neg} \subseteq Q_2$.
Let $\alpha \in Q_{1\neg}$.

$$\Rightarrow \neg\alpha \notin Q_1 \Rightarrow \neg\alpha \notin P_1$$
$$\Rightarrow \neg\alpha \notin P_\neg = \{\beta : \neg\beta \notin P\}$$
$$\Rightarrow \neg\neg\alpha \in P.$$

But we have for any $\beta \in \mathcal{F}$ either $\neg\beta \notin P$ or $\neg\neg\beta \notin P$. Hence we have $\neg\alpha \notin P$.

$$\alpha \in P_\neg \text{ and } PR_{c\neg}Q_2, \text{ hence } \alpha \in Q_2.$$

Hence we have $Q_{1\neg} \subseteq Q_2$.

The enhanced united kite of negations with Stone and dual Stone negations can
be seen in Fig. 21.

Theorem 44. *(Regularity)* $\alpha \wedge \neg\alpha \vdash \beta \vee \sim \beta$ *is valid in a K_- frame $(W, R, \leq)$*
if and only if the K_- frame satisfies the following first order property:

$$\forall x ((\forall y (xR_\neg y \rightarrow x \leq y)) \vee (\forall z (xR_\sim z \rightarrow z \leq x))). \qquad (*)$$

Moreover, it is canonical.

Proof. Let $(*)$ hold in any K_- frame $(W, R, \leq)$, and let $x \in W$.
Assume $\forall y (xR_\neg y \rightarrow x \leq y)$ is true. Let us show that $x \nvDash \alpha \wedge \neg\alpha$. Assume $x \vDash \alpha$.
Let $xR_\neg y$, then by our assumption $x \leq y$. Hence using hereditary property of $\vDash$,
$y \vDash \alpha$. So $x \nvDash \neg\alpha$, whereby $x \nvDash \alpha \wedge \neg\alpha$.
Now let $\forall z (xR_\sim y \rightarrow z \leq x)$ be true. Let us show that $x \vDash \beta \vee \sim \beta$. Let $x \nvDash \beta$.
Let $xR_\sim z$, then by our assumption $z \leq x$. Using hereditary property of $\vDash$ again,
we have $z \nvDash \beta$, hence $x \vDash \sim \beta$. We have $x \vDash \beta \vee \sim \beta$.

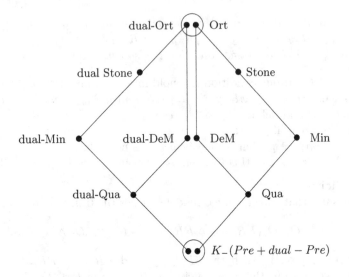

Fig. 21. Enhanced Dunn's United Kite of Negations

In either case, if $x \vDash \alpha \wedge \sim \alpha$ then $x \vDash \beta \vee \sim \beta$ holds, i.e. $\alpha \wedge \neg\alpha \vdash \beta \vee \sim \beta$ is valid.

Now, let (*) not hold. Then $\exists x((\exists y_1(xR_\neg y_1 \wedge x \nleq y_1)) \wedge (\exists y_2(xR_\sim y_2 \wedge y_2 \nleq x)))$. Let us define $\vDash$ as:

$$y \vDash p \text{ if and only if } x \leq y,$$

$$z \vDash q \text{ if and only if } z \nleq x.$$

Let us show that $\vDash$ is a well defined consequence relation.

(1) Let $y \vDash p$ and $y \leq y'$. Then $x \leq y \leq y'$. Hence $y' \vDash p$.

(2) Let $z \vDash q$ and $z \leq z'$. If $z' \nvDash q$ then by definition of $\vDash$ we would have $z' \leq x$. Hence $z \leq z' \leq x$, which implies $z \nvDash q$ contradicting our assumption.

Now let us show that $x \vDash p \wedge \neg p$. $x \vDash p$ using the definition of $\vDash$. As (*) not hold, we have y_1 in W such that $xR_\neg y_1$ and $x \nleq y_1$. By definition $y_1 \nvDash p$. Hence $x \vDash \neg p$.

Let us show that $x \nvDash q \vee \sim q$. As $x \leq x$, hence $x \nvDash q$. We also have an element y_2 in W such that $xR_\sim y_2$ and $y_2 \nleq x$. Hence $x \nvDash \sim q$. So $x \nvDash q \vee \sim q$.

Canonicity:

Let us show that the canonical frame $(W_c, R_c, \subseteq)$ also satisfies the frame condition (*), i.e., it satisfies:

$$\forall P(\forall Q(PR_{c\neg}Q \to P \subseteq Q) \vee (\forall Q'(PR_{c\sim}Q' \to Q' \subseteq P))).$$

So let $P \in W_c$ and suppose there exists a prime theory Q such that $Q(PR_{c\neg}Q \wedge P \nsubseteq Q)$. Let us show that $\forall Q'(PR_{c\sim}Q' \to Q' \subseteq P)$.

$P \nsubseteq Q$ implies there exists a formula $\alpha \in P$ and $\alpha \notin Q$. $PR_{c\neg}Q$ implies existence of a prime theory Q_1 such that PR_cQ_1 and $Q_1 \subseteq Q$. Hence $\alpha \notin Q_1$. So, $\alpha \notin P_\neg$, but then by definition of $P_\neg$, $\neg\alpha \in P$. Hence $\alpha \wedge \neg\alpha \in P$. By our assumption $\alpha \wedge \neg\alpha \vdash \beta \vee \sim \beta$. Hence for any formula β we have either $\beta \in P$ or $\sim \beta \in P$.

Now let $PR_{c\sim}Q'$. Our claim is $Q' \subseteq P$. So, let $\gamma \in Q'$. By definition, $PR_{c\sim}Q'$ if and only if $P(R_c \circ \subseteq^{-1})Q'$ which implies existence of a prime theory Q_1' such that PR_cQ_1' and $Q_1' \supseteq Q'$. Hence $\gamma \in Q_1'$. But $Q_1' \subseteq P_\sim$. $\gamma \in P_\sim$ implies $\sim \gamma \notin P$. Hence $\gamma \in P$. Finally we have established $Q' \subseteq P$.

Hence canonicity holds.

Now, let $\mathcal{L}_{RDSA}$ be the logic which contains all axioms and postulates of the logic K_- along with the following rules and postulates.

(i) $\alpha \wedge \sim \alpha \vdash \perp$.

$$\frac{\alpha \wedge \beta \vdash \gamma}{}$$

(ii) $\alpha \wedge \sim \gamma \leq \sim \beta$.

(iii) $\top \vdash \sim \alpha \vee \sim\sim \alpha$.

(iv) $\alpha \vee \neg\alpha \vdash \perp$.

$$\frac{\gamma \vdash \alpha \vee \beta}{}$$

(v) $\neg\beta \vdash \alpha \vee \neg\gamma$.

(vi) $\neg\alpha \wedge \neg\neg\alpha \vdash \perp$.

(vii) $\alpha \wedge \neg\alpha \vdash \beta \vee \sim \beta$ (Regularity).

Now, let us denote by $\mathcal{A}_{RDSA}$ the class of all regular double Stone algebras, $\mathcal{R}_{RDSA}$ the class of all regular double Stone algebras of the form $B^{[2]}$, where B is a Boolean algebra and $\mathcal{RS}_{RDSA}$ the class of all $\mathcal{RS}$ considered as regular double Stone algebras.

Definition 35. *We call a K_- frame $(W, R, \leq)$ a regular double Stone frame if it satisfies the following first order conditions.*

1. $\forall x \forall y (x R_\sim y \rightarrow y R_\sim x)$.
2. $\forall x (x R_\sim x)$.
3. $\forall x \forall y_1 \forall y_2 (x R_\sim y_1 \wedge x R_\sim y_2 \rightarrow (y_1 R_\sim y_2 \wedge y_2 R_\sim y_1))$.
4. $\forall x \forall y (x R_\neg y \rightarrow y R_\neg x)$.
5. $\forall x (x R_\neg x)$.
6. $\forall x \forall y_1 \forall y_2 (x R_\neg y_1 \wedge x R_\neg y_2 \rightarrow (y_1 R_\neg y_2 \wedge y_2 R_\neg y_1))$.
7. $\forall x ((\forall y (x R_\neg y \rightarrow x \leq y)) \vee (\forall z (x R_\sim z \rightarrow z \leq x)))$.

Let us denote by $\mathcal{F}_{RDSA}$ the class of all regular double Stone frames.

Theorem 45. *For any $\alpha, \beta \in \mathcal{F}$, the following are equivalent.*

1. $\alpha \vdash_{\mathcal{L}_{RDSA}} \beta$.
2. $\alpha \vDash_{\mathcal{A}_{RDSA}} \beta$.
3. $\alpha \vDash_{\mathcal{R}_{RDSA}} \beta$.
4. $\alpha \vDash_{\mathcal{RS}_{RDSA}} \beta$.
5. $\alpha \vDash_{\mathcal{F}_{RDSA}} \beta$.

So it appears that a negation having the Stone and dual Stone with regularity property would occupy a node in the 'intersection' of the lopsided kite and its dual.

4   Discrete Dualities for Kleene, Stone, Double Stone and Regular double Stone Algebras

Stone's representation theorem for Boolean algebras led to the emergence of duality theory (cf. e.g. [10]). Some well known duality results are the following.

1. Jónsson-Tarski duality between modal algebras and Kripke frames.
2. Esakia duality between Heyting algebras and Esakia spaces.
3. Priestley duality between distributive lattices (Heyting algebras, Topological Boolean algebra) and Priestley spaces.

Duality results present advantages in both the fields of logic and algebra. On the one hand, these provide representations of various algebras in terms of set lattices. On the other hand, one obtains either of the algebraic or relational semantics of some logics from the other.

Note that the above-mentioned duality results are topological. Orłowska and Rewitzky in [55–57] have discussed various duality results in which topology is not involved. They have termed these results as 'discrete dualities'. Let us discuss the steps to establish discrete duality, following [56,57]. Let Alg be a class of algebras and Frm a class of frames.

1. For each $\mathcal{A} \in Alg$, associate a frame $CFrame(\mathcal{A}) \in Frm$ (the *canonical frame* of $\mathcal{A}$).
2. For each $\mathcal{F} \in Frm$, associate an algebra $CAlg(\mathcal{F}) \in Alg$ (the *complex algebra* of $\mathcal{F}$).
3. Prove the following representation theorems.
   (a) For each $\mathcal{A} \in Alg$, there exists an embedding from $\mathcal{A}$ to $CAlg(CFrame)(\mathcal{A})$.
   (b) For each $\mathcal{F} \in Frm$, there exists an embedding from $\mathcal{F}$ to $CFrame(CAlg)(\mathcal{F})$.

In this section, we establish duality results between the following.

(i)   Class $\mathcal{A}_K$ of all Kleene algebras and class $\mathbb{F}_K$ of all Kleene frames.
(ii) Class $\mathcal{A}_{K'}$ of all dual Kleene algebras and class $\mathbb{F}_{DK}$ of all dual Kleene frames.
(iii) Class $\mathcal{A}_S$ of all Stone algebras and class $\mathbb{F}_S$ of all Stone frames.
(iv) Class $\mathcal{A}_{DS}$ of all double Stone algebras and class $\mathbb{F}_{DS}$ of all dual Stone frames.
(v)   Class $\mathcal{A}_{K'}$ of all double Kleene algebras and class $\mathbb{F}_{DK}$ of all double Kleene frames.
(vi) Class $\mathcal{A}_{RDSA}$ of all regular double Stone Algebras and class $\mathcal{F}_{RDSA}$ of all regular double Stone frames.

It is worth mentioning here that in [31], Düntsch and Orłowska obtained duality results between double Stone and regular double Stone algebras and certain classes of partially ordered sets. In this section, we also present duality results for double Stone and regular double Stone algebras and certain classes of partially ordered sets, but those involving an additional relation, namely the classes of compatibility, exhaustive and K_- frames. One must also mention that Dunn obtained various representation results of lattice based algebras (with negation) in [24,26,29], induced by perp semantics. We obtain the representations in the framework of discrete duality introduced by Orłwska and Rewitzky. In the discrete duality representation results of certain classes of frames are obtained as well.

In this section, by an embedding between algebras, we mean a one-one and structure preserving map. Let $(W, R, \leq)$ and $(W', R', \leq')$ be frames. A map $\phi : W \to W'$ is called an *embedding* if for any $a, b \in W$,

(a) $a \leq b$ if and only if $\phi(a) \leq' \phi(b)$.
(b) aRb if and only if $\phi(a)R'\phi(b)$.

Let us recall the following for a given partially ordered set $(U, \leq)$.

(i) $X \subseteq U$ is called an *upward closed* set if $x \in X$ and $x \leq y$ then $y \in X$.
(ii) $X \subseteq U$ is called a *downward closed* set if $x \in X$ and $y \leq x$ then $y \in X$.

We provide complete details in the proofs of results in Sect. 4.1. Proofs of results in Sects. 4.2 and 4.3 can be obtained similarly.

4.1   Dualities Arising from Compatibility Frames

Let $(U, \leq)$ be a poset and let K_U be the collection of all upward closed subsets of U. Then it is well known that K_U is a bounded distributive lattice, where join and meet are given by set theoretic union and intersection respectively. $\emptyset$ and U are the bottom and top elements respectively.

It is interesting to see how a compatibility frame $(U, C, \leq)$ enhances the structure
$(K_U, \cup, \cap, \emptyset, U)$. Let $A \in K_U$ and define $\sim A := \{x \in U : \forall y(xCy \to y \notin A)\}$. Using the definition of compatibility frame, one can show that $\sim$ is a well defined operation on K_U. In fact $\sim$ satisfies the following properties. For all $A, B \in K_U$,

1. $A \subseteq B \Rightarrow \sim B \subseteq \sim A$.
2. $\sim (A \cup B) = \sim A \cap \sim B$.
3. $\sim \emptyset = U$.

Hence $(K_U, \cup, \cap, \emptyset, U)$ is a $K_i - algebra$.
Now, let $(K, \vee, \wedge, \sim, 0, 1)$ be a $K_i - algebra$. Let us consider the set U_K which is the collection of all prime filters of K. Define a binary relation C_K on U_K as follows. For $P, Q \in U_K$,

$$PC_KQ \text{ if and only if for all } a \in K, \sim a \in P \Rightarrow a \notin Q.$$

It is shown by Dunn [24] that the tuple $(U_K, C_K, \subseteq)$ is a compatibility frame.

Definition 36. [24]

1. Let $(K, \vee, \wedge, \sim, 0, 1)$ be a $K_i - algebra$. The structure $(U_K, C_K, \subseteq)$ is called the canonical frame for K.
2. Let $(U, C, \leq)$ be a compatibility frame. The structure $(K_U, \cup, \cap, \sim, \emptyset, U)$ is called the complex algebra of the frame $(U, C, \leq)$.

Now, let us establish the duality results for the $K_i - algebras$.

Theorem 46. Let $\mathcal{K} := (K, \vee, \wedge, \sim, 0, 1)$ be a $K_i - algebra$. There exists a compatibility frame $(W, C, \leq)$ such that $\mathcal{K}$ can be embedded into the complex algebra of $(W, C, \leq)$.

The proof of this theorem uses the usual technique of prime filters (cf. e.g. [24, 29]). Let us sketch the proof to clarify the context.

Proof. Let us consider the set U_K of all prime filters of the distributive lattice K. We have mentioned that the tuple $(U_K, C_K, \subseteq)$ is a compatibility frame. Now take the complex algebra of this canonical frame $(K_{U_K}, \cup, \cap, \emptyset, U_K)$. Let us define a map $h : K \to K_{U_K}$ as follows. For $a \in K$,

$$h(a) := \{P \in U_K : a \in P\}.$$

h is proved in literature to be well defined, one-one, and to preserve the join, meet operations of K. Let us prove that $h(\sim a) = \sim h(a)$. By definition $h(\sim a) := \{P \in U_K : \sim a \in P\}$ and $\sim h(a) = \{P : \forall Q(PC_cQ \to Q \notin h(a))\}$.
Let $P \in h(\sim a)$, i.e. $\sim a \in P$. Let PC_KQ. But then $\sim a \in P$ implies $a \notin Q$, i.e. $Q \notin h(a)$.
Now let $P \in \sim h(a)$. We want to show that $\sim a \in P$. On the contrary, let us assume $\sim a \notin P$. Consider the filter $a \uparrow$. Define the set $Q^* := \{x \in K : \sim x \in P\}$. We show that Q^* is an ideal. Let $x \in Q^*$ and $b \leq x$. Then $\sim x \leq \sim b$. As P is a filter, $b \in Q^*$. Now let $x, b \in Q^*$, we have $\sim (x \vee b) \leq \sim x \wedge \sim b$. Hence $\sim (x \vee b) \in P$. So, $x \vee b \in Q^*$.
We also have $a \uparrow \cap Q^* = \emptyset$ as, if $x \in a \uparrow \cap Q^*$ then $a \leq x$ and $\sim x \in P$. But $a \leq x \Rightarrow \sim x \leq \sim a$ hence $\sim a \in P$, which is a contradiction. Now by prime filter theorem, there exists a prime filter Q such that $a \uparrow \subseteq Q$ and $Q \cap Q^* = \emptyset$.
    Hence we have shown the existence of a prime filter Q such that PC_KQ and $Q \in h(a)$, which is a contradiction.

Theorem 47. Let $(W, C, \leq)$ be a compatibility frame. Then there exists a $K_i - algebra$ $\mathcal{K} := (K, \vee, \wedge, \sim, 0, 1)$ such that $(W, C, \leq)$ can be embedded into the canonical frame of $\mathcal{K}$.

Proof. Consider the complex algebra $(K_W, \cup, \cap, \emptyset, W)$ of the compatibility frame $(W, C, \leq)$. K_W is a $K_i - algebra$ as mentioned earlier. Let us define a map $\phi : W \to W_{K_W}$ as:

$$\phi(a) := \{C \in K_W : a \in C\}, \ a \in W.$$

It has been proved in literature that ϕ is an order embedding (cf. e.g. [56]). Let us prove that aCb if and only if $\phi(a)C_{K_W}\phi(b)$.

Let aCb and $\sim C \in \phi(a)$. $a \in\sim C$ implies $b \notin C$, i.e., $C \notin \phi(b)$. Hence $\phi(a)C_{K_W}\phi(b)$.

Now, let $not(aCb)$. Consider the set $C := \{z \in W : not(aCz)\}$. It can be shown that C is an upward closed set. Clearly, $\sim C \in \phi(a)$ and $C \in \phi(b)$. So, $not(\phi(a)C_{K_W}\phi(b))$.

Like the correspondence results presented in the earlier sections, one can easily establish correspondence results between classes of compatibility frames and classes of complex algebras. For the class of Kleene frames, we obtain the following, by mimicking the proofs of Theorems 25 and 26.

Proposition 9. *Let* $(W,C,\leq)$ *be a compatibility frame. Its complex algebra is a Kleene algebra if and only if* $(W,C,\leq)$ *is a Kleene frame.*

Lemma 5. *Let* $\mathcal{K} := (K,\vee,\wedge,\sim,0,1)$ *be a Kleene algebra. Then its canonical frame is a Kleene frame.*

Now, we obtain the duality result for Kleene algebras.

Theorem 48. *Let* $\mathcal{K} := (K,\vee,\wedge,\sim,0,1)$ *be a Kleene algebra. Then there exists a compatibility frame* $(W,C,\leq)$ *such that* $\mathcal{K}$ *can be embedded into the complex algebra of* $(W,C,\leq)$.

Proof. Let $\mathcal{K}$ be a Kleene algebra. Consider the canonical frame of $\mathcal{K}$. Then using Lemma 5, $(U_K,C_K,\subseteq)$ is a Kleene frame. Hence using Proposition 9, its complex algebra is a Kleene algebra. The mapping h in the proof of Theorem 46 is the required embedding.

Now, let us analyze the above result from the perspective of rough set theory. By definition, assuming (U,R) is an approximation space, a rough set is an ordered pair $(\mathsf{L}X,\mathsf{U}X)$ for some $X \subseteq U$. Theorem 48 says that elements of a Kleene algebra can be also looked upon as sets, where Kleene negation is defined by a compatibility relation. Let us illustrate this through an example.

Example 9. Let $U := \{a,b,c\}$, define a relation R on U as, $aRa, aRb, bRb,$ bRa and cRc. Then (U,R) is an equivalence relation. $\mathcal{RS} = \{(\emptyset,\emptyset),$ $(\emptyset,ab),(ab,ab),(c,c),(c,U),(U,U)\}$. Recall from Sect. 2, $\mathcal{RS}$ is a Kleene algebra, where $\sim(\mathsf{L}A,\mathsf{U}A) := ((\mathsf{U}A)^c,(\mathsf{L}A)^c)$. The Hasse diagram of $\mathcal{RS}$ is given in Fig. 21. Prime filters of $\mathcal{RS}$ are given by:

$P_1 = \{(\emptyset,ab) =\sim (c,U),(ab,ab) =\sim (c,c),(c,U) =\sim (\emptyset,ab),(U,U) =\sim (\emptyset,\emptyset)\}$.
$P_2 = \{(c,c) =\sim (ab,ab),(c,U) =\sim (\emptyset,ab),(U,U) =\sim (\emptyset,\emptyset)\}$.
$P_3 = \{(ab,ab) =\sim (c,c),(U,U) =\sim (\emptyset,\emptyset)\}$.

By definition of C, we have: $P_1CP_3, P_2CP_2, P_3CP_3, P_3CP_1$. Now, $h(\emptyset,\emptyset) = \emptyset$, $h(c,c) = \{P_2\}$, $h(\emptyset,ab) = \{P_1\}$, $h(c,U) = \{P_1,P_2\}$, $h(ab,ab) = \{P_1,P_3\}$, $h(U,U) = \{P_1,P_2,P_3\}$.

Illustration of the isomorphism is given in Fig. 22.

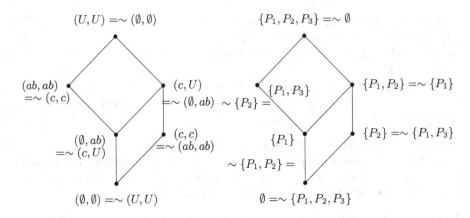

Fig. 22. $\mathcal{RS} \hookrightarrow K_{W_K}$

Theorem 49. *Let $(W, C, \leq)$ be a Kleene frame. Then there exists a Kleene algebra $\mathcal{K} := (K, \vee, \wedge, \sim, 0, 1)$ such that $(W, C, \leq)$ can be embedded into the canonical frame of $\mathcal{K}$.*

Proof. Let $(W, C, \leq)$ be a Kleene frame. Then using Proposition 9, the complex algebra $(K_W, \cup, \cap, \sim, \emptyset, U)$ is a Kleene algebra. Now, Lemma 5 states that $(W_{K_W}, C_{K_W}, \subseteq)$ is a Kleene frame. The map ϕ in the proof of Theorem 47 is the required embedding.

Let us illustrate this theorem through an example.

Example 10. Let us recall the Kleene frame $(W, C, \leq)$ given in Example 5. Let $W := \{a, b, c\}$. The partial order $\leq$ on U is defined as:

$$a \leq a, c \leq a, c \leq c \text{ and } b \leq b.$$

The compatibility relation C on U is defined as:

$$aCc, bCb, cCc, cCa.$$

Let us list all the upward closed sets of the poset $(W, \leq)$.
$C_0 = \emptyset, C_1 = \{a\}, C_2 = \{b\}, C_3 = \{a, c\}, C_4 = \{a, b\}, C_5 = W$.
$K_W = \{C_0, C_1, C_2, C_3, C_4, C_5\}$. Let us now turn to the operator $\sim$, which is defined as:

$$\sim C_i := \{x \in W : \forall y(xCy \rightarrow y \notin C_i)\}.$$

$\sim C_0 = C_5, \sim C_1 = C_4, \sim C_2 = C_3, \sim C_3 = C_2, \sim C_4 = C_1, \sim C_5 = C_0$.
The Hasse diagram of the Kleene algebra $(K_W, \cup, \cap, \sim, \emptyset, W)$ is given in Fig. 23.

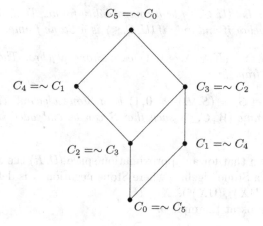

$$C_5 =\sim C_0$$

$$C_4 =\sim C_1 \qquad C_3 =\sim C_2$$

$$C_2 =\sim C_3 \qquad C_1 =\sim C_4$$

$$C_0 =\sim C_5$$

Fig. 23. $(K_W, \cup, \cap, \sim, \emptyset, W)$

Now the map, $\Phi : W \to W_{K_W}$ given as

$$\Phi(a) := \{C_1, C_3, C_4, C_5\}.$$
$$\Phi(b) := \{C_2, C_4, C_5\}.$$
$$\Phi(c) := \{C_3, C_5\}.$$

is the embedding of the structure $(W, C, \leq)$ into $(W_{K_W}, C_{K_W}, \subseteq)$. The embedding Φ can be depicted as in Fig. 24.

We get similar results for Stone algebras and frames.

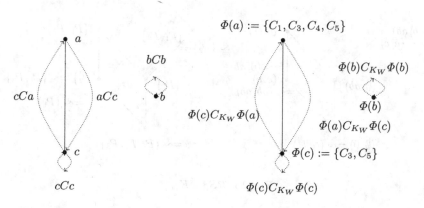

$$\Phi(a) := \{C_1, C_3, C_4, C_5\}$$

a

bCb

$cCa \qquad aCc \qquad \bullet b$

$\Phi(b) C_{K_W} \Phi(b)$

$\Phi(c) C_{K_W} \Phi(a)$

$\Phi(b)$

$\Phi(a) C_{K_W} \Phi(c)$

c

$\Phi(c) := \{C_3, C_5\}$

$cCc \qquad\qquad \Phi(c) C_{K_W} \Phi(c)$

Fig. 24. $(W, C, \leq) \hookrightarrow (W_{K_W}, C_{K_W}, \subseteq)$

Proposition 10. *Let $(U, C, \leq)$ be a compatibility frame. Then its complex algebra is a Stone algebra if and only if $(U, C, \leq)$ is a Stone frame.*

Lemma 6. *Let $\mathcal{K} := (K, \vee, \wedge, \sim, 0, 1)$ be a Stone algebra. Then its canonical frame is a Stone frame.*

Theorem 50. *Let $\mathcal{S} := (S, \vee, \wedge, \sim, 0, 1)$ be a Stone algebra. Then there exists a compatibility frame $(W, C, \leq)$ such that $\mathcal{S}$ can be embedded into the complex algebra of $(W, C, \leq)$.*

It is already known that for an approximation space (U, R) the collection $\mathcal{RS}$ of rough sets forms a Stone algebra, where Stone negation $\sim$ is defined as:
$\sim (LX, UX) := ((UX)^c, (UX)^c), \ X \subseteq U.$
Let us take a re-look at Example 9.

Example 11. Consider the approximation space of Example 9. Let us re-write the prime filters.
$P_1 = \{(\emptyset, ab), (ab, ab) =\sim (c, c), (c, U), (U, U) =\sim (\emptyset, \emptyset)\}.$
$P_2 = \{(c, c) =\sim (ab, ab) =\sim (\emptyset, ab), (c, U), (U, U) =\sim (\emptyset, \emptyset)\}.$
$P_3 = \{(ab, ab) =\sim (c, c), (U, U) =\sim (\emptyset, \emptyset)\}.$
By definition of C, we have: $P_1 C P_1, \ P_1 C P_3, P_2 C P_2, P_3 C P_3, P_3 C P_1.$
Now, $h(\emptyset, \emptyset) = \emptyset, h(c, c) = \{P_2\}, h(\emptyset, ab) = \{P_1\}, h(c, U) = \{P_1, P_2\}, h(ab, ab) = \{P_1, P_3\}, h(U, U) = \{P_1, P_2, P_3\}.$
Illustration of the isomorphism is given in Fig. 25.

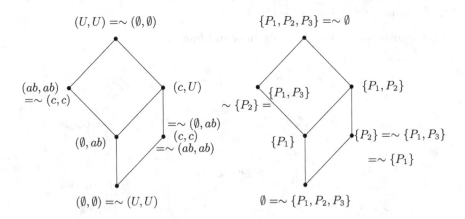

Fig. 25. $\mathcal{RS} \hookrightarrow K_{W_K}$

Theorem 51. *Let* $(W, C, \leq)$ *be a Stone frame. Then there exists a Stone algebra* $\mathcal{S} := (S, \vee, \wedge, \sim, 0, 1)$ *such that* $(W, C, \leq)$ *can be embedded into the canonical frame of* $\mathcal{S}$.

Example 12. Let us re-consider the compatibility frame $(W, C, \leq)$ of Example 6. The upward closed sets of the poset $(W, \leq)$ are as given there.
Now $K_W = \{C_0, C_1, C_2, C_3, C_4, C_5\}$. The negation operator $\sim$ defined as:

$$\sim C_i := \{x \in W : \forall y(xCy \to y \notin C_i)\},$$

gives $\sim C_0 = C_5$, $\sim C_1 = C_2$, $\sim C_2 = C_3$, $\sim C_3 = C_2$, $\sim C_4 = C_0$, $\sim C_5 = C_0$.
The Hasse diagram of the Stone algebra $(K_W, \cup, \cap, \sim, \emptyset, W)$ is given in Fig. 26.

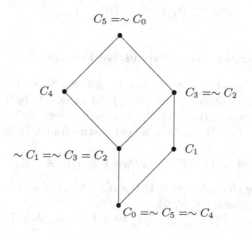

Fig. 26. $(K_W, \cup, \cap, \sim, \emptyset, W)$

The map $\Phi : W \to W_{K_W}$ given as:

$$\Phi(a) := \{C_1, C_3, C_4, C_5\}.$$
$$\Phi(b) := \{C_2, C_4, C_5\}.$$
$$\Phi(c) := \{C_3, C_5\}.$$

is the embedding of the structure $(W, C, \leq)$ into $(W_{K_W}, C_{K_W}, \subseteq)$. The pictorial representation of this embedding is given in Fig. 27.

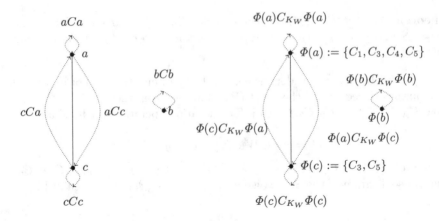

Fig. 27. $(W, C, \leq) \hookrightarrow (W_{K_W}, C_{K_W}, \subseteq)$

4.2   Dualities Arising from Exhaustive Frames

This section is very much similar to the previous one. So, we just state the results and omit the proofs. Given a poset $(W, \leq)$, let us denote by K_W the collection of all downward closed sets of $(W, \leq)$. It is well known that K_W is a bounded distributive lattice. Let $(W, R, \leq)$ be an exhaustive frame. Define $\neg : K_W \to K_W$ as:

$$\neg A := \{x \in U : \exists y (xRy \land y \notin A)\}, \ A \in K_W.$$

Then the structure $(K_W, \lor, \land, \neg, 0, 1)$ is a $K_u - algebra$, and is called the canonical algebra of the frame $(W, R, \leq)$.

On the other hand, let $(K, \lor, \land, \neg, 0, 1)$ be a $K_u - algebra$. Let us consider the set W_K which is the collection of all prime filters of K. Define a binary relation R_K on W_K as follows. For $P, Q \in W_K$,

$$PR_KQ \text{ if and only if for all } a \in K \text{ if } \neg a \notin P \text{ then } a \in Q.$$

It has been shown by Dunn [29] that the structure $(W_K, R_K, \supseteq)$ is an exhaustive frame. We call the structure $(W_K, R_K, \supseteq)$ the canonical frame of the $K_u - algebra$ $(K, \lor, \land, \neg, 0, 1)$.

Let us state the duality results for classes of K_u-algebras and exhaustive frames.

Theorem 52.

(a) Let $\mathcal{K} := (K, \lor, \land, \neg, 0, 1)$ be a $K_u - algebra$. Then there exists an exhaustive frame $(W, R, \leq)$ such that $\mathcal{K}$ can be embedded into the complex algebra of $(W, R, \leq)$.

(b) Let $(W, R, \leq)$ be an exhaustive frame. Then there exists a $K_u - algebra$ $\mathcal{K} := (K, \lor, \land, \neg, 0, 1)$ such that $(W, R, \leq)$ can be embedded into canonical frame of $\mathcal{K}$.

Proof. Let us simply mention the embeddings.

(a) The map $\phi : K \to K_{W_K}$ defined as

$$\phi(a) := \{P \in W_K : a \in P\}, \ a \in K,$$

is the required embedding.

(b) The map $\psi : W \to W_{K_W}$ defined as

$$\psi(x) := \{D \in K_W : x \notin D\}, \ x \in W,$$

is the required embedding.

Proposition 11. *Let* $(W, R, \leq)$ *be an exhaustive frame. Then its complex algebra is a dual Kleene algebra if and only if the exhaustive frame* $(W, R, \leq)$ *is dual Kleene.*

Theorem 53.

(i) *Let* $\mathcal{K} := (K, \vee, \wedge, \neg, 0, 1)$ *be a dual Kleene algebra. Then there exists an exhaustive frame* $(W, C, \leq)$ *such that* $\mathcal{K}$ *can be embedded into the complex algebra of* $(W, R, \leq)$.

(ii) *Let* $(W, R, \leq)$ *be a dual Kleene frame. Then there exists a dual Kleene algebra* $\mathcal{K} := (K, \vee, \wedge, \neg, 0, 1)$ *such that* $(W, R, \leq)$ *can be embedded into the canonical frame of* $\mathcal{K}$.

For an illustration of part (i) of this theorem, we refer to Example 9, and note that $\sim$ and $\neg$ are the same.

Proposition 12. *Let* $(W, R, \leq)$ *be an exhaustive frame. Then its complex algebra is a dual Stone algebra if and only if the frame* $(W, R, \leq)$ *is a dual Stone frame.*

Theorem 54. *Let* $\mathcal{D} := (D, \vee, \wedge, \neg, 0, 1)$ *be a dual Stone algebra. Then there exists an exhaustive frame* $(W, R, \leq)$ *such that* $\mathcal{D}$ *can be embedded into the complex algebra of* $(W, R, \leq)$.

The collection $\mathcal{RS}$ of rough sets for an approximation space (U, R) forms a dual Stone algebra, the dual Stone negation $\neg$ being defined as:
$\neg(LX, UX) := ((LX)^c, (LX)^c), \ X \subseteq U.$

Example 13. Let us consider the approximation space of Example 6. The prime filters are given as follows. (We are re-writing the prime filters to emphasize the negation $\neg$).
$P_1 = \{(\emptyset, ab), (ab, ab) = \neg(c, c) = \neg(c, U), (c, U), (U, U) = \neg(\emptyset, \emptyset) = \neg(\emptyset, ab)\}$.
$P_2 = \{(c, c) = \neg(ab, ab), (c, U), (U, U) = \neg(\emptyset, \emptyset) = \neg(\emptyset, ab)\}$.
$P_3 = \{(ab, ab) = \neg(c, c) = \neg(c, U), (U, U) = \neg(\emptyset, \emptyset) = \neg(\emptyset, ab)\}$.
By definition of R, we have: $P_1 R P_1$, $P_1 R P_3$, $P_2 R P_2$, $P_3 R P_3$, $P_3 R P_1$. Now,
$h(\emptyset, \emptyset) = \emptyset$, $h(c, c) = \{P_2\}$, $h(\emptyset, ab) = \{P_1\}$, $h(c, U) = \{P_1, P_2\}$, $h(ab, ab) = \{P_1, P_3\}$, $h(U, U) = \{P_1, P_2, P_3\}$. Pictorial representation of the isomorphism is given in Fig. 28.

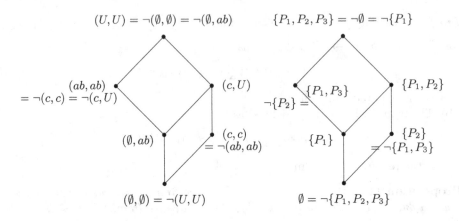

Fig. 28. $\mathcal{D} \hookrightarrow K_{W_D}$

Theorem 55. *Let* $(W, C, \leq)$ *be a dual Stone frame. Then there exists a dual Stone algebra* $\mathcal{D} := (D, \vee, \wedge, \neg, 0, 1)$ *such that* $(W, R, \leq)$ *can be embedded into the canonical frame of* $\mathcal{D}$.

4.3 Dualities Arising from K_- Frames

This section is also very similar to Sect. 4.1, and we just state the results without proofs. Let $(U, R, \leq)$ be a K_- frame. Let us recall that K_U denotes the collection of all upward closed sets. Define unary maps $\sim, \neg : K_U \to K_U$ as:

$$\sim A := \{x \in U : \forall y(xRy \to y \notin A)\}$$

$$\neg A := \{x \in U : \exists y(xRy \wedge y \notin A)\}.$$

That the operators $\sim, \neg$ are well defined can be seen using the properties of K_- frames. Moreover, it can be easily seen that the structure $(K_U, \cup, \cap, \sim, \neg, \emptyset, U)$ is a K_--algebra. In fact the structure $(K_U, \cup, \cap, \sim, \neg, \emptyset, U)$ is called the complex algebra of the K_- frame $(U, R, \leq)$. The other way around, let $(K, \vee, \wedge, \sim, \neg, 0, 1)$ be a K_--algebra. Let U_K be the collection of all prime filters of the distributive lattice K. Let $P \in U_K$, and let us denote by $P_\sim$, the set $\{a \in K : \sim a \notin P\}$, and $P_\neg$, the set $\{a \in K : \neg a \notin P\}$. Define a relation R_c on U_K as: PR_cQ if and only if $P_\neg \subseteq Q \subseteq P_\sim$. Then $(U_K, R_c, \subseteq)$ is a K_- frame and is called the canonical frame of the K_--algebra K. Before proceeding to the duality results, let us recall Lemma 4. Although this lemma was proved by Dunn in the context of logic, one can easily provide the proof in the algebraic context.

Lemma 7. *Let* P *and* Q *be prime filters such that* $P_\neg \subseteq Q$. *Then there exists a prime filter* S *such that* PR_cS *and* $S \subseteq Q$.

Lemma 8. *Let* P *and* Q *be prime filters such that* $Q \subseteq P_\sim$. *Then there exists a prime filter* S *such that* PR_cS *and* $Q \subseteq S$.

Theorem 56. *Let* $\mathcal{K} := (K, \vee, \wedge, \sim, \neg, 0, 1)$ *be a* K_-*-algebra. Then there exists a frame* $(W, R, \leq)$ *such that* $\mathcal{K}$ *can be embedded into the complex algebra of* $(W, R, \leq)$.

Proof. Let us consider the set U_K of all prime filters of the distributive lattice K. The tuple $(U_K, R_c, \subseteq)$ is a K_- frame. Now consider the complex algebra $(K_{U_K}, \cup, \cap, \emptyset, U_K)$ of this canonical frame. Take the usual map $h : K \to K_{U_K}$ defined for any $a \in K$ as

$$h(a) := \{P \in U_K : a \in P\}.$$

Let us prove that

1. $h(\sim a) = \sim h(a)$,
2. $h(\neg a) = \neg h(a)$.

1. By definition $h(\sim a) := \{P \in U_K :\sim a \in P\}$ and $\sim h(a) = \{P : \forall Q(PR_cQ \to Q \notin h(a))\}$. Let $P \in h(\sim a)$. This imply $\sim a \in P$, but then $\sim a \in P$ imply $a \notin P_\sim$. Let PR_cQ then using definition of R_c we have $Q \notin h(a)$.

Let $P \in\sim h(a)$. We want to show that $\sim a \in P$. On the contrary, let us assume $\sim a \notin P$. Consider the filter $a \uparrow$. Consider the set $P_\sim^c$. As done for Theorem 46, it can be shown that $P_\sim^c$ is an ideal.

Also $a \uparrow \cap P_\sim^c = \emptyset$ as, if $x \in a \uparrow \cap P_\sim^c$ then $a \leq x$ and $\sim x \in P$. But $a \leq x \Rightarrow \sim x \leq \sim a$ and hence $\sim a \in P$ – which is a contradiction.

Hence we have $a \uparrow \cap P_\sim^c = \emptyset$. Now using the prime filter theorem, there exists a prime filter Q such that $a \uparrow \subseteq Q$ and $Q \cap P_\sim^c = \emptyset$. Thus $Q \subseteq P_\sim$. Now using Lemma 8, we have the existence of a prime filter S such that PR_cS and $Q \subseteq S$.

Hence we have shown the existence of a prime filter S such that PR_cS and $S \in h(a)$, which is a contradiction.

2. Let $P \in \neg h(a)$. This implies $\exists Q(PR_cQ \wedge a \notin Q)$. $a \notin Q \Rightarrow a \notin P_\neg \Rightarrow \neg a \in P$. Conversely, let $P \in h(\neg a)$, which means $\neg a \in P$. It can be easily shown that $P_\neg$ is a filter and $a \notin P_\neg$. Hence there exists a prime filter Q such that $a \notin Q$ and $P_\neg \subseteq Q$. So, using Lemma 8, we have the required prime filter.

Theorem 57. *Let* $(W, R, \leq)$ *be a* K_- *frame. Then there exists a* K_-*-algebra,* $\mathcal{K} := (K, \vee, \wedge, \sim, \neg, 0, 1)$ *such that* $(W, R, \leq)$ *can be embedded into the canonical frame of* $\mathcal{K}$.

Proof. Consider the complex algebra $(K_W, \cup, \cap, \sim, \neg, \emptyset, W)$ of the K_- frame $(W, R, \leq)$. Then K_W is a K_--algebra. Let us define a map $\phi : W \to W_{K_W}$ as:

$$\phi(a) := \{C \in K_W : a \in C\}, \ a \in W.$$

Let us show that aRb if and only if $\phi(a)R_c\phi(b)$.
Let aRb. $C \in \phi(a)_\neg$ implies $\neg C \notin \phi(a)$. But $a \notin \neg C$ implies $b \in C$. Now, let $C \notin \phi(a)_\sim$. This implies $a \in\sim C$ and $b \notin C$. Hence $\phi(a)_\neg \subseteq \phi(b) \subseteq \phi(a)_\sim$.
Let $not(\phi(a)R_c\phi(b))$. Assume $\phi(a)_\neg \not\subseteq \phi(b)$ and $C \in \phi(a)_\neg$, $C \notin \phi(b)$. $C \in \phi(a)_\neg$ implies $a \notin \neg C$. So if aRb then $b \in C$ which is a contradiction. Hence $not(aRb)$. Similarly, one can show $not(aRb)$, if $\phi(b) \not\subseteq \phi(a)_\sim$.

The correspondence and completeness results proved in Theorems 37 and 38 can be re-phrased in terms of set algebras.

Theorem 58.

1. Let $(U, R, \leq)$ be a K_- frame. Then its complex algebra is a double Kleene algebra if and only if $(U, R, \leq)$ is a double Kleene frame.
2. Let $\mathcal{K} := (K, \vee, \wedge, \sim, \neg, 0, 1)$ be a double Kleene algebra. Then its canonical frame is a double Kleene frame.

Now, the duality results can be established via the following.

Theorem 59.

1. Let $\mathcal{K} := (K, \vee, \wedge, \sim, \neg, 0, 1)$ be a double Kleene algebra. Then there exists a K_- frame $(W, R, \leq)$ such that $\mathcal{K}$ can be embedded into the complex algebra of $(W, R, \leq)$.
2. Let $(W, R, \leq)$ be a double Kleene frame. Then there exists a double Kleene algebra $\mathcal{K} := (K, \vee, \wedge, \sim, \neg, 0, 1)$ such that $(W, R, \leq)$ can be embedded into the canonical frame of $\mathcal{K}$.

Similar duality results can be established in case of regular double Stone algebras.

Theorem 60.

1. Let $(U, R, \leq)$ be a K_- frame. Then its complex algebra is a regular double Stone algebra if and only if $(U, R, \leq)$ is a regular double Stone frame.
2. Let $\mathcal{K} := (K, \vee, \wedge, \sim, \neg, 0, 1)$ be a regular double Stone algebra. Then its canonical frame is a regular double Stone frame.

Duality results can be stated as follows.

Theorem 61.

1. Let $\mathcal{K} := (K, \vee, \wedge, \sim, \neg, 0, 1)$ be a regular double Stone algebra. Then there exists a K_- frame $(W, R, \leq)$ such that $\mathcal{K}$ can be embedded into the complex algebra of $(W, R, \leq)$.
2. Let $(W, R, \leq)$ be a regular double Stone frame. Then there exists a regular double Stone algebra $\mathcal{K} := (K, \vee, \wedge, \sim, \neg, 0, 1)$ such that $(W, R, \leq)$ can be embedded into canonical frame of $\mathcal{K}$.

$\mathcal{RS}$ for an approximation space (U, R) forms a regular double Stone algebra as well, where Stone negation $\sim$ and dual stone negation $\neg$ are defined as:
$\sim (LX, UX) := ((UX)^c, (UX)^c)$, and $\neg(LX, UX) := ((LX)^c, (LX)^c), X \subseteq U$.

Example 14. Let us consider the approximation space of Example 9. Note that for a prime filter P, $P_\neg := \{a : \neg a \notin P\}$ and $P_\sim := \{a :\sim a \notin P\}$.
$P_{1\neg} = \{(ab, ab), (U, U)\}$ and $P_{1\sim} = \{(\emptyset, ab), (ab, ab), (c, U), (U, U)\}$.
$P_{2\neg} = \{(c, c), (c, U), (U, U)\} = P_{2\sim}$ and
$P_{3\neg} = \{(ab, ab), (U, U)\}$ and $P_{3\sim} = \{(\emptyset, ab), (ab, ab), (c, U), (U, U)\}$.
The relation R can be obtained as:

$$P_1 R P_1, \ P_1 R P_3, \ P_2 R P_2, \ P_3 R P_1 \text{ and } P_3 R P_3.$$

Pictorial representation of the isomorphism is as in Fig. 22.

5 Granule-Based Rough Sets from Quasi Order-Generated Covering-Based Approximation Spaces, Algebras and Representations

The essence of classical rough set theory is to approximate a set X in an approximation space (U, R) with the help of 'granules', where a granule is an equivalence class in the domain U. It is thus that, as mentioned in previous sections, the lower and upper approximation operators are defined with the help of these granules:

$$\mathsf{L}X := \bigcup \{[x] : [x] \subseteq X\},$$
$$\mathsf{U}X := \bigcup \{[x] : [x] \cap X \neq \emptyset\} = \bigcup \{[x] : x \in X\}, \qquad (*)$$

where $[x]$ is the equivalence class containing the element x of U. Moreover, unions of the granules give the definable sets. These sets are termed so, because they are exactly describable by the two approximation operators: X is definable, if and only if $X = \mathsf{L}X = \mathsf{U}X$. The simplest definable sets are the equivalence classes. One can represent L and U in terms of definable sets as well. Let $\mathcal{D}$ denote the collection of all definable sets.

$$\mathsf{L}X := \bigcup \{D \in \mathcal{D} : D \subseteq X\},$$
$$\mathsf{U}X := \bigcap \{D \in \mathcal{D} : X \subseteq D\}. \qquad (**)$$

Hence, the lower approximation is the largest definable set contained in X, while the upper approximation of X is the smallest definable set which contains X. The representation in $(**)$ shows the topological importance of L and U. In fact, L is an interior operator on the power set $\mathcal{P}(U)$, while U is a closure operator on $\mathcal{P}(U)$. Moreover, L and U are dual to each other, hence they generate the same (clopen) topology. Definable sets are the only open (closed) sets in this topology.

In practice however, granules may not be disjoint, or may not arise from an equivalence relation. It is interesting to see how a set may be approximated by granules in these general situations. A lot of ground has been covered in the study of generalized approximation spaces (cf. [69]), and we have a number of different notions of approximations of sets, for instance, by Yao [77–80]. In this work, we ask the following question. Consider a set U and a collection of granules $\{O_i : i \in \Lambda\}$ such that $\bigcup_{i \in \Lambda} O_i = U$. How should a set $X (\subseteq U)$ be approximated in the (generalized) approximation space $(U, \{O_i : i \in \Lambda\})$? We put forth the following natural requirements from a pair of operators $\mathsf{L}, \mathsf{U} : \mathcal{P}(U) \rightarrow \mathcal{P}(U)$ that we would like to call (respectively) the lower and upper approximations in $(U, \{O_i : i \in \Lambda\})$.

1. For $X \subseteq U$, $\mathsf{L}X \subseteq X \subseteq \mathsf{U}X$.
2. If O is a granule, then the lower and upper approximations of O are O itself, i.e., $\mathsf{L}O = O$ and $\mathsf{U}O = O$.

3. For $X \subseteq U$, further approximations of its lower and upper approximations do not lead to anything new, i.e.

$$LLX = LX, UUX = UX, \text{ and}$$
$$ULX = LX, LUX = UX.$$

In the classical case (cf. (*), (**)), L and U do, in fact, satisfy 1 2, 3.

Our interest in this section lies in the algebraic studies of generalized rough set theory, where the set approximations are defined through granules in the approximation space, and satisfy at least properties 1-3 above. There are at least two ways to generalize classical rough set theory: replacing the equivalence relation by a different binary relation (e.g. a tolerance), or replacing the partition due to the equivalence relation by a different collection of subsets of the domain (e.g. a covering). A lot of work has been done on algebraic structures of rough sets based on binary relations, a bulk of it by Järvinen (e.g. [40]). On the other hand, there is work, for instance by Bonikowski [12], on algebraic structures formed in covering-based approximation spaces. Ours is an amalgamation of these two lines of work.

We have organized this section as follows. In Sect. 5.1 we introduce a pair of approximation operators L,U in 'quasi order-generated covering-based approximation spaces (QOCAS)', and study their algebraic and topological properties. In Sect. 5.2 we study the algebraic structures of definable sets in QOCAS, the rough set theoretic view of some classical results in this regard, and finally, a representation theorem for the class of completely distributive lattices in which the set of completely join irreducible elements is join dense. In the context of completely distributive lattices, we should mention the work in [22,84], with the observation that the approximation operators studied in these papers are different from the ones considered here. In Sect. 5.3 we establish relationships between the collection $\mathcal{RS}$ of rough sets and the collection $\mathcal{R}$ of generalized rough sets in QOCAS. We study the algebraic behavior of the collections of rough sets in Sect. 5.4. This study culminates in a representation result for completely distributive Heyting algebras in which the set of completely join irreducible elements is join dense. We end the section by observing connections of the notions presented here, to the dominance-based rough set approach.

The content of this section is based on the articles [48,49].

5.1 Granule-Based Definition of Rough Sets in QOCAS

A pair $(U, \mathcal{C})$ is called a *covering-based approximation space*, when U is a non-empty set and $\mathcal{C}$ is a covering of U, i.e. it is a non-empty subset of $\mathcal{P}(U)$ such that $\bigcup \mathcal{C} = U$. Now elements of $\mathcal{C}$ may be considered as granules, and we would like to have lower and upper approximation operators L, U on $(U, \mathcal{C})$ that satisfy the properties 1-3 mentioned in the previous section. L and U have been defined in several ways in literature (cf. [69], or [53]), and it is interestingly observed by Samanta and Chakraborty in [70] that none of them capture all of 1, 2 and 3 together.

In this section, we consider coverings generated by quasi orders. Formally, let U be a set and R a quasi order on U. The family $\{R(x) : x \in U\}$ forms a covering of U, where $R(x) := \{y \in U : xRy\}$, i.e. $R(x)$ is the R-neighbourhood of x in U. So, given (U, R), one gets a covering-based approximation space $(U, \{R(x) : x \in U\})$. Let us call such covering-based approximation spaces *quasi order-generated covering-based approximation spaces* (QOCAS), and for simplicity, use the denotation (U, R) in place of $(U, \{R(x) : x \in U\})$. It is clear that one also has the converse: a covering-based approximation space $(U, \mathcal{C})$ gives rise to a QOCAS $(U, \{R_{\mathcal{C}}(x) : x \in U\})$, considering $R_{\mathcal{C}}(x) := \cap\{C \in \mathcal{C} : x \in C\}$. In line with the expressions of L, U in (*), treating R-neighbourhoods as granules, one obtains definitions of the lower and upper approximation operators in a QOCAS as follows.

Definition 37. For any $X \subseteq U$, $\mathsf{L}, \mathsf{U} : \mathcal{P}(U) \to \mathcal{P}(U)$ are such that

$$\mathsf{L}X := \bigcup\{R(x) : R(x) \subseteq X\},$$
$$\mathsf{U}X := \bigcup\{R(x) : x \in X\}. \qquad (***)$$

As in the classical case, $\mathcal{RS} := \{(\mathsf{L}X, \mathsf{U}X) : X \subseteq U\}$ then gives the collection of *rough sets in the QOCAS* (U, R).

We observe that some of the properties of the classical approximation operators hold here as well. Before proceeding, let us note the following lemma.

Lemma 9. (cf. [42,85]). R *is a quasi order on a set* U *if and only if* $y \in R(x)$ *implies* $R(y) \subseteq R(x)$, *for any* $x, y \in U$.

Let (U, R) be a QOCAS.

Proposition 13. *For any* $X, Y \subseteq U$, L *and* U *satisfy the following properties.*

1. $\mathsf{L}(U) = U$ *and* $\mathsf{L}(\emptyset) = \emptyset$.
2. $\mathsf{U}(U) = U$ *and* $\mathsf{U}(\emptyset) = \emptyset$.
3. $\mathsf{L}(X) \subseteq X \subseteq \mathsf{U}(X)$.
4. $\mathsf{LL}(X) = \mathsf{L}(X)$ *and* $\mathsf{LU}(X) = \mathsf{U}(X)$.
5. $\mathsf{UU}(X) = \mathsf{U}(X)$ *and* $\mathsf{UL}(X) = \mathsf{L}(X)$.
6. $\mathsf{L}(X \cap Y) = \mathsf{L}(X) \cap \mathsf{L}(Y)$ *and* $\mathsf{U}(X \cup Y) = \mathsf{U}(X) \cup \mathsf{U}(Y)$.
7. $\mathsf{L}(X) \cup \mathsf{L}(Y) \subseteq \mathsf{L}(X \cup Y)$ *and* $\mathsf{U}(X \cap Y) \subseteq \mathsf{U}(X) \cap \mathsf{U}(Y)$.
8. $\mathsf{L}(X) = X$ *if and only if* $\mathsf{U}(X) = X$.
9. *The pair* (L, U) *forms a Galois connection on the poset* $(\mathcal{P}(U), \subseteq)$: $\mathsf{U}A \subseteq B$ *if and only if* $A \subseteq \mathsf{L}B$, *for any* $A, B \subseteq U$.

Proof. 1. Let $x \in U$, so $R(x) \subseteq U$. Hence $\mathsf{L}(U) = U$. For each $x \in U$, $R(x) \neq \emptyset$, Hence $\mathsf{L}(\emptyset) = \emptyset$.
2. As R is reflexive, hence $\mathsf{U}(U) = U$. Clearly using the definition, we have $\mathsf{U}(\emptyset) = \emptyset$.
3. From the definition of L, clearly we have $\mathsf{L}(X) \subseteq X$. Reflexivity of R shows that $X \subseteq \mathsf{U}(X)$.

4. We already have $LL(X) \subseteq L(X)$. Let $x \in L(X)$, hence there exists y such that $x \in R(y)$ and $R(y) \subseteq X$ but then $R(x) \subseteq R(y)$. Hence $R(x) \subseteq L(X)$. So, $LL(X) = L(X)$.

We have $LU(X) \subseteq U(X)$. Let $x \in U(X)$. Using the definition, there exists y such that $x \in R(y)$ and $R(y) \subseteq U(X)$. $x \in R(y)$ implies $R(x) \subseteq R(y) \subseteq U(X)$. Hence $LU(X) = U(X)$.

5. $UU(X) \supseteq U(X)$. Now, let $x \in UU(X)$, hence $x \in R(x) \subseteq U(X)$. Hence $UU(X) = U(X)$.

We already have $UL(X) \supseteq L(X)$. Now, let $x \in UL(X)$, hence $x \in R(x) \subseteq L(X)$. Hence $UL(X) = L(X)$.

6. Let $x \in L(X) \cap L(Y)$, hence $R(x) \subseteq X \cap Y$. Now let $x \in U(X \cup Y)$ Using the definition, there exists y such that $x \in R(y)$ and $y \in X \cup Y$. Other directions are trivial.

7. Using monotone property of L and U, we have the desired result.

8. Let $LX = X$. Hence we have $X = LX = ULX = UX$.

9. Let us show that $UA \subseteq B$ if and only if $A \subseteq LB$. We have:

$$UA \subseteq B \Rightarrow LUA \subseteq LB \Rightarrow UA \subseteq LB \Rightarrow A \subseteq UA \subseteq LB.$$

$$A \subseteq LB \Rightarrow UA \subseteq ULB \Rightarrow UA \subseteq LB \Rightarrow UA \subseteq LB \subseteq B.$$

That L and U so defined, satisfy all the properties 1-3 required of approximation with granules, follows from the above proposition. Let us illustrate the definitions and properties through an example.

Example 15. Let $U := \{a, b, c, d\}$ and $\mathcal{C} := \{\{a, b, c\}, \{b\}, \{c\}, \{c, d\}\}$ be a covering of U. Then $R_{\mathcal{C}}(a) = \{a, b, c\}$, $R_{\mathcal{C}}(b) = \{b\}$, $R_{\mathcal{C}}(c) = \{c\}$ and $R_{\mathcal{C}}(d) = \{c, d\}$. For $X \subseteq U$, $L(X)$, $U(X)$ are then given by Table 1.

Table 1. .

X	$L(X)$	$U(X)$	X	$L(X)$	$U(X)$
$\emptyset$	$\emptyset$	$\emptyset$	ac	c	abc
a	$\emptyset$	abc	ad	$\emptyset$	U
b	b	b	bd	b	bcd
c	c	c	abc	abc	abc
d	$\emptyset$	cd	bcd	bcd	bcd
ab	b	abc	abd	b	U
bc	bc	bc	acd	c	U
cd	cd	cd	U	U	U

Observation 1. *Consider $X := \{b\}$ in Example 15, then $U(X) = \{b\}$, but $\{b\} \neq \{a, b\} = (L(\{a, c, d\}))^c = (L(\{b\}^c))^c$, hence (L, U) are not dual operators.*

As we have seen, classical approximation operators have granule-based as well as topological representations. The operators defined through (***) also have a topological representation. This is due to the well-known one-one correspondence between Alexandrov topologies and quasi orders on any set U. Indeed, the family $\{R(x) : x \in U\}$ for a quasi order R on U forms a minimal basis for an Alexandrov topology on U. Let us denote this topology by $\mathcal{T}_R$. When R is a quasi order, so is R^{-1}. So, $\mathcal{T}_{R^{-1}}$ also generates an Alexandrov topology.

Corollary 2. L *is the interior operator in the topology* $\mathcal{T}_R$, *while* U *is the closure operator in* $\mathcal{T}_{R^{-1}}$. *Hence, alternatively,* L *and* U *may be defined, for any* $X \subseteq U$, *as*

$$\mathsf{L}X := \bigcup \{D \in \mathcal{T}_R : D \subseteq X\},$$

$$\mathsf{U}X := \bigcap \{D \in \mathcal{T}_R : X \subseteq D\}.$$

Note the analogy with the topological definitions (**) of L, U in classical rough set theory. The above representation of L, U indicates that we are approximating subsets of U in the *bitopological* space $(U, \mathcal{T}_R, \mathcal{T}_{R^{-1}})$. For $D \in \mathcal{T}_R$, we have $\mathsf{L}D = D$ and $\mathsf{U}D = D$. Further, using the fact that L and U generate dual topologies, we have $\mathsf{L}X = X$ if and only if $\mathsf{U}X = X$, for any subset X of U (cf. Proposition 13). So, we take definable sets here to be exactly the open sets in the topological space $(U, \mathcal{T}_R)$, or closed sets in the topological space $(U, \mathcal{T}_{R^{-1}})$. Following our notational convention, $\mathcal{D}$ is the set of all definable sets in the QOCAS (U, R), but note that $\mathcal{D}$ is just the same as $\mathcal{T}_R$. It is also clear that when R is an equivalence relation on U, $R = R^{-1}$, whence $\mathcal{T}_R = \mathcal{T}_{R^{-1}}$, and L and U coincide with the classical lower and upper approximation operators.

Now, we give a comparison between L, U, and other operators in literature that are defined using a quasi order. The operators in 1-4 below are from [64], while those in 5 are from [85]. Let $(U, \mathcal{C})$ be a covering based approximation space.

1. $L_1(X) := \{x \in U : R_{\mathcal{C}}(x) \subseteq X\}$,
   $U_1(X) := \{x \in U : R_{\mathcal{C}}(x) \cap X \neq \emptyset\}$.
2. $L_2(X) := \{x \in U : \exists u(u \in R_{\mathcal{C}}(x) \wedge R_{\mathcal{C}}(u) \subseteq X)\}$,
   $U_2(X) := \{x \in U : \forall u(u \in R_{\mathcal{C}}(x) \rightarrow R_{\mathcal{C}} \cap X \neq \emptyset)\}$.
3. $L_3(X) := \{x \in U : \forall u(x \in R_{\mathcal{C}}(u) \rightarrow R_{\mathcal{C}}(u) \subseteq X)\}$,
   $U_3(X) := \cup\{R_{\mathcal{C}}(x) : R_{\mathcal{C}}(x) \cap X \neq \emptyset\}$.
4. $L_4(X) := \{x \in U : \forall u(x \in R_{\mathcal{C}}(u) \rightarrow u \in X)\}$,
   $U_4(X) := \cup\{R_{\mathcal{C}}(x) : x \in X\}$.

$(L_1, U_1) - (L_4, U_4)$ are dual operator pairs. There is a non-dual operator pair defined using $R_{\mathcal{C}}(x)$ also.

5. $L_5(X) := \cup\{C \in \mathcal{C} : C \subseteq X\}$,
   $U_5(X) := L_5(X) \cup (\cup\{R_{\mathcal{C}}(x) : x \in (X \setminus L_5(X))\})$
   $\qquad = \cup\{R_{\mathcal{C}}(x) : x \in X\}$.

Proposition 14. *Let (U, C) be a covering-based approximation space and $X \subseteq U$.*

1. $L_1 = L$.
2. $L \subseteq L_2$.
3. $L_3(X) \subseteq L(X)$ *and* $U(X) \subseteq U_3(X)$.
4. $U_4 = U$.
5. $L_5(X) \subseteq L(X)$ *and* $U_5 = U$.

Let us show that containment of L_5 inside L can be proper, i.e., L_5 is different from L.

Example 16. Let $U := \{a, b, c, d\}$ and $C := \{\{a, b, c\}, \{b, c\}, \{c, d\}\}$. Then $R_C(a) = \{a, b, c\}$, $R_C(b) = \{b, c\}$, $R_C(c) = \{c\}$, $R_C(d) = \{c, d\}$. Let $X = \{c\}$, then $L_5 X = \emptyset$ but $LX = X = \{c\}$.

We show through the following example that U_1, U_2 and L_4 are not comparable with U and L respectively.

Example 17. Let $U := \{a, b, c, d\}$ and $C := \{\{a, b, c\}, \{b, c\}, \{c\}, \{c, d\}\}$. Then $R_C(a) = \{a, b, c\}$, $R_C(b) = \{b, c\}$, $R_C(c) = \{c\}$, $R_C(d) = \{c, d\}$.
     Let $X := \{a, b, c\}$. Then $U(\{a, b, c\}) = \{a, b, c\}$ and $U_1(\{a, b, c\}) = U = U_2(\{a, b, c\})$. So, here $U(X) \subseteq U_1(X) = U_2(X)$. But if we consider $X := \{a, d\}$, then $U(\{a, d\}) = U$, while $U_1(\{a, d\}) = \{a, d\}$, and $U_2(\{a, d\}) = \emptyset$. Hence, in this case, $U_1(X) \subseteq U(X)$, as well as $U_2(X) \subseteq U(X)$. So U and U_1, U_2 are not comparable.
     Again consider $X := \{a, b, c\}$. Then $L(\{a, b, c\}) = \{a, b, c\}$ and $L_4(\{a, b, c\}) = \{a, b\}$. So, $L_4(X) \subseteq L(X)$. But taking $X := \{a, d\}$ as before, we find $L(\{a, d\}) = \emptyset$, while $L_4(\{a, d\}) = \{a, d\}$. So, $L(X) \subseteq L_4(X)$, in this case. Hence L and L_4 are not comparable either.

Observation 2. Thus (L, U) is comparable only with (L_3, U_3) and (L_5, U_5), and is better than either as a pair of approximations.

5.2   Classical Algebraic Structures Represented Through the Collection of Definable Sets in QOCAS

We focus now on the algebraic structures formed by definable sets in a QOCAS, and representation results obtained in terms of definable sets for certain classes of lattices. In other words, we see how some classical algebraic structures may be viewed as definable sets in some QOCAS.
     Recall that, for a QOCAS (U, R), definable sets forming the collection $\mathcal{D}$, and open sets in the topology $\mathcal{T}_R$ are identical. Hence each representation result presented in this section has a topological interpretation. We shall also be using $\mathcal{D}$ and $\mathcal{T}_R$ interchangeably in the following. In this context, it should be noted that even though some of the results given here may be known already, we are re-presenting them in the context of rough sets.
     Let us begin with a well-known result about algebras of open sets in a topological space.

Theorem 62. [65] *The lattice of open sets of a topological space is a Heyting algebra. Conversely, any Heyting algebra is embeddable into the Heyting algebra of open sets of some topological space.*

In Theorem 64 below, we note that any Heyting algebra is, in fact, embeddable into the Heyting algebra of open sets of some Alexandrov topological space. Observe that, in particular, $\mathcal{D}(= \mathcal{T}_R)$ is a Heyting algebra.

When $\mathcal{C} := \{R(x) : x \in U\}$ is a partition of U, we know that $(\mathcal{D}, \cup, \cap, ^c, \emptyset, U)$ forms a complete atomic Boolean algebra in which each equivalence class is an atom, and any definable set in (U, R) is the union of some atoms (^{c} denotes the set-theoretic complement in $\mathcal{D}$). Moreover, every complete atomic Boolean algebra is isomorphic to the Boolean algebra formed by the collection of definable sets in some classical approximation space (based on an equivalence relation). In this context, we address a natural question for any QOCAS (U, R): under what conditions on granules, is $(\mathcal{D}, \cup, \cap, ^c, \emptyset, U)$ a Boolean algebra? One can prove the following.

Proposition 15. $(\mathcal{D}, \cup, \cap, ^c, \emptyset, U)$ *forms a Boolean algebra, if and only if* $\{R(x) : x \in U\}$ *forms a partition of* U.

Proof. Suppose $\{R(x) : x \in U\}$ forms a partition of U. As definable sets are unions of some $R(x)$'s, $\mathcal{D}$ is just the classical collection of definable sets, and as is well-known, $(\mathcal{D}, \cup, \cap, ^c, \emptyset, U)$ is a Boolean algebra.

Conversely, let $(\mathcal{D}, \cup, \cap, ^c, \emptyset, U)$ be a Boolean algebra, and suppose that the collection $\{R(x) : x \in U\}$ is not a partition of U. Then there exist $x, y \in U$ such that $x \neq y$, $R(x) \neq R(y)$ and $R(x) \cap R(y) \neq \emptyset$. So there is z in U such that $z \in R(x) \cap R(y)$. Thus $R(z) \subseteq R(x)$ and $R(z) \subseteq R(y)$. As $R(x) \neq R(y)$, we must have one of $R(z) \subsetneq R(x)$ and $R(z) \subsetneq R(y)$. Suppose $R(z) \subsetneq R(x)$. Consider the pseudo-complement $R(z)^+$ and *dual* pseudo-complement $R(z)^-$ of $R(z)$, viz.

$$R(z)^+ := \cup\{D \in \mathcal{D} : D \cap R(z) = \emptyset\},$$
$$R(z)^- := \cap\{D \in \mathcal{D} : D \cup R(z) = U\}.$$

Let $D \in \mathcal{D}$ such that $D \cap R(z) = \emptyset$. This implies $x \notin D$. Indeed, if $x \in D$, $R(x) \subseteq D$. So, $R(z) \subseteq R(x)$ implies that $D \cap R(z) = R(z) \neq \emptyset$, a contradiction. Hence $x \notin R(z)^+$.

Now let $D \in \mathcal{D}$ such that $D \cup R(z) = U$. As $R(z) \subsetneq R(x)$, $x \notin R(z)$. So $x \in D$, whence $x \in R(z)^-$.

Thus $R(z)^+ \neq R(z)^-$. But as $(\mathcal{D}, \cup, \cap, ^c, \emptyset, U)$ is a Boolean algebra, the pseudo-complement and dual pseudo-complement of an element are the same (as its complement). So this gives a contradiction.

We now proceed to investigate what structure, in general, $\mathcal{D}$ forms here. First, we observe from [65] that elements of any distributive lattice or Heyting algebra, may be regarded as definable sets in some QOCAS. As mentioned in Sect. 5.1, if $\mathcal{C}$ is an arbitrary covering of U, we get a QOCAS $(U, R_\mathcal{C})$ considering the relation $R_\mathcal{C}$ such that $R_\mathcal{C}(x) = \cap\{C \in \mathcal{C} : x \in C\}$, $x \in U$.

Theorem 63. *Let* $\mathcal{L} := (L, \vee, \wedge, 0, 1)$ *be a distributive lattice. There exists a* $QOCAS\ (U, R)$ *such that* $\mathcal{L}$ *can be embedded into the lattice of its definable sets.*

Let us give a sketch of the proof, as we shall use the constructions involved in it in the sequel.

Proof. Let U be the set of all prime filters of $\mathcal{L}$, and for each $a \in L$, let $\mathcal{C}_a := \{\Delta \in U : a \in \Delta\}$. Then, clearly, $\mathcal{C} := \{\mathcal{C}_a : a \in L\}$ is a covering for U. So $(U, R_\mathcal{C})$ is a QOCAS. Now for $\Delta \in U$, $R_\mathcal{C}(\Delta) = \bigcap\{\mathcal{C}_a \in \mathcal{C} : \Delta \in \mathcal{C}_a\} = \bigcap\{\mathcal{C}_a \in \mathcal{C} : a \in \Delta\}$. Thus definable sets, and in particular, $\mathcal{C}_a, a \in L$, are open sets in the Alexandrov topology $\mathcal{T}_{R_\mathcal{C}}$. Define a map $h : L \longrightarrow \mathcal{D}$, as $h(a) := \mathcal{C}_a$. h is shown to be a lattice embedding.

Moreover, we have the following for any Heyting algebra.

Theorem 64. *For any Heyting algebra* $\mathcal{L} := (L, \vee, \wedge, \rightarrow, 0, 1)$, *there exists a* $QOCAS$ *such that* $\mathcal{L}$ *can be embedded into the Heyting algebra of its definable sets.*

Proof. We take $(U, R_\mathcal{C})$, as in the proof of Theorem 63. Apart from the Alexandrov topology $\mathcal{T}_{R_\mathcal{C}}$ on U generated by $\{R_\mathcal{C}(\Delta) : \Delta \in U\}$ as subbasis, there is a topology $\mathcal{T}$ on U considering $\{\mathcal{C}_a : a \in L\}$ as subbasis. Note that in the topological space $(U, \mathcal{T}_{R_\mathcal{C}})$, $\mathcal{C}_a$ is an open set for each $a \in L$. Hence $\mathcal{T}_{R_\mathcal{C}}$ is finer than $\mathcal{T}$. Now we know that any topological space on U, and hence $\mathcal{T}$, forms a Heyting algebra with set union and intersection, and the operation $\rightarrow$ defined as

$$X \rightarrow Y := I(X^c \cup Y),\ X, Y \subseteq U,$$

I being the topological interior. It is shown in [65] that given a Heyting algebra $\mathcal{L}$, the map h as in the proof of Theorem 63, i.e. $h : L \longrightarrow \mathcal{D}$ such that $h(a) := \mathcal{C}_a$, $a \in L$, is an embedding into the Heyting algebra of open sets of the topology $\mathcal{T}$. In other words, it is proved that $h(a \rightarrow b) = h(a) \rightarrow h(b) = I_\mathcal{T}(h(a)^c \cup h(b))$, where $I_\mathcal{T}$ is the interior with respect to $\mathcal{T}$. In the same lines, we can show here for the topology $\mathcal{T}_{R_\mathcal{C}}$, that $h(a \rightarrow b) = h(a) \rightarrow h(b) = I_{\mathcal{T}_{R_\mathcal{C}}}(h(a)^c \cup h(b))$. So h provides a (Heyting) embedding into $\mathcal{D}$.

So what kind of algebras are *exactly* determined by the collection $\mathcal{D}$ of definable sets? Theorem 65 below answers the question. To get to the result, we first recall the following.

Definition 38. *Let* $\mathcal{L} := (L, \vee, \wedge, 0, 1)$ *be a complete lattice. A filter* Δ *of a complete lattice* $\mathcal{L}$ *is said to be* complete, *if and only if* $a_i \in \Delta$, *for all* $i \in I$, *implies* $\wedge_{i \in I} a_i \in \Delta$. *Further, a complete filter is said to be* completely prime, *provided* $\vee_{i \in I} a_i \in \Delta$ *implies* $a_i \in \Delta$, *for some* $i \in I$.

Recall from Sect. 1 that $\mathcal{J}_L$ denotes the set of all completely join irreducible elements of a complete lattice L.

The granules $R(x)$, $x \in U$, of a QOCAS (U, R) form a minimal basis for the topology $\mathcal{T}_R(= \mathcal{D})$. Hence the completely join irreducible elements of $\mathcal{D}$ are the

$R(x)$'s. So each element in $\mathcal{D}$, i.e. any definable set, is the join of completely join irreducible elements. In other words, $\mathcal{D}$ forms a completely distributive lattice in which $\mathcal{J}_\mathcal{D}$ is join dense.

One can prove the following.

Proposition 16. *Let* $\mathcal{L} := (L, \vee, \wedge, 0, 1)$ *be a completely distributive lattice in which* $\mathcal{J}_L$ *is join dense. Then there is a set of complete and completely prime filters which separates the points of L.*

Proof. Let $a \in \mathcal{J}_L$ and consider $a \uparrow := \{b \in L : a \leq b\}$. Let $\{b_i\}_i \in I \subseteq a \uparrow$, where I is some index set. Then $a \leq b_i$, for all $i \in I$. This implies that $a \leq \wedge b_i$. Hence $\wedge b_i \in a \uparrow$. Now let $\vee b_i \in a \uparrow$. This implies $a \leq \vee b_i$. Using complete join irreducibility of a, we get $a \leq b_i$, for some i. Hence $b_i \in a \uparrow$, for some i. Let $a, b \in L$ such that $a \nleq b$. By the assumption of the proposition, $a = \vee \{a_i : a_i \in \mathcal{J}_L\}$ and $b = \vee \{b_i : b_i \in \mathcal{J}_L\}$. So $a_i \nleq b$, for some i, and thus $a \in a_i \uparrow$ but $b \notin a_i \uparrow$. $\{c \uparrow : c \in \mathcal{J}_L\}$ is then the required set of filters.

Proposition 16 applies, in particular, to the lattice formed by $\mathcal{D}$. We now get a representation of these kinds of completely distributive lattices.

Theorem 65. *Let* $\mathcal{L} := (L, \vee, \wedge, 0, 1)$ *be a completely distributive lattice in which* $\mathcal{J}_L$ *is join dense. Then there exists a QOCAS (U, R) such that $\mathcal{L}$ is isomorphic to the lattice of its definable sets.*

Proof. By Proposition 16, there exists a set of complete and completely prime filters which separates the points of L. So, let U be such a set of filters. The proof now follows the lines of that for Theorem 63, considering this set U as the domain of the approximation space. The map $h : L \longrightarrow \mathcal{D}$ is a lattice homomorphism. Using the fact that U contains filters which are complete and completely prime, the collection $\{\mathcal{C}_a : a \in L\}$ forms a complete lattice with respect to set union and intersection. Hence $\mathcal{D} = \{\mathcal{C}_a : a \in L\}$, and it follows that h is onto. That h is one-one follows from the fact that U separates the points of L. Moreover, $h(0) = \emptyset$ and $h(1) = U$. Hence h is the required isomorphism.

5.3 The Collections $\mathcal{R}$ and $\mathcal{RS}$ of Rough Sets in a QOCAS: Relationships

In this section, we study the collection $\mathcal{RS} := \{(\mathsf{L}X, \mathsf{U}X) : X \subseteq U\}$ of rough sets, and the collection $\mathcal{R} := \{(D_1, D_2) : D_1 \subseteq D_2, \ D_1, D_2 \in \mathcal{D}\}$ of 'generalized' rough sets in a QOCAS (U, R). Let us recall that $\mathcal{R}$ has been considered in the classical rough set context in [4]. As the collection $\mathcal{D}$ of definable sets in (U, R) is the same as the collection of open sets of the topology $\mathcal{T}_R$, $\mathcal{R}$ may be viewed as a collection of ordered pairs of open sets of $\mathcal{T}_R$, i.e., $\mathcal{R} \subseteq \mathcal{T}_R \times \mathcal{T}_R$. Now, we already have $\mathcal{RS} \subseteq \mathcal{R}$. When $\mathcal{C}$ is a partition, it is observed in [5] that $\mathcal{R}$ and $\mathcal{RS}$ are equivalent if and only if each equivalence class of $(U, \mathcal{C})$ contains at least two elements. In other words, for any pair (D_1, D_2) of definable sets in $\mathcal{R}$, there is $X(\subseteq U)$ such that the lower approximation of X, $\mathsf{L}(X) = D_1$ and the upper approximation of X, $\mathsf{U}(X) = D_2$, if and only if $|D_2 \setminus D_1| \neq 1$. On investigating the situation here, in [48] we find that the same condition works here as well.

Theorem 66. $\mathcal{R} = \mathcal{RS}$, *if and only if* $|D_2 \setminus D_1| \neq 1$ *for every* $(D_1, D_2) \in \mathcal{R}$.

Proof. First let us assume $\mathcal{R} = \mathcal{RS}$.

Suppose the condition of the theorem does not hold. Then there exist $D_1, D_2 \in \mathcal{D}$ such that $D_1 \subseteq D_2$, and $|D_2 \setminus D_1| = 1$. Let $D_1 = \cup_{y \in Z_1} R(y)$ and $D_2 = \cup_{z \in Z_2} R(z)$, where $Z_1, Z_2 \subseteq U$. Now, $D_1 \subseteq D_2$, and $|D_2 \setminus D_1| = 1$ imply that there is $x \in D_2$ such that $x \notin D_1$. $x \in D_2$ means $x \in R(z)$ for some $z \in Z_2$. If $x \neq z$, then $z \in D_1$, and so $R(z) \subseteq D_1$. This implies that $x \in D_1$, a contradiction to our assumption. Hence if $x \in R(z) \subseteq D_2$, $z = x$. Now suppose there is $X \subseteq U$ such that $\mathsf{L}(X) = D_1$ and $\mathsf{U}(X) = D_2$. $\mathsf{U}(X) = D_2$ implies $x \in X$: otherwise, $x \in \mathsf{U}(X) = \cup\{R(z) : z \in X\} = D_2$ implies there is $z \in X \subseteq D_2$ such that $x \in R(z)$, a contradiction. Hence $x \in X$. We also have $\mathsf{L}(X) = D_1$, hence $\mathsf{L}(X) = D_1 \subseteq X \subseteq \mathsf{U}(X) = D_2$ reduces to $\mathsf{L}(X) = D_1 = X = \mathsf{U}(X) = D_2$, a contradiction to the assumption that $|D_2 \setminus D_1| = 1$.

Conversely, let us assume that the condition of the theorem holds, and let $(D_1, D_2) \in \mathcal{R}$. Let $D_2 = \cup_{y \in W} R(y)$, where $W \subseteq U$ is such that, for all $y, z \in W$ with $y \neq z$, neither of $R(y) \subseteq R(z)$ or $R(z) \subseteq R(y)$ holds. Note that we can always find such $W \subseteq U$. Let $Z := \{y \in W : y \notin D_1\}$. Now, let us consider $X := D_1 \cup Z$. We claim that $\mathsf{L}(X) = D_1$ and $\mathsf{U}(X) = D_2$.

Let us first prove that $\mathsf{L}(X) = D_1$. As $D_1 \subseteq X$, $\mathsf{L}(D_1) = D_1 \subseteq \mathsf{L}(X)$. Now, let $x \in \mathsf{L}(X)$. Then $R(x) \subseteq X$.

Case 1: If $R(x) \subseteq D_1$, we have nothing to prove.

Case 2: Assume $R(x) \not\subseteq D_1$. Then we have two subcases.

Subcase 1: If $R(x) \subseteq Z$, then $x = y$ for some $y \in Z$. Observe that the condition of the theorem implies that

$$|R(w)| \geq 2, \text{ for all } w \in U.$$

Hence, $|R(y)| \geq 2$, which implies that there exists some $z \neq y$ such that $z \in R(y)$. This implies $R(z) \subseteq R(y)$, which is a contradiction. Hence this subcase is ruled out.

Subcase 2: Now, assume there are two disjoint non-empty sets A and B such that $A \cup B = R(x)$, and $A \subseteq D_1$, $B \subseteq Z$. Clearly $x \notin A$, otherwise $x \in A \subseteq D_1$ would imply $R(x) \subseteq D_1$, a contradiction to our assumption. Hence $x \in B \subseteq Z$ and so $x = y$ for some $y \in Z$. As D_1 and Z are disjoint, $R(x) \setminus D_1 = B \subseteq Z$. By the condition of the theorem, $B = R(x) \setminus D_1 = R(x) \setminus (D_1 \cap R(x))$, has at least two elements. Hence there is $z \in B$ and $z \neq x$. This implies that $z \in R(x)$, and hence $R(z) \subseteq R(x)$, which is a contradiction to the fact that $z, x \in Z$.

Hence case 2 is ruled out and we have $R(x) \subseteq D_1$. Thus $\mathsf{L}(X) = D_1$.

Let us prove that $\mathsf{U}(X) = D_2$. We already have $X \subseteq D_2$, which implies $\mathsf{U}(X) \subseteq \mathsf{U}(D_2) = D_2$. Now let $x \in D_2$. Then there exists a y where $y \in W$, such that $x \in R(y)$. If $y \in D_1$, then $x \in R(y) \subseteq D_1$. If $y \notin D_1$, then $y \in Z$. This implies that $R(y) \subseteq \mathsf{U}(Z) \subseteq \mathsf{U}(X)$, and hence $x \in \mathsf{U}(X)$. So, $\mathsf{U}(X) = D_2$.

Let us illustrate the above theorem through an example.

Example 18. Let us recall Example 15. It is clear that the condition of the theorem is violated: $(\emptyset, R_{\mathcal{C}}(b)) = (\emptyset, b) \in \mathcal{R}$. So there is no $X \subseteq U$, such that $L(X) = \emptyset$ and $U(X) = \{b\}$.

We can make another interesting point through this example. Let us consider $X := \{a, b\}$ and $Y := \{c, d\}$. Then $L(X) \cup L(Y) = \{b, c, d\}$, $U(X) \cup U(Y) = \{a, b, c, d\}$, and $(L(X) \cup L(Y), U(X) \cup U(Y)) \in \mathcal{R}$. This pair also violates the condition of the theorem, and so there does not exist any $Z(\subseteq U)$ such that $L(X) \cup L(Y) = L(Z)$ and $U(X) \cup U(Y) = U(Z)$. Note that this is contrary to the case in classical rough set theory, when $\mathcal{C}$ is a partition of U (cf. [4,5]).

As we noted while proving Theorem 66, the condition of the theorem implies that $|R(x)| \geq 2$, for all $x \in U$. Let us show through an example that the converse need not hold, i.e. $|R(x)| \geq 2$, for all $x \in U$, does not imply the condition of Theorem 66.

Example 19. Let $U := \{a, b, c, d, e\}$ and $\mathcal{C} := \{\{a, b, c\}, \{b, c\}, \{d, e\}\}$. Then $R_{\mathcal{C}}(a) = \{a, b, c\}$, $R_{\mathcal{C}}(b) = \{b, c\}$, $R_{\mathcal{C}}(c) = \{b, c\}$, $R_{\mathcal{C}}(d) = \{d, e\}$ and $R_{\mathcal{C}}(e) = \{d, e\}$. Note that for all $x \in U$ $R_{\mathcal{C}}(x) \geq 2$, but it does not satisfy the condition of the theorem: consider the pair $(R_{\mathcal{C}}(b), R_{\mathcal{C}}(a))$ in $\mathcal{R}$.

We consider the natural order $\leq$ on $\mathcal{RS}$ and $\mathcal{R}$ inherited from $\mathcal{P}(U) \times \mathcal{P}(U)$, viz. for any $(D_1, D_2), (D_3, D_4) \in \mathcal{R}$ (or $\mathcal{RS}$), $(D_1, D_2) \leq (D_3, D_4)$ if and only if $D_1 \subseteq D_3$ and $D_2 \subseteq D_4$. Even though $\mathcal{R}$ and $\mathcal{RS}$ may not coincide for an approximation space (U, R), we now proceed to prove that there exists an approximation space (U', R') such that $\mathcal{R}$ and $\mathcal{R}'$ (corresponding to (U', R')) are order isomorphic and $\mathcal{R}' = \mathcal{RS}'$.

Theorem 67. *Let (U, R) be an approximation space such that there are definable sets $D_1, D_2 \in \mathcal{D}$ with $D_1 \subset D_2$ and $|D_2 \setminus D_1| = 1$. Then there exists an approximation space (U', R') such that the collection $\mathcal{D}'$ of its definable sets is lattice isomorphic to $\mathcal{D}$ and, for all $D_1', D_2' \in \mathcal{D}'$ with $D_1' \subset D_2'$, $|D_2' \setminus D_1'| \neq 1$.*

Proof. Let $\mathcal{S} := \{(D_1, D_2) : D_1, D_2 \in \mathcal{D}$ and $D_1 \subset D_2$ and $|D_2 \setminus D_1| = 1\}$. Now, corresponding to $\mathcal{S}$, we consider another set

$$\mathcal{S}' := \{x \in U : (D_1, D_2) \in \mathcal{S} \text{ and } x \in D_2 \text{ but } x \notin D_1\}.$$

Let $\mathcal{S}''$ be a disjoint copy of $\mathcal{S}'$: $\mathcal{S}'' := \{x' : x \in \mathcal{S}'\}$, and $U' := U \cup \mathcal{S}''$. We next define a quasi order on U'. For each $x' \in U'$,

$$R'(x') := \begin{cases} R(x') \cup \{y' \in \mathcal{S}'' : y \in R(x') \cap \mathcal{S}'\}, & \text{if } x' \in U \\ R(x) = R(x) \cup \{y' \in \mathcal{S}'' : y \in R(x) \cap \mathcal{S}'\}, & \text{if } x' \in \mathcal{S}'' \text{ for } x \in \mathcal{S}'. \end{cases}$$

Let us show R' is a quasi order. For that, it is enough to show that $y' \in R'(x')$ if and only if $R'(y') \subseteq R'(x')$, for all $x' \in U'$. If $R'(y') \subseteq R'(x')$, by definition of R', $y' \in R'(x')$.

Conversely, let $y' \in R'(x')$. If $x' \in U$, there are two cases.

Case 1: $y' \in R(x')$. Then $R(y') \subseteq R(x')$ and also $R(y') \cap \mathcal{S}' \subseteq R(x') \cap \mathcal{S}'$. Hence $R'(y') \subseteq R'(x')$.

Case 2: $y' \in S''$ such that $y \in R(x') \cap S'$. Then $R'(y') = R'(y) = R(y) \cup \{z' \in S'' : z \in R(y) \cap S'\}$ and $R(y) \subseteq R(x')$. Hence $R'(y') \subseteq R'(x')$.
If $x' \in S''$, $R'(x') = R'(x)$ for $x \in S'$. It can be proved similarly as above that $R'(y') \subseteq R'(x')$.
Then clearly, $(U', R' := \{R'(x') : x' \in U'\})$ is a QOCAS.

Let $\mathcal{D}'$ denote the collection of definable sets corresponding to (U', R'). Observe that if $D' \in \mathcal{D}'$ then $D' \setminus S'' \in \mathcal{D}$. Now suppose, if possible, there are $D_1', D_2' \in \mathcal{D}'$ such that $D_1' \subset D_2'$ and $|D_2' \setminus D_1'| = 1$. Then there is $x' \in U'$ such that $x' \in D_2'$, and $x' \notin D_1'$. Let $x' \in U$. Then $x' \in D_2' \setminus S''$, $x' \notin D_1' \setminus S''$ and $D_1' \setminus S'' \subseteq D_2' \setminus S''$ with $|(D_2' \setminus S'') \setminus (D_1' \setminus S'')| = 1$. Hence $x' \in S'$. But then by the definition of R', there is $x'' \in S''$ such that $R'(x') = R'(x'')$, which means that $x'', x' \in D_2'$ and $x'', x' \notin D_1'$. So $|D_2' \setminus D_1'| \neq 1$, a contradiction to our assumption. Now if $x' \in S''$ then there is an $x \in S'$ such that $R'(x) = R'(x')$ and so $x, x' \in D_2'$ and $x, x' \notin D_1'$. Hence $|D_2' \setminus D_1'| \neq 1$, again a contradiction. Consider the collections $\mathcal{J}_{\mathcal{D}}, \mathcal{J}_{\mathcal{D}'}$ of the completely join irreducible elements of $\mathcal{D}$ and $\mathcal{D}'$ respectively. Then $\mathcal{J}_{\mathcal{D}} = \{R(x) : x \in U\}$ and $\mathcal{J}_{\mathcal{D}'} = \{R'(x') : x' \in U'\}$. Let us define a map $\phi : \mathcal{J}_{\mathcal{D}} \longrightarrow \mathcal{J}_{\mathcal{D}'}$ such that

$$\phi(R(x)) := R'(x), \ x \in U.$$

We show that ϕ is an order isomorphism.
Let $R(x), R(y) \in \mathcal{J}_{\mathcal{D}}$ and $R(x) \subseteq R(y)$. Then $R(x) \cap S' \subseteq R(y) \cap S'$, and $\{y' \in S'' : y \in R(x) \cap S'\} \subseteq \{y' \in S'' : y \in R(y) \cap S'\}$. By definition of R', $R'(x) \subseteq R'(y)$.
If $R'(x'), R'(y') \in \mathcal{J}_{\mathcal{D}'}$ and $R'(x') \subseteq R'(y')$ then with a similar argument as above, we can prove that $R(x'') \subseteq R(y'')$, where x'', y'' are x or x' and y or y' respectively according as $x', y' \in U$, or $x', y' \in S''$: in the former case $x'' = x', y'' = y'$ and in the latter, $x'' = x, y'' = y$ where x', y' are copies of $x, y \in S'$ respectively.
ϕ is onto: let $R'(x') \in \mathcal{J}_{\mathcal{D}'}$. If $x' \in U$, then $\phi(R(x')) = R'(x')$. If $x' \in S''$, $R'(x') = R'(x)$, where x' is the copy of $x \in S'$, and $\phi(R(x)) = R'(x)$.
Now, every element of $\mathcal{D}$ and $\mathcal{D}'$ can be written as a join of elements from $\mathcal{J}_{\mathcal{D}}, \mathcal{J}_{\mathcal{D}'}$ respectively. So we consider the natural extension of $\phi : \mathcal{J}_{\mathcal{D}} \longrightarrow \mathcal{J}_{\mathcal{D}'}$ to the map $\overline{\phi} : \mathcal{D} \longrightarrow \mathcal{D}'$ defined as

$$\overline{\phi}(D) := \cup\{\phi(R(x)) : R(x) \subseteq D\},$$

where $D := \cup\{R(x) : R(x) \subseteq D\}$. Hence, using Lemma 1, $\overline{\phi}$ is a lattice isomorphism.

Using Theorems 66 and 67, we get the following corollary.

Corollary 3. *For every approximation space (U, R), there exists an approximation space (U', R') with the following properties.*

(i) $R' = \mathcal{R}S'$, *i.e. for each $(D'_1, D'_2) \in R'$, there exists an $X \subseteq U'$ with $LX = D'_1$ and $UX = D'_2$.*

(ii) R *and R' are order isomorphic.*

Let us illustrate Theorem 67 and Corollary 3 through the following example.

Example 20. Let $U := \{a, b\}$. Let us define a relation R on U as: $R(a) := ab = U$ and $R(b) := b$. Then R is a quasi order on U.

$$\mathcal{D} = \{\emptyset, b, U\}.$$
$$\mathcal{R} = \{(\emptyset, \emptyset), (\emptyset, b), (\emptyset, U), (b, b), (b, U), (U, U)\}.$$

The Hasse diagram of $(\mathcal{D}, \subseteq)$ and $(\mathcal{R}, \leq)$ is given in Fig. 29.

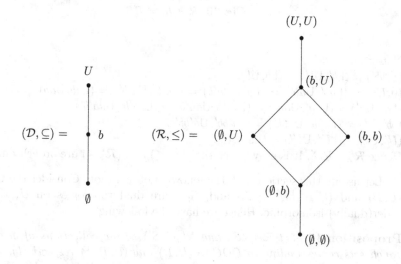

Fig. 29. Hasse diagram of ordered structures $(\mathcal{D}, \subseteq)$ and $(\mathcal{R}, \leq)$

$\mathcal{S} = \{(\emptyset, b), (b, U)\}$, $\mathcal{S}' = \{a, b\}$ and $\mathcal{S}'' := \{a', b'\}$. So $U' := U \cup \{a', b'\} = \{a, b, a', b'\}$. $R'(a) = aa'bb' = R'(a') = U, R'(b) = bb' = R'(b')$.

$$\mathcal{D}' = \{\emptyset, bb', U'\}.$$
$$\mathcal{R}' = \{(\emptyset, \emptyset), (\emptyset, bb'), (\emptyset, U'), (bb', bb'), (bb', U'), (U', U')\}.$$

A pictorial representation of the ordered structures $\mathcal{R}$ and $\mathcal{R}'$ is given in Fig. 30.

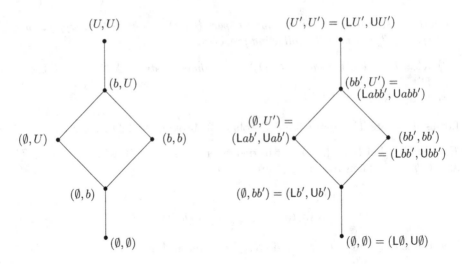

Fig. 30. $\mathcal{R} \cong \mathcal{R}' = \mathcal{RS}'$

Note that
$(\emptyset, bb') = (Lb', Ub') = (Lb, Ub)$,
$(\emptyset, U') = (Lab', Uab') = (La'b, Ua'b) = (La'b', Ua'b') = (Lab, Uab) = (La, Ua) = (La', Ua') = (Laa', Uaa') = (Laa'b, Uaa'b) = (Laa'b', Uaa'b')$,
$(bb', U') = (Labb', Uabb') = (La'bb', Ua'bb')$,
$(U', U') = (LU', UU') = (LU, UU)$.
Hence $\mathcal{R}' = \mathcal{RS}'$. It is easy to see that $(\mathcal{R}, \leq)$ and $(\mathcal{R}', \leq)$ are order isomorphic.

Let us end this section with the following observation. Consider the QOCAS (U, R) and (U, R^{-1}). As $\mathcal{T}_R$ and $\mathcal{T}_{R^{-1}}$ are dual topologies on U, they are order(dually) isomorphic. Hence we have the following.

Proposition 17. *Let $\mathcal{R}$ ($\mathcal{RS}$) and $\mathcal{R}'$ ($\mathcal{RS}'$) be the collections of generalized rough sets corresponding to QOCAS (U, R) and (U, R^{-1}) respectively. Then $\mathcal{R}$ ($\mathcal{RS}$) and $\mathcal{R}'$ ($\mathcal{RS}'$) are dual isomorphic via the map $(X, Y) \hookrightarrow (Y^c, X^c)$.*

5.4   Algebraic Structures of $\mathcal{R}$ and $\mathcal{RS}$

It has been shown earlier that $\mathcal{D}$ forms a Heyting algebra and completely distributive lattice. These structures can also be induced in $\mathcal{R}$.

Proposition 18. *$(\mathcal{R}, \cup, \cap, (\emptyset, \emptyset), (U, U))$ is a complete lattice, where, for $(D_1, D_2), (D_1', D_2') \in \mathcal{R}$,*

$$(D_1, D_2) \cup (D_1', D_2') := (D_1 \cup D_1', D_2 \cup D_2'),$$

$$(D_1, D_2) \cap (D_1', D_2') := (D_1 \cap D_1', D_2 \cap D_2').$$

In fact, $\mathcal{R}$ forms a complete sublattice of $\mathcal{P}(U) \times \mathcal{P}(U)$, and so a completely distributive lattice, whence a Heyting algebra. More explicitly, we have

Proposition 19. $(\mathcal{R}, \cup, \cap, \Rightarrow, (\emptyset, \emptyset), (U, U))$ *is a Heyting algebra, with*
$(D_1, D_2) \Rightarrow (D_1', D_2') := \cup \{(D, D') \in \mathcal{R} : (D_1, D_2) \cap (D, D') \subseteq (D_1', D_2')\},$
for $(D_1, D_2), (D_1', D_2') \in \mathcal{R}.$
In particular, $\mathcal{R}$ *is a pseudo-complemented lattice, where the pseudo-complement of* $(D_1, D_2) \in \mathcal{R}$ *is given by*

$$\sim_s (D_1, D_2) := (\sim D_2, \sim D_2),$$

where $\sim D_2$ *is the pseudo-complement of* D_2 *in the Heyting algebra* $\mathcal{D}$.

So in the case when the condition of Theorem 66 is satisfied, $\mathcal{RS}$ forms a completely distributive lattice. But we can show through an example that, in some cases, $\mathcal{RS}$ may not even form a lattice with respect to $\leq$.

Example 21. Let $U := \{a, b, c, d, e\}$, $R(a) := \{a, d, e\}$, $R(b) := \{b, d, e\}$, $R(c) := \{c, d\}$, $R(d) := \{d\}$ and $R(e) := \{d, e\}$. $(U, \{R(x) : x \in U\})$ is a QOCAS, and $\mathcal{RS}$ for this approximation space is given in Table 2. Now

Table 2. $\mathcal{RS}$ for QOCAS (U, R) in Example 21

X	$(\mathrm{L}X, \mathrm{U}X)$	X	$(\mathrm{L}X, \mathrm{U}X)$
$\emptyset$	$(\emptyset, \emptyset)$	abc	$(\emptyset, U)$
a	$(\emptyset, ade)$	bcd	$(cd, bcde)$
b	$(\emptyset, bde)$	cde	(cde, cde)
c	$(\emptyset, cd)$	abd	$(d, abde)$
d	(d, d)	abe	$(\emptyset, abde)$
e	$(\emptyset, de)$	acd	$(cd, acde)$
ab	$(\emptyset, abde)$	ade	(ade, ade)
bc	$(\emptyset, bcde)$	bce	$(\emptyset, bcde)$
cd	(cd, cd)	bde	(bde, bde)
de	(de, de)	ace	$(\emptyset, acde)$
ac	$(\emptyset, acde)$	$abcd$	(cd, U)
ad	(d, ade)	$bcde$	$(bcde, bcde)$
ae	$(\emptyset, ade)$	$abce$	$(\emptyset, U)$
bd	(d, bde)	$abde$	$(abde, abde)$
be	$(\emptyset, bde)$	$acde$	$(acde, acde)$
ce	$(\emptyset, cde)$	U	(U, U)

consider $(cd, cd), (\emptyset, de) \in \mathcal{RS}$. One may observe from the table that the set $\{(cde, cde), (cd, acde), (cd, bcde), (cd, U), (bcde, bcde), (acde, acde)\}$ gives all the upper bounds of $\{(cd, cd), (\emptyset, de)\}$, and that it does not have a least element in $\mathcal{RS}$. Hence $(\mathcal{RS}, \leq)$ is not a lattice.

Table 3. $\mathcal{RS}$ for QOCAS (U, R) in Example 22

X	(LX, UX)	X	(LX, UX)
$\emptyset$	$(\emptyset, \emptyset)$	ac	(c, abc)
a	$(\emptyset, abc)$	ad	$(\emptyset, U)$
b	(b, b)	bd	(b, bcd)
c	(c, c)	abc	(abc, abc)
d	$(\emptyset, cd)$	bcd	(bcd, bcd)
ab	(b, abc)	abd	(b, U)
bc	(bc, bc)	acd	(cd, U)
cd	(cd, cd)	U	(U, U)

One can, in fact, provide a condition different from that in Theorem 66, under which $\mathcal{RS}$ becomes a lattice.

Proposition 20. *Let* (U, R) *be a QOCAS such that the operators* L, U *have the following property:*

$$\text{if } LA = LB \text{ and } UA = UB \text{ then } A = B, \text{ for all } A, B \subseteq U. \qquad (\bowtie)$$

Then $(\mathcal{RS}, \vee, \wedge, (\emptyset, \emptyset), (U, U))$ *is a distributive lattice, where,*

$$(LA, UA) \vee (LB, UB) := (L(A \cup B), U(A \cup B)),$$
$$(LA, UA) \wedge (LB, UB) := (L(A \cap B), U(A \cap B)).$$

Proof. That $\vee$ and $\wedge$ are well-defined operators, follows from the property $(\bowtie)$. Now using Definition 3, it is easy to establish that $(\mathcal{RS}, \vee, \wedge, (\emptyset, \emptyset), (U, U))$ is a distributive lattice.

The following example uses Proposition 20 to show that the condition of Theorem 66 is not necessary for $\mathcal{RS}$ to form a (completely distributive) lattice.

Example 22. Let $U := \{a, b, c, d\}$ and let $R(a) = \{a, b, c\}$, $R(b) = \{b\}$, $R(c) = \{c\}$ and $R(d) = \{c, d\}$. Then $(U, \{R(x) : x \in U\})$ is a QOCAS. $\mathcal{RS}$ for the approximation space (U, R) is then given by Table 3. From the table, it is easy to check that whenever $LA = LB$ and $UA = UB$ then $A = B$. Hence $(\mathcal{RS}, \vee, \wedge, (\emptyset, \emptyset), (U, U))$ is a distributive lattice, where $\vee$ and $\wedge$ are defined as in Proposition 20. As $\mathcal{RS}$ is finite, it is completely distributive. But $\mathcal{R} \neq \mathcal{RS}$.

Now, let **R** and **RS** denote respectively the collections of sets $\mathcal{R}$ and $\mathcal{RS}$ for different approximation spaces. Then Corollary 3 gives a well-defined and one-one correspondence between **R** and **RS** quotiented by order isomorphism. However, note that this correspondence is not onto, due to the observation that $(\mathcal{RS}, \leq)$ may fail to be a lattice (cf. Example 21). So one may think of **R** as being properly embedded in **RS**, up to order isomorphism.

Now, let us abstract the construction generalized rough sets. It is well known that, if $\mathcal{A}$ is an abstract algebraic structure then $\mathcal{A} \times \mathcal{A}$ is also an algebraic structure of the same type as $\mathcal{A}$. We have seen that the collection $\mathcal{D}$ of definable sets in a QOCAS (U, R) forms a Heyting algebra and completely distributive lattice in which the set $\mathcal{J}_{\mathcal{D}} = \{R(x) : x \in U\}$ of completely join irreducible elements is join dense. From $\mathcal{D}$, we have the set $\mathcal{R} := \{(D_1, D_2) : D_1 \subseteq D_2, \ D_1, D_2 \in \mathcal{D}\}$.

Now, let us abstract this construction to the cases when $\mathcal{L} := (L, \vee, \wedge, 0, 1)$ is a

1. Heyting algebra, and
2. completely distributive lattice in which $\mathcal{J}_L$ is join dense.

Consider the set

$$L^{[2]} := \{(a, b) \in L \times L : a \leq b\}.$$

Let us define join and meet on $L^{[2]}$ componentwise as,

$$(a, b) \vee (c, d) = (a \vee c, b \vee d),$$

$$(a, b) \wedge (c, d) = (a \wedge c, b \wedge d),$$

where $(a, b), (c, d) \in L^{[2]}$. Also, $(0, 0), (1, 1)$ are bounds of $L^{[2]}$.
Hence $\mathcal{R}_L := (L^{[2]}, \vee, \wedge, (0, 0), (1, 1))$ is a bounded distributive lattice. In this section we study the enhanced structure of $\mathcal{R}_L$, when L is considered to be Heyting algebra, and also when it is a completely distributive lattice in which the set of join irreducible elements is join dense. Note that in the case when B is a Boolean algebra, $B^{[2]}$ may not form a Boolean algebra. But it can be shown to form a 3-valued Łukasiewicz algebra, pre-rough algebra and double Stone Algebra [11]. In case $\mathcal{L}$ is a Heyting algebra, $\mathcal{R}_L$ also forms a Heyting algebra. This is established by the following theorem.

Theorem 68. *Let* $\mathcal{H} := (H, \vee, \wedge, \rightarrow, 0, 1)$ *be a Heyting algebra. Then* $\mathcal{R}_{\vec{H}} := (\mathcal{R}_H, \rightarrow)$, *where for all* $(a, b), (c, d) \in H^{[2]}$,

$$(a, b) \rightarrow (c, d) := ((a \rightarrow c) \wedge (b \rightarrow d), b \rightarrow d),$$

is a Heyting algebra.

Proof. Let us show that $(a, b) \rightarrow (c, d) = max\{(e, f) \in H^{[2]} : (a, b) \wedge (e, f) \leq (c, d)\} = ((a \rightarrow c) \wedge (b \rightarrow d), b \rightarrow d)$.
Let $(e, f) \in H^{[2]}$ such that $(a, b) \wedge (e, f) \leq (c, d)$. Then $a \wedge e \leq c$ and $b \wedge f \leq d$. These imply $e \leq a \rightarrow c$ and $f \leq b \rightarrow d$. But we already have $e \leq f$, hence $e \leq (a \rightarrow c) \wedge (b \rightarrow d)$. So $(e, f) \leq ((a \rightarrow c) \wedge (b \rightarrow d), b \rightarrow d)$.
Now, $((a \rightarrow c) \wedge (b \rightarrow d), b \rightarrow d) \wedge (a, b) = ((a \rightarrow c) \wedge (b \rightarrow d) \wedge a, (b \rightarrow d) \wedge b)$.
But $(a \rightarrow c) \wedge a \leq c$ and $(b \rightarrow d) \wedge b \leq d$. Hence $(a \rightarrow c) \wedge (b \rightarrow d) \wedge a \leq c$ and $((a \rightarrow c) \wedge (b \rightarrow d) \wedge a, (b \rightarrow d) \wedge b) \leq (c, d)$.
Thus $(a, b) \rightarrow (c, d) = ((a \rightarrow c) \wedge (b \rightarrow d), b \rightarrow d)$ and $\mathcal{R}_{\vec{H}} = (\mathcal{R}_H, \rightarrow)$ is a Heyting algebra.

Proposition 21. *Let $\mathcal{H} := (H, \vee, \wedge, \rightarrow, 0, 1)$ be a Heyting algebra. Then $\mathcal{H}$ can be embedded in $\mathcal{R}_{\vec{H}}$ via the map $a \hookrightarrow (a, a)$.*

Proof. Let $\phi : H \rightarrow H^{[2]}$, defined as, $\phi(a) := (a, a)$, $a \in H$. Let us show that $\phi(a \rightarrow b) = (a \rightarrow b, a \rightarrow b) = (a, a) \rightarrow (b, b)$. Now, $(a, a) \rightarrow (b, b) = \bigvee\{(c, d) : (a, a) \wedge (c, d) \le (b, b)\}$. We already have $a \wedge (a \rightarrow b) \le b$. Moreover, if $(a, a) \wedge (c, d) \le (b, b)$ then $a \wedge c \le b$ and $a \wedge d \le b$. Hence $c \le a \rightarrow b$ and $d \le a \rightarrow b$, i.e. $(c, d) \le (a \rightarrow b, a \rightarrow b)$.

So, $\phi(a \rightarrow b) = (a \rightarrow b, a \rightarrow b) = (a, a) \rightarrow (b, b) = \phi(a) \rightarrow \phi(b)$.

In other words, the Heyting algebra $\mathcal{H}$ is isomorphic to the subalgebra of the Heyting algebra $\mathcal{R}_{\vec{H}}$ formed by the set $H' := \{(a, a) : a \in H\}$.
Moreover, we also have

Lemma 10.

1. Let $\mathcal{H}_1$ be embeddable into Heyting algebra $\mathcal{H}_2$. Then $\mathcal{R}_{H_1}$ is embeddable into $\mathcal{R}_{H_2}$.
2. Let $\mathcal{H}_1$ and $\mathcal{H}_2$ be isomorphic Heyting algebras. Then $\mathcal{R}_{H_1}$ and $\mathcal{R}_{H_2}$ are isomorphic as Heyting algebras.

Proof. Let ϕ be the given embedding (isomorphism). Then the map $\Phi : H_1^{[2]} \rightarrow H_2^{[2]}$ given as
$$\Phi(a, b) := (\phi(a), \phi(b))$$
is the required embedding (isomorphism).

Similar results can be proved for a completely distributive lattice in which the set of join irreducible elements is join dense.

Proposition 22. $\mathcal{R}_L := (L^{[2]}, \vee, \wedge, (0, 0), (1, 1))$ *is a completely distributive lattice in which the set $\mathcal{J}_{\mathcal{R}_L}$ of its completely join irreducible elements is join dense. $\mathcal{J}_{\mathcal{R}_L}$ is the collection of pairs of the form $(0, a)$, (a, a), $a \in \mathcal{J}_L$.*

Proof. Let us show that for $a \in \mathcal{J}_L$, $(0, a)$ is a completely join irreducible element of $\mathcal{R}_L$. Indeed, let $(0, a) = \bigvee_{i \in \Lambda}(a_i, b_i)$. This implies that for all $i \in \Lambda, a_i = 0$, and $a = \bigvee_{i \in \Lambda} b_i$ implies that $a = b_j$, for some $j \in \Lambda$. Hence $(0, a) = (0, b_j) = (a_j, b_j)$. Let us next show that for each $a \in \mathcal{J}_L$, (a, a) is a completely join irreducible element of $\mathcal{R}_L$. Let $(a, a) = \bigvee_{i \in \Lambda}(a_i, b_i)$. Then $a = \bigvee_{i \in \Lambda} a_i$, implying that $a = a_j$, for some $j \in \Lambda$. We also have $a = \bigvee_{i \in \Lambda} b_i$. So for each $i \in \Lambda, b_i \le a$, but this means $a = a_j \le b_j \le a$. Hence $(a, a) = (a_j, b_j)$, for some $j \in \Lambda$.

Moreover, these are the only completely join irreducible elements of $\mathcal{R}_L$. Now, let $(a, b) \in L^{[2]}$. $a, b \in L$ imply $a = \bigvee_{i \in \Lambda} a_i$ and $b = \bigvee_{j \in \Lambda'} b_j$, for some $a_i, b_j \in \mathcal{J}_L$. Hence, $(a, b) = \bigvee_{i \in \Lambda}(a_i, a_i) \vee \bigvee_{j \in \Lambda'}(0, b_j)$.

Observation 3.

(i) $\mathcal{L}$ is order embeddable into $\mathcal{R}_L$ via the map $a \hookrightarrow (a, a)$, $a \in L$.
(ii) Clearly, for any QOCAS (U, R), the set of completely join irreducible elements of $\mathcal{R}$ is given by $\mathcal{J}_{\mathcal{R}} = \{(\emptyset, R(x)), (R(x), R(x)) : x \in U\}$.

That the above abstractions of $\mathcal{R}$ are correct, is established by the following result.

Theorem 69.

1. Let $\mathcal{H} := (H, \vee, \wedge, \to, 0, 1)$ be a Heyting algebra. Then there exists a QOCAS (U, R) such that $\mathcal{R}_{\vec{H}}$ is embeddable into $\mathcal{R}$.
2. Let $\mathcal{L} := (L, \vee, \wedge, 0, 1)$ be a completely distributive lattice in which $\mathcal{J}_L$ is join dense. Then there exists a QOCAS (U, R) such that $\mathcal{R}_L$ is isomorphic to $\mathcal{R}$.

Proof. 1. Let (U, R) be the approximation space as in the proof of Theorem 64. We have seen in the proof of the theorem that the map $h : H \longrightarrow D$ defined as $h(a) := C_a = \{\Delta \in U : a \in \Delta\}$ is an embedding. Now, let $\mathcal{R}$ be the corresponding collection of rough sets in (U, R). Define a map $\phi : \mathcal{R}_H \longrightarrow \mathcal{R}$ as $\phi(a, b) := (h(a), h(b))$. We have seen in Theorem 64 that h is an embedding. Let us show that ϕ is also an embedding.
As h distributes over join and meet, so does ϕ. Now $\phi((a, b) \to (c, d)) = \phi((a \to c) \wedge (b \to d), (b \to d)) = (h(a \to c) \wedge h(b \to d), h(b \to d))$. As h distributes over $\to$, we have $\phi((a, b) \to (c, d)) = ((h(a) \to h(c)) \wedge (h(b) \to h(d)), (h(b) \to h(d)) = ((h(a), h(b)) \to (h(c), h(d)) = \phi(a, b) \to \phi(c, d)$.
Now let $\phi(a, b) = \phi(c, d)$. This implies that $h(a) = h(c), h(b) = h(d)$ and as h is one-one, we have $a = c, b = d$. So $(a, b) = (c, d)$, whence ϕ is one-one.
2. Take the approximation space and map h as in the proof of Theorem 65. As h is an isomorphism, it is not difficult to see that the map ϕ defined as in part 1 above, is also an isomorphism.

In fact, using Corollary 3, we have

Corollary 4.

1. $\mathcal{R}_{\vec{H}}$ is isomorphic to a subalgebra of $\mathcal{RS}'$ for some approximation space (U', R').
2. $\mathcal{R}_L$ is isomorphic to $\mathcal{RS}'$ for an approximation space (U', R').

For a Heyting algebra $\mathcal{H} := (H, \vee, \wedge, \to, 0, 1)$, let us consider $H_1 := \{(0, a) : a \in H\}$ and $H_2 := \{(a, a) : a \in H\}$.

Proposition 23.

1. H_1 forms a Heyting algebra $\mathcal{H}_1$ that is not a subalgebra of $\mathcal{R}_{\vec{H}}$.
2. H_2 forms a Heyting subalgebra $\mathcal{H}_2$ of $\mathcal{R}_{\vec{H}}$.
3. $\mathcal{H}_1 \cong \mathcal{H}_2$.
4. $\mathcal{R}_{\vec{H}} \cong \mathcal{R}_{H_1} \cong \mathcal{R}_{H_2}$.

Proof. 1. Let $(0, a), (0, b) \in H_1$. $(0, a) \to_{H_1} (0, b) = max\{(0, c) \in H_1 : (0, c) \wedge (0, a) \leq (0, b)\} = (0, a \to b)$. But in $\mathcal{R}_{\vec{H}}$, $(0, a) \to (0, b) = ((0 \to 0) \wedge (a \to b), a \to b) = (a \to b, a \to b)$. Hence the Heyting algebra $\mathcal{H}_1$ induced on the set H_1 by $\mathcal{H}$ is not a subalgebra of $\mathcal{R}_{\vec{H}}$.
2. Let $(a, a), (b, b) \in H_2$. $(a, a) \to_{H_2} (b, b) = max\{(c, c) \in H_2 : (c, c) \wedge (a, a) \leq (b, b)\} = (a \to a, b \to b)$. In $\mathcal{R}_{\vec{H}}$ also, $(a, a) \to (b, b) = ((a \to b) \wedge (a \to b), a \to$

$b) = (a \rightarrow b, a \rightarrow b)$. Hence $\mathcal{H}_2$, the Heyting algebra induced on the set H_2 by $\mathcal{H}$, is a subalgebra of $\mathcal{R}_{\vec{H}}$.

3. The map $(0, a) \hookrightarrow (a, a)$ is the required isomorphism.

4. The map $(a, b) \hookrightarrow ((a, a), (b, b))$, from $\mathcal{R}_{\vec{H}}$ to $\mathcal{R}_{H_2}$ is the required isomorphism. $\mathcal{R}_{H_1} \cong \mathcal{R}_{H_2}$ follows from part 3 and Lemma 10.

Now, let us discuss the variants of the above results in the context of a completely distributive lattice in which the set of completely join irreducible elements is join dense.

For any completely distributive lattice $\mathcal{L} := (L, \vee, \wedge, 0, 1)$ in which $\mathcal{J}_L$ is join dense, let $\mathcal{J}_{\mathcal{R}_1} := \{(0, a) : a \in \mathcal{J}_L\}$ and $\mathcal{J}_{\mathcal{R}_2} := \{(a, a) : a \in \mathcal{J}_L\}$. So $\mathcal{J}_{\mathcal{R}_L} = \mathcal{J}_{\mathcal{R}_1} \cup \mathcal{J}_{\mathcal{R}_2}$. Let $\mathcal{L}_1 := (L_1 := \{(0, a) : a \in L\}, \vee, \wedge, (0, 0), (0, 1))$ be the sublattice generated by $\mathcal{J}_{\mathcal{R}_1}$, and $\mathcal{L}_2 := (L_2 := \{(a, a) : a \in L\}, \vee, \wedge, (0, 0), (1, 1))$ that generated by $\mathcal{J}_{\mathcal{R}_2}$. $\mathcal{L}_1$ and $\mathcal{L}_2$ are isomorphic via the map $(0, a) \hookrightarrow (a, a)$, $a \in L$. Further, $\mathcal{L}_1$ and $\mathcal{L}_2$ are complete sublattices of $\mathcal{R}_L$, hence they are also completely distributive. The sets of completely join irreducible elements of $\mathcal{L}_1$ and $\mathcal{L}_2$ are just $\mathcal{J}_{\mathcal{R}_1}$ and $\mathcal{J}_{\mathcal{R}_2}$ respectively, and these are also join dense in the respective lattices.

Now we can construct the lattices $\mathcal{R}_{L_1}$ and $\mathcal{R}_{L_2}$ on the sets $L_1^{[2]}$ and $L_2^{[2]}$ as before, and obtain the following result. Observe that any pair $(a, b) \in L^{[2]}$ can be identified with the pair $((a, a), (b, b)) \in L_2^{[2]}$.

Theorem 70.

(a) Let $\mathcal{L} := (L, \vee, \wedge, 0, 1)$ be a completely distributive lattice in which $\mathcal{J}_L$ is join dense. Then $\mathcal{R}_L \cong \mathcal{R}_{L_2} \cong \mathcal{R}_{L_1}$.

(b) Let $\mathcal{L} := (L, \vee, \wedge, 0, 1)$ be a complete lattice. Then the following are equivalent.

   (i) There exists a completely distributive sublattice $\mathcal{L}_1$ of $\mathcal{L}$ in which $\mathcal{J}_{L_1}$ is join dense such that $\mathcal{L} \cong \mathcal{R}_{L_1}$.

   (ii) There exists an approximation space (U, R) such that $\mathcal{L} \cong \mathcal{R}$.

Proof. Let us sketch the proof of (b). (i) implies (ii), by using Theorem 69. Now, assume (ii) and let $\phi : \mathcal{R} \longrightarrow \mathcal{L}$ be the given isomorphism. Applying part (a), we have $\mathcal{R} = \mathcal{R}_{\mathcal{D}} \cong \mathcal{R}_{L_1} \cong \mathcal{R}_{L_2}$, where $L_1 := \{(\emptyset, D) : D \in \mathcal{D}\}$ and $L_2 := \{(D, D) : D \in \mathcal{D}\}$. Hence through the isomorphism ϕ, we have $\mathcal{L} \cong \mathcal{R}_{\phi(L_1)} \cong \mathcal{R}_{\phi(L_2)}$.

We have seen that in an approximation space $(U, \mathcal{C})$ definable sets $\mathcal{D}$ forms a Heyting algebra. Further any Heyting algebra can be embedded into definable sets $\mathcal{D}$ for some approximation space $(U, \mathcal{C})$ Theorem 64. And from Observation 3, definable set $\mathcal{D}$ is isomorphic to a sublattice of set of rough sets $\mathcal{R}$ via map $D \hookrightarrow (D, D)$, hence also we have the following representation result for Heyting algebras.

Theorem 71. *For any Heyting algebra $\mathcal{L} := (L, \vee, \wedge, \rightarrow, 0, 1)$, there exists a QOCAS (U', R') such that $\mathcal{L}$ is isomorphic to a subalgebra of the Heyting algebra $\mathcal{RS}'$.*

Proof. Let us take the approximation space (U, R) as in the proof of Theorem 64. Note that the map in question is $\phi : L \longrightarrow \mathcal{R}$ given by

$$\phi(a) := (h(a), h(a)), \ a \in L.$$

We just need to verify that ϕ is a Heyting homomorphism, i.e. $\{(h(a), h(a)) : a \in L\}$ forms a Heyting subalgebra of $\mathcal{R}$. In particular, we show that $\phi(a \to b) = \phi(a) \to \phi(b)$. For that, consider
$(h(a), h(a)) \to (h(b), h(b))$

$$= \bigcup\{(D, D') \in \mathcal{R} : D \cap h(a) \subseteq h(b), D' \cap h(a) \subseteq h(b)\}$$
$$= \bigcup\{(D', D') \in \mathcal{R} : D' \cap h(a) \subseteq h(b)\}.$$

But $D' \cap h(a) \subseteq h(b)$ implies $D' \subseteq h(a) \to h(b)$. Hence, we have
$(h(a), h(a)) \to (h(b), h(b))$

$$= \bigcup\{(D, D') \in \mathcal{R} : D \cap h(a) \subseteq h(b), D' \cap h(a) \subseteq h(b)\}$$
$$= \bigcup\{(D', D') \in \mathcal{R} : D' \cap h(a) \subseteq h(b)\}$$
$$= (h(a) \to h(b), h(a) \to h(b))$$
$$= (h(a \to b), h(a \to b)).$$

Now, by Corollary 3, there exists an approximation space (U', R') such that $\mathcal{R}$ is lattice isomorphic to $\mathcal{R}'$ and $\mathcal{R}' = \mathcal{RS}'$. Further, as $\mathcal{R}$ and $\mathcal{R}'$ are Heyting algebras, they are Heyting isomorphic also. Hence we have the result.

Note that if $\mathcal{L}$, in particular, is a *complete* Heyting algebra, Theorem 71 does not guarantee that the image of $\mathcal{L}$ under the map ϕ in the proof would be a complete Heyting subalgebra of $\mathcal{R}$. This is because the set $\{h(a) : a \in L\}$ may not form a complete lattice. However, we can obtain a representation result for the class of all completely distributive Heyting algebras $\mathcal{L}$ in which $\mathcal{J}_L$ is join dense. Before proceeding further, let us note that Lemma 1 and Corollary 3 can easily be extended to the case of completely distributive Heyting algebras, because relative pseudo-complement is defined via the order of the algebras.

Theorem 72. *Let $\mathcal{L} := (L, \vee, \wedge, \to, 0, 1)$ be a completely distributive Heyting algebra in which $\mathcal{J}_L$ is join dense. Then there is a QOCAS (U', R') such that $\mathcal{L}$ is isomorphic to a complete subalgebra of the Heyting algebra $\mathcal{RS}'$.*

Proof. Define a relation R on the set $\mathcal{J}_L$ as,

$$xRy \text{ if and only if } y \leq x, x, y \in \mathcal{J}_L.$$

Clearly R is a quasi order on $\mathcal{J}_L$, so that $\{R(x) : x \in \mathcal{J}_L\}$ is a covering for $\mathcal{J}_L$. Consider the approximation space $(\mathcal{J}_L, \{R(x) : x \in \mathcal{J}_L\})$. Hence completely join irreducible elements of $\mathcal{R}$ for the approximation space $(\mathcal{J}_L, \{R(x) : x \in \mathcal{J}_L\})$ are just the elements in the sets $\mathcal{J}_1 := \{(\emptyset, R(x)) : x \in \mathcal{J}_L\}$ and

$\mathcal{J}_2 := \{(R(x), R(x)) : x \in \mathcal{J}_L\}$. Let us define a map $\phi : \mathcal{J}_L \longrightarrow \mathcal{J}_2$ as

$$\phi(x) := (R(x), R(x)), \ x \in \mathcal{J}_L.$$

It is easy to see that ϕ is an order-preserving bijection. Now consider the subalgebra $\mathcal{R}_2 := \{(D, D) : D \in \mathcal{D}\}$ of $\mathcal{R}$ generated by $\mathcal{J}_2$ in the approximation space $(\mathcal{J}_L, \{R(x) : x \in \mathcal{J}_L\})$. $\mathcal{R}_2$ is, in fact, a complete Heyting subalgebra of $\mathcal{R}$; $\mathcal{J}_2$ is the set of completely join irreducible elements of $\mathcal{R}_2$.

Extend the map ϕ to $\overline{\phi} : \mathcal{L} \longrightarrow \mathcal{R}_2$ as in Lemma 1:

$$\overline{\phi}(x) := \bigvee_{y \in \mathcal{J}_L(x)} \phi(y), \ x \in L.$$

Then $\overline{\phi}$ is a Heyting isomorphism. Finally, using Corollary 3, we have the result.

Note that in the above proof, if instead of the map $\phi : \mathcal{J}_L \longrightarrow \mathcal{J}_2$, we define the map $\psi : \mathcal{J}_L \to \mathcal{J}_1$ as

$$\psi(x) := (\emptyset, R(x)), \ x \in \mathcal{J}_L,$$

then the corresponding extension map $\overline{\psi} : \mathcal{L} \longrightarrow \mathcal{R}_1 := \{(\emptyset, D) : D \in \mathcal{D}\}$ is also a lattice isomorphism. But it is easy to check that $\mathcal{R}_1$ may not be a Heyting subalgebra of $\mathcal{R}$.

5.5   Connections with Dominance-Based Rough Set Approach

We end this section by observing a connection between the basic notions of the dominance-based rough set approach (DRSA) [35] and our granule-based approach to generalized rough sets. Consider an information table (U, Q, V, f), where U is a finite set of objects, Q a finite set of attributes or criteria, $V := \bigcup_{q \in Q} V_q$, the set of attribute-values with V_q giving the values for the criterion q. $f : U \times Q \to V$ is the assignment function such that $f(x, q) \in V_q$. Q is taken to consist of a set C of condition criteria and a decision criterion d. It is assumed that the domain V_q of any criterion $q \in Q$ is a set of real numbers. Each $q \in Q$ induces an 'outranking' relation $\succeq_q$ on U such that the following relation holds:

$$x \succeq_q y, \text{ if and only if } f(x, q) \geq f(y, q).$$

$x \succeq_q y$ means that x is at least as good as y with respect to the criterion q. For any set $P(\subseteq C)$ of conditional criteria and objects $x, y \in U$, we say x dominates y with respect to P, denoted by $x D_P y$, if x is better than y on every criterion belonging to P, i.e., $x \succeq_q y$, for all $q \in P$. This dominance relation D_P is reflexive and transitive. For any $x \in U$, let

$$D_P^+(x) := \{y \in U : y D_P x\}, \text{ and}$$
$$D_P^-(x) := \{y \in U : x D_P y\}.$$

$D_P^+(x)$ and $D_P^-(x)$ represent the sets of $P - dominating$ and $P - dominated$ objects with respect to $x \in U$, respectively.

Now, let us look at the above from our perspective. Firstly, D_P^+ is a quasi order on U, and hence (U, D_P^+) is a QOCAS. Similarly, D_P^- defines the QOCAS $(U, \mathcal{C}_P^-)$, and $D_P^-(x), x \in U$, are granules therein. Thus an information table (U, Q, V, f) gives rise to QOCAS $(U, D_P^+), (U, D_P^-)$ through the dominance relations D_P, $P \subseteq C$. The algebraic structures formed by the definable sets in these QOCAS, therefore, are given by the study here. It should be mentioned that our approach to the study of algebraic structures is quite different from that in [37].

The notion of definable sets in a QOCAS gets a significant interpretation in DRSA. Consider (U, D_P^+) as above, and assume that $V_d := \{1, \ldots, n\}$, so that it induces a partition **Cl** on U, viz. **Cl** $:= \{Cl_t : t \in V_d\}$, where $Cl_t := \{x \in U : f(x,d) = t\}$. The equivalence classes Cl_t may also be assumed to be preference-ordered according to the increasing order of class indices, i.e., for all $r, s \in V_d$ such that $r \geq s$, the objects from Cl_r are strictly preferred to the objects from Cl_s. One then defines upward unions and downward unions of equivalence classes respectively as:

$$Cl_t^{\geq} := \cup_{s \geq t} Cl_s, \; Cl_t^{\leq} := \cup_{s \leq t} Cl_s, t = 1, \ldots, n.$$

Objects in $Cl_t^{\geq}$ are those belonging to the class Cl_t or to a more preferred class, while $Cl_t^{\leq}$ is the set of objects belonging to Cl_t or to less preferred classes. Now there is a notion of *inconsistency* defined in DRSA, when an object $x \in U$ belongs to the upward union of classes $Cl_t^{\geq}, t = 2, \ldots, n$, with some *ambiguity*, viz. when one of the following two conditions hold:

1. x belongs to class Cl_t or better, but it is $P - dominated$ by an object y belonging to a class worse than Cl_t, i.e., $x \in Cl_t^{\geq}$ but $D_P^+(x) \cap Cl_{t-1}^{\leq} \neq \emptyset$.
2. x belongs to a class worse than Cl_t, but it $P-dominates$ an object y belonging to class Cl_t or better, i.e., $x \notin Cl_t^{\geq}$ but $D_P^-(x) \cap Cl_{t-1}^{\leq} \neq \emptyset$.

One finds that the lower and upper approximations of $Cl_t^{\geq}$ in $(U, \mathcal{C}_P^+)$, viz.

$$\mathsf{L}(Cl_t^{\geq}) := \{x \in U : D_P^+(x) \subseteq Cl_t^{\geq}\} \text{ and}$$
$$\mathsf{U}(Cl_t^{\geq}) := \{x \in U : D_P^-(x) \cap Cl_t^{\geq} \neq \emptyset\} = \cup\{D_P^+(x) : x \in Cl_t^{\geq}\},$$

give, respectively, the sets of all objects belonging to $Cl_t^{\geq}$ without any ambiguity, and that of all those objects belonging to $Cl_t^{\geq}$ with or without ambiguity. Then it can be seen that the classes $Cl_t^{\geq}$ that are definable in (U, D_P^+), are exactly those classes such that no object of U belongs to them with any ambiguity (i.e. the ones with $\mathsf{L}(Cl_t^{\geq}) = Cl_t^{\geq}$.

6   Negations and Logics of Rough Sets from QOCAS

Let us recall the examples of 3-valued LM algebras constructed by Moisil (cf. [17]). Let B be a Boolean algebra and define operators $\neg$ and L on $B^{[2]}$ respectively as, $\neg(a, b) := (b^c, a^c)$ and $L(a, b) := (a, a)$, $a, b \in B$, where ^{c} is the Boolean negation in B. Then the structure $B^{[2]}$ enhanced with $\neg$ and L is a 3-valued LM algebra. Let us also recall that such operators on $B^{[2]}$ lead to examples of *pre-rough algebras* studied by Banerjee and Chakraborty [4]. We have seen in Sect. 5 that for a QOCAS, the collection $\mathcal{D}$ of definable sets forms a distributive lattice, Heyting algebra and completely distributive lattice in which the set $\mathcal{J}_{\mathcal{D}}$ of completely join irreducible elements is join dense. Moreover, the collection $\mathcal{R}$ also inherits these structures. It is interesting to see how the unary operator L defined on the set $\mathcal{R}$ behaves, when $\mathcal{R}$ is considered as a distributive lattice, Heyting algebra or completely distributive lattice in which $\mathcal{J}_{\mathcal{R}}$ is join dense. Abstraction of these properties leads us to introduce the classes of *rough lattices*, *rough Heyting algebras* and *complete rough lattices*. Moreover, we provide the representation of these structures in terms of rough sets. We discuss corresponding logics and present a rough set semantics for them.

We also pursue another line of work here. The Boolean negation induces various negations on the set $B^{[2]}$, such as De Morgan, Kleene, intuitionistic and dual intuitionistic negation. It is natural to ask the following questions. What kind of unary operators are induced by intuitionistic and dual intuitionistic negations on the set $L^{[2]}$, when L is considered to be a distributive pseudo and dual pseudo complemented lattice respectively? Can these unary operators be called 'negations'? To answer the latter, we again follow Dunn's approach to the study of negations. We introduce some unary operators via the intuitionistic and dual intuitionistic negations, study their properties, and then characterize them in the classes of compatibility frames and exhaustive frames. One is able to demonstrate that these negations occupy new positions in Dunn's kite and its dual kite.

We have organized this section as follows. In Sect. 6.1 we introduce two unary operators I, C and study their interactions with distributive lattices, Heyting algebras and completely distributive lattices in which the set of join irreducible elements is join dense. Moreover, we prove rough set representation of the classes of rough lattices, rough Heyting algebras and complete rough lattices. In Sect. 6.2, we discuss two logics, one which is based on the distributive lattice logic (DLL) extended with two modal operators, and another based on $MIPC$ introduced by Prior (cf. [7]). Further, due to the representation results proved earlier, we obtain a rough set semantics for these logics. In the final section (Sect. 6.3), we study the new negations defined with the help of intuitionistic and dual intuitionistic negations.

6.1 $\mathcal{R}$ and $\mathcal{RS}$ with Operators

Let $\mathcal{B} := (B, \vee, \wedge, ^c, 0, 1)$ be a Boolean algebra. Although the set $B^{[2]}$ was first studied by Moisil in the context of 3-valued LM algebras, as mentioned in Sect. 1, extensive research on algebras based on the set $B^{[2]}$ in case B is complete and atomic, has been done by Banerjee and Chakraborty [3–5] in the context of rough set theory.

Let (U, R) be a classical approximation space, and consider the collection $\mathcal{R} := \{(D_1, D_2) : D_1 \subseteq D_2, \ D_1, D_2 \in \mathcal{D}\}$, i.e. the set $\mathcal{D}^{[2]}$. Define operators $L, M : \mathcal{R} \longrightarrow \mathcal{R}$ as

$$L(D_1, D_2) := (D_1, D_1), M(D_1, D_2) := (D_2, D_2).$$

The motivation of defining L, M in the context of rough set theory is that when $\mathcal{R} = \mathcal{RS} := \{(LX, UX) : X \subseteq U\}$, the operators L and M extract respectively, the lower and upper approximations of the set concerned: $L(LX, UX) = (LX, LX)$; $M(LX, UX) = (UX, UX)$, $X \subseteq U$.

It is shown by Banerjee and Chakraborty in [4] that the structure $(\mathcal{R}, \vee, \wedge, \neg, L, (\emptyset, \emptyset), (U, U))$ is a pre-rough algebra, where $\neg$ is the De Morgan negation on $\mathcal{R}$ (cf. Sect. 1). In the same paper, representation of pre-rough algebras has also been proved. Further, Banerjee and Chakraborty extended the structure of a pre-rough algebra to define a *rough algebra*, and obtained its rough set representation. Later, it was observed by Banerjee that the rough set representation of pre-rough algebras can also be obtained.

In this section we follow the same line of work, in the generalized scenario of a QOCAS.

Let (U, R) be a QOCAS. Taking cues from the work of Moisil as well as Banerjee and Chakraborty, we define two unary operators I and C on $\mathcal{R}$ as follows.

$$I(D_1, D_2) := (D_1, D_1); \ C(D_1, D_2) := (D_2, D_2), \ D_1, D_2 \in \mathcal{D}.$$

In Sect. 5, we found that in the general set-up of a Heyting algebra and an arbitrary completely distributive lattice $\mathcal{L}$ having join-dense $\mathcal{J}_L$, $\mathcal{R}_L$ and $\mathcal{R}_L^{\rightarrow}$ abstract $\mathcal{R}$. Now let L be a bounded distributive lattice. One may define operators I and C on $\mathcal{R}_L$ as

$$I(a, b) := (a, a); \ C(a, b) := (b, b), \ a, b \in L.$$

Let us denote by $\mathcal{R}_L^{IC}$ and $\mathcal{R}^{IC}$ respectively, the lattices $\mathcal{R}_L$ and $\mathcal{R}$ enhanced with the operators I, C. To represent an algebra in terms of $\mathcal{RS}$, we need the following theorem, which is obtained as an easy extension of Theorem 67 and Corollary 3 to the case of $\mathcal{R}^{IC}$.

Theorem 73. *Let (U, R) be a QOCAS. There exists a QOCAS (U', R') such that $R' = \mathcal{RS}'$ and $\mathcal{R}^{IC}$ corresponding to (U, R) is isomorphic to $\mathcal{R}'^{IC}$ corresponding to (U', R').*

In the following section we study the enhanced algebraic structure $\mathcal{R}_L^{IC}$, where L is a bounded distributive lattice.

Now, recall that when L is a bounded distributive lattice, $\mathcal{R}_L := (L^{[2]}, \vee, \wedge, (0, 0), (1, 1))$ is also a bounded distributive lattice, where, for all $(a, b), (c, d) \in L^{[2]}$ we have,

$$(a, b) \vee (c, d) := (a \vee c, b \vee d),$$
$$(a, b) \wedge (c, d) := (a \wedge c, b \wedge d),$$

and $(0, 0)$ and $(1, 1)$ are top and bottom elements respectively.
Let us first state some properties of I and C on $\mathcal{R}_L$.

Proposition 24. *For* $(a, b), (c, d) \in L^{[2]}$, *$I$ and C satisfy the following properties.*

1. $I((a, b) \wedge (c, d)) = I(a, b) \wedge I(c, d),\ C((a, b) \vee (c, d)) = C(a, b) \vee C(c, d)$.
2. $I((a, b) \vee (c, d)) = I(a, b) \vee I(c, d),\ C((a, b) \wedge (c, d)) = C(a, b) \wedge C(c, d)$.
3. $I(a, b) \leq (a, b),\ (a, b) \leq C(a, b)$.
4. $I(1, 1) = (1, 1),\ C(0, 0) = (0, 0)$.
5. $II(a, b) = I(a, b)$ *and* $CC(a, b) = C(a, b)$.
6. $IC(a, b) = C(a, b)$ *and* $CI(a, b) = I(a, b)$.
7. $I(a, b) \leq I(c, d)$ *and* $C(a, b) \leq C(c, d)$ *imply* $(a, b) \leq (c, d)$.

In particular, I and C are homomorphisms on the lattice $\mathcal{R}_L$. Now, let us abstract the above properties to define a new algebraic structure, viz. a *rough lattice*.

Definition 39. $\mathcal{L} := (L, \vee, \wedge, I, C, 0, 1)$ *is said to be a* rough lattice, *provided* $(L, \vee, \wedge, 0, 1)$ *is a bounded distributive lattice and* I, C *are unary operators on* L *that satisfy the following properties.*

1. $I(a \wedge b) = Ia \wedge Ib, \qquad C(a \vee b) = Ca \vee Cb$.
2. $I(a \vee b) = Ia \vee Ib, \qquad C(a \wedge b) = Ca \wedge Cb$.
3. $Ia \leq a,\ a \leq Ca$.
4. $I1 = 1,\ C0 = 0$.
5. $IIa = Ia$ *and* $CCa = Ca$.
6. $ICa = Ca$ *and* $CIa = Ia$.
7. $Ia \leq Ib$ *and* $Ca \leq Cb$ *imply* $a \leq b$.

Note that in particular, any pre-rough algebra is a rough lattice. Moreover, for any bounded distributive lattice L, the structure $\mathcal{R}_L^{IC}$ is a rough lattice. In particular, for any $QOCAS$ $\mathcal{R}^{IC}$ is a rough lattice.

We now show that any rough lattice $\mathcal{L} := (L, \vee, \wedge, I, C, 0, 1)$ gives rise to another rough lattice, denoted $\mathcal{IC}_L^{I'C'}$ as follows.

Consider the set $IC_L := \{(Ia, Ca) : a \in L\}$. Following the theme of this section, let us define operators I', C' on the sublattice $\mathcal{IC}_L := (IC, \vee, \wedge, (0, 0), (1, 1))$ of $L \times L$ as

$$I'(Ia, Ca) := (Ia, Ia);\ C'(Ia, Ca) := (Ca, Ca),\ \text{for any } (Ia, Ca) \in IC_L.$$

As $(Ia, Ia) = (IIa, CIa)$ and $(Ca, Ca) = (ICa, CCa)$, we obtain that I', C' are well-defined. Then we have

Proposition 25. *Let* $\mathcal{L} := (L, \vee, \wedge, I, C, 0, 1)$ *be a rough lattice.* $\mathcal{IC}_L^{I'C'}$ *is a rough lattice, where* $\mathcal{IC}_L^{I'C'}$ *denotes the lattice* $\mathcal{IC}_L$ *enhanced with the operators* I' *and* C'.

Proof. Let $(Ia, Ca), (Ib, Cb) \in IC_L$.

1. $I'((Ia, Ca) \wedge (Ib, Cb)) = I'(Ia \wedge Ib, Ca \wedge Cb) = I'(I(a \wedge b), C(a \wedge b))$
   $= (I(a \wedge b), I(a \wedge b)) = I'(Ia, Ca) \wedge I'(Ib, Cb)$.
   $C'((Ia, Ca) \vee (Ib, Cb)) = C'(Ia \vee Ib, Ca \vee Cb) = C'(I(a \vee b), C(a \vee b))$
   $= (C(a \vee b), C(a \vee b)) = C'(Ia, Ca) \vee C'(Ib, Cb)$
2. $I'((Ia, Ca) \vee (Ib, Cb)) = I'(Ia \vee Ib, Ca \vee Cb) = I'(I(a \vee b), C(a \vee b))$
   $= (I(a \vee b), I(a \vee b)) = I'(Ia, Ca) \vee I'(Ib, Cb)$. $C'((Ia, Ca) \wedge (Ib, Cb))$
   $= C'(Ia \wedge Ib, Ca \wedge Cb) = C'(I(a \wedge b), C(a \wedge b))$
   $= (C(a \wedge b), C(a \wedge b)) = C'(Ia, Ca) \wedge I'(Ib, Cb)$.
3. $I'(Ia, Ca) = (Ia, Ia) \leq (Ia, Ca)$ and $(Ia, Ca) \leq (Ca, Ca) = C'(Ia, Ca)$.
4. $I'(1, 1) = (I1, I1) = (1, 1)$ and $C'(0, 0) = (C0, C0) = (0, 0)$.
5. $I'I'(Ia, Ca) = I'(Ia, Ia) = (Ia, Ia) = I'(Ia, Ca)$ and $C'C'(Ia, Ca)$
   $= C'(Ca, Ca) = (Ca, Ca) = C'(Ia, Ca)$.
6. $I'C'(Ia, Ca) = I'(Ca, Ca) = (Ca, Ca) = C'(Ia, Ca)$ and $C'I'(Ia, Ca)$
   $= C'(Ia, Ia) = (Ia, Ia) = I'(Ia, Ca)$.
7. Let $I'(Ia, Ca) \leq I'(Ib, Cb)$ and $C'(Ia, Ca) \leq C'(Ib, Cb)$. These imply
   $(Ia, Ia) \leq (Ib, Ib)$ and $(Ca, Ca) \leq (Cb, Cb)$. Hence, $Ia \leq Ib$ and $Ca \leq Cb$,
   but this means $(Ia, Ca) \leq (Ib, Cb)$.

Further, we prove the following.

Proposition 26. *Every rough lattice* $\mathcal{L} := (L, \vee, \wedge, I, C, 0, 1)$ *is isomorphic to the lattice* $\mathcal{IC}_L^{I'C'}$.

Proof. Let us define a map $\phi : L \longrightarrow IC_L$ as

$$\phi(a) := (Ia, Ca), \ a \in L.$$

Let us show that ϕ is the required map.
(1) $\phi(a \vee b) = (I(a \vee b), C(a \vee b)) = (Ia \vee Ib, Ca \vee Cb) = (Ia, Ca) \vee (Ib, Cb) = \phi(a) \vee \phi(b)$.
(2) $\phi(a \wedge b) = (I(a \wedge b), C(a \wedge b)) = (Ia \wedge Ib, Ca \wedge Cb) = (Ia, Ca) \wedge (Ib, Cb) = \phi(a) \wedge \phi(b)$.
(3) $\phi(Ia) = (IIa, CIa) = (Ia, Ia) = I'(Ia, Ca) = I'\phi(a)$.
(4) $\phi(Ca) = (ICa, CCa) = (Ca, Ca) = C'(Ia, Ca) = C'\phi(a)$.
(5) $\phi(a) = \phi(b)$ implies $(Ia, Ca) = (Ib, Cb)$. Hence $Ia = Ib$ and $Ca = Cb$.
Using 7, we have $a = b$. Clearly ϕ is an onto map, making ϕ a bijection.
Hence ϕ is the required isomorphism.

Let us provide another construction of a rough lattice from a given rough lattice $\mathcal{L} := (L, \vee, \wedge, I, C, 0, 1)$. Consider the set $D_L := \{a \in L : Ia = a\}$. As $CIa = Ia$ and $ICa = Ca$, $D_L = \{a \in L : Ca = a\}$ also. Moreover, as $IIa = Ia$ and $CCa = Ca$, $D_L = \{Ia : a \in L\} = \{Ca : a \in L\}$. Thus $IC_L \subseteq D_L^{[2]}$.
Note that in case of a pre-rough algebra, the set D_L forms a Boolean algebra. Here, one observes the following.

Proposition 27. *For a rough lattice* $\mathcal{L} := (L, \vee, \wedge, I, C, 0, 1)$, D_L *forms a bounded distributive sublattice of* $\mathcal{L}$.

Proof. Let $a, b \in D_L$. Now $a \vee b \in L$. Let us show that $a \vee b \in D_L$. $I(a \vee b) = Ia \vee Ib = a \vee b$. Hence $a \vee b \in D_L$. With similar lines $a \wedge b \in D_L$. As, $I(0) = 0$ and $I(1) = 1$, we have 0 and 1 are lower and upper bounds of D_L respectively.

So $\mathcal{R}_{D_L}^{I'C'}$ is a rough lattice. The following proposition is analogous to (representation) Theorem 2 which was in the context of pre-rough algebras.

Proposition 28. *Every rough lattice* $\mathcal{L} := (L, \vee, \wedge, I, C, 0, 1)$ *is embeddable into* $\mathcal{R}_{D_L}^{I'C'}$.

Proof. Let us consider the composition of the map ϕ in Proposition 26 and the inclusion map from IC_L into $D_L^{[2]}$. This gives the required embedding.

Now, let us take a re-look at rough sets from a QOCAS, in the context of rough lattices.

Observation 4. *Let* (U, R) *be a QOCAS. Then* $\mathcal{D}$ *is a bounded distributive lattice. Hence* $\mathcal{R} = \mathcal{R}_\mathcal{D}$, *and as mentioned earlier,* $\mathcal{R}^{IC}$ *is a rough lattice. Now, let us see how the rough lattices* $\mathcal{R}_{\mathcal{D}_\mathcal{R}}^{I'C'}$ *and* $\mathcal{IC}_\mathcal{R}^{I'C'}$ *behave: what are the respective base sets* $\mathcal{D}_\mathcal{R}$, $IC_\mathcal{R}$?
$\mathcal{D}_\mathcal{R} = \{(D, D) : D \in \mathcal{D}\}$ *and* $\mathcal{D}_\mathcal{R}^{[2]} = \{((D, D), (D', D)') : D \subseteq D'\}$. *Moreover,*
$IC_\mathcal{R} = \{(I(D_1, D_2), C(D_1, D_2)) : D_1 \subseteq D_2\} = \{((D_1, D_1), (D_2, D_2)) : D_1 \subseteq D_2\}$.
So note that, $\mathcal{D}_\mathcal{R}^{[2]} = IC_\mathcal{R}$. *Observe also that* $\mathcal{R}^{IC}$ *is embeddable into* $\mathcal{R}_{\mathcal{D}_\mathcal{R}}^{I'C'}$ *via the map*
$$(D_1, D_2) \hookrightarrow ((D_1, D_1), (D_2, D_2)).$$

Let us end this section with the rough set representation of rough lattices. Note that Lemma 10 can easily be extended to the case of distributive lattices.

Theorem 74 (*Rough Set Representation*). *Let* $\mathcal{L} := (L, \vee, \wedge, I, C, 0, 1)$ *be a rough lattice.*

1. *There exists an approximation space* (U, R) *such that* $\mathcal{L}$ *can be embedded in* $\mathcal{R}^{IC}$.
2. *There exists an approximation space* (U, R) *such that* $\mathcal{L}$ *can be embedded in* $\mathcal{RS}^{IC}$.

Proof. 1. Let us consider the distributive sublattice D_L of $\mathcal{L}$. Following Theorem 63, there exists an approximation space (U, R) such that D_L is embeddable into the lattice $\mathcal{D}$ of definable sets (via $a \hookrightarrow h(a)$). Hence using the note above, $\mathcal{R}_{D_L}$ is embeddable into generalized rough sets $\mathcal{R}$ (via $(a, b) \hookrightarrow (h(a), h(b))$). Now, the map: $\Phi : L \to \mathcal{R}^{IC}$ defined as, $\Phi(a) := (h(Ia), h(Ca))$ is the desired embedding. In particular, let us show that Φ preserves I: $\Phi(Ia) = (h(IIa), h(CIa)) = (h(Ia), h(Ia)) = I\Phi(a)$.
2. Using Theorem 73, we have the desired result.

Now, let $\mathcal{H} := (H, \vee, \wedge, \rightarrow, 0, 1)$ be a Heyting algebra. In this section we study the algebraic structure of the set $H^{[2]}$. In Theorem 68, we observed that $\mathcal{R}_H$ also forms a Heyting algebra, where relative pseudo complement '$\rightarrow$' on $H^{[2]}$ is defined as follows for $(a, b), (c, d) \in H^{[2]}$:

$$(a, b) \rightarrow (c, d) := ((a \rightarrow c) \wedge (b \rightarrow d), b \rightarrow d).$$

As $\mathcal{H}$ is a Heyting algebra, it is a bounded distributive lattice. So, all the results of the previous section (Sect. 6.1) apply to the Heyting algebra $\mathcal{H}$. It is interesting to see how I and C interact with $\rightarrow$ of $\mathcal{R}_H$. The following proposition is an extension of Proposition 24.

Proposition 29. *For $(a, b), (c, d) \in H^{[2]}$, I and C satisfy the following properties.*

1. $I((a, b) \rightarrow (c, d)) = (I(a, b) \rightarrow I(c, d)) \wedge (C(a, b) \rightarrow C(c, d))$.
2. $C((a, b) \rightarrow (c, d)) = C(a, b) \rightarrow C(c, d)$

Proof. 1. $I((a, b) \rightarrow (c, d)) = I((a \rightarrow c) \wedge (b \rightarrow d), (b \rightarrow d)) = ((a \rightarrow c) \wedge (b \rightarrow d), (a \rightarrow c) \wedge (b \rightarrow d)) = (a \rightarrow c, a \rightarrow c) \wedge (b \rightarrow d, b \rightarrow d) = (I(a, b) \rightarrow I(c, d)) \wedge (C(a, b) \rightarrow C(c, d))$.
2. $C((a, b) \rightarrow (c, d)) = C((a \rightarrow c) \wedge (b \rightarrow d), (b \rightarrow d)) = (b \rightarrow d, b \rightarrow d) = (b, b) \rightarrow (d, d) = C(a, b) \rightarrow C(c, d)$.

Let us abstract these properties to an arbitrary Heyting algebra, and introduce the notion of a *rough Heyting algebra*.

Definition 40. $\mathcal{H} := (H, \vee, \wedge, \rightarrow, I, C, 0, 1)$ *is said to be a* rough Heyting algebra, *provided* $(H, \vee, \wedge, \rightarrow, 0, 1)$ *is a rough lattice and* I, C *satisfy the following properties.*

1. $I(a \rightarrow b) = (Ia \rightarrow Ib) \wedge (Ca \rightarrow Cb)$.
2. $C(a \rightarrow b) = Ca \rightarrow Cb$.

Observe that for any Heyting algebra $\mathcal{H} := (H, \vee, \wedge, \rightarrow, 0, 1)$, $\mathcal{R}_H^{IC}$ is a rough Heyting algebra. In particular for any QOCAS, $\mathcal{R}^{IC}$ is a rough Heyting algebra.

Let us digress a little at this point, and show the connection between rough Heyting algebras and $T-rough$ algebras for any finite poset $\mathbf{T} := (T, \leq)$, the latter introduced by Sanjuan [71] in the context of classical rough set theory.

Definition 41 [71]. *Given a finite poset* $\mathbf{T} := (T, \leq)$, *an abstract algebra* $(H, \vee, \wedge, \rightarrow, (\pi_t)_{t \in T}, 0, 1)$ *is called a* T-rough algebra *provided the following conditions are satisfied.*

1. $(H, \vee, \wedge, \rightarrow, 0, 1)$ *is a Heyting algebra.*
2. $\pi_t : H \rightarrow H$ *such that, for any* $x, y \in H$ *and* $t, u, v \in T$
   (a) $\pi_t(x \vee y) = \pi_t x \vee \pi_t y$,
   (b) $\pi_t(x \wedge y) = \pi_t x \wedge \pi_t y$,
   (c) $\pi_t \pi_u x = \pi_u x$,
   (d) $\pi_t(0) = 0$,

(e) $\pi_t(x) \vee \neg\pi_t(x) = 1$, *where* $\neg x := x \to 0$,

(f) $\pi_t(x \to y) = \bigwedge_{v \geq t}(\pi_v(x) \to \pi_v(y))$,

(g) $\bigwedge_{v \geq t} \pi_v(x) \vee x = x$.

Now, let us consider a QOCAS (U, R) and consider the rough Heyting algebra $\mathcal{R}_H^{IC}$. With the ordering $I \leq C$, the structure $\mathbf{T} := (\{I, C\}, \leq)$ is a poset (in fact a chain). The rough Heyting algebra $\mathcal{R}_H^{IC}$ satisfies all the axioms of $\{I, C\}$-rough algebra except the axioms g and Boolean property e of I and C.

Returning to rough Heyting algebras, we prove the following.

Lemma 11. *Let* $\mathcal{H}$ *and* $\mathcal{H}'$ *be isomorphic Heyting algebras. Then* $\mathcal{R}_H^{IC}$ *and* $\mathcal{R}_H'^{IC}$ *are isomorphic.*

Proof. As, $\mathcal{H}$ and $\mathcal{H}'$ are isomorphic, and let ϕ be that isomorphism. Hence using Lemma 10, $\mathcal{R}_\mathcal{H}$ and $\mathcal{R}_{\mathcal{H}'}$ are isomorphic as Heyting algebras via the map $\Phi(a, b) := (\phi(a), \phi(b))$. Φ can be extended to the required isomorphism, as $\Phi(I(a, b)) = \Phi(a, a) = (\phi(a), \phi(a)) = I\Phi(a, b)$.

Example 23. Let $H := \{0, a, b, c, d, 1\}$, and consider the lattice $\mathcal{H}$ (with domain H), depicted in Fig. 31.

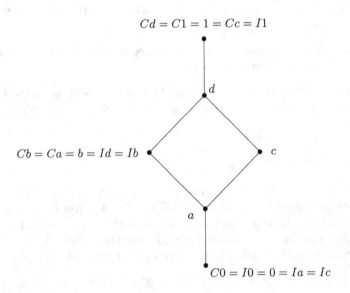

Fig. 31. $\mathcal{H}$

Being a finite distributive lattice, H is also a Heyting algebra. Let us define the unary operators I and C as:

$$I0 = 0, Ia = 0, Ib = b, Ic = 0, Id = b \text{ and } 1 = 1.$$
$$C0 = 0, Ca = b, Cb = b, Cc = 1, Cd = 1 \text{ and } C1 = 1.$$

Now let us prove that the structure $(H, \vee, \wedge, \rightarrow, I, C, 0, 1)$ is a rough Heyting algebra.

1. $I(x \vee y) = Ix \vee Iy,\ I(x \wedge y) = Ix \wedge Iy$:
   Let $y = 0$, then $I0 = 0$. $I(x \vee 0) = I(0 \vee x) = Ix = I0 \vee Ix$. $I(x \wedge 0) = I(0 \wedge x) = I0 = 0 = I0 \wedge Ix$ Now, let $y = a$, then $Ia = 0$. For $a \leq x$ we have: $I(a \vee x) = Ix = Ia \vee Ix$ and $I(a \wedge x) = Ia = Ia \wedge Ix$.
   Let $y = b$, then $Ib = b$. For $b \leq x$: $I(b \vee x) = Ix = b \vee Ix$ and $I(b \wedge x) = Ib = b \wedge Ix$. Moreover, $I(b \vee c) = Id = b = b \vee b = Ic \vee Ib$ and $I(b \wedge c) = Ia = 0 = b \vee 0 = Ib \wedge Ic$.
   Let $y = c$, then $Ic = 0$. For $c \leq x$: $I(c \vee x) = Ix = 0 \vee Ix$ and $I(c \wedge x) = Ic = 0 = Ic \wedge Ix$. As above, $I(b \vee c) = Ic \vee Ib$ and $I(b \wedge c) = Ic \wedge Ib$.
   Let $y = d$, then $Id = b$. $I(d \vee 1) = I1 = Id \vee I1$ and $I(d \wedge 1) = Id = b = Id \wedge I1$.

2. $C(x \vee y) = Cx \vee Cy,\ C(x \wedge y) = Cx \wedge Cy$:
   Let $y = 0$, then $C0 = 0$. $C(x \vee 0) = C(0 \vee x) = C(x) = C0 \vee Cx$. $C(x \wedge 0) = C(0 \wedge x) = C(0) = 0 = C0 \wedge Cx$.
   Now, let $y = a$, then $Ca = b$. For $a \leq x$ we have: $C(a \vee x) = Cx = Ca \vee Cx$ and $C(a \wedge x) = Ca = b = Ca \wedge Cx$.
   Let $y = b$, then $Cb = b$. For $b \leq x$: $C(b \vee x) = Cx = b \vee Cx$ and $C(b \wedge x) = Cb = b \wedge Cx$. Moreover, $C(b \vee c) = Cd = 1 = 1 \vee b = Cc \vee Cb$ and $C(b \wedge c) = Ca = b = b \vee 1 = Cb \wedge Cc$.
   Let $y = c$, then $Cc = 1$. For $c \leq x$: $C(c \vee x) = Cx = 1 = Cc \vee Cx$ and $C(c \wedge x) = Cc = 1 = 1 \wedge 1 = Cc \wedge Cx$. As above, $C(b \vee c) = Cb \vee Cc$ and $C(b \wedge c) = Cc \wedge Cb$.
   Let $y = d$, then $Cd = 1$. $C(d \vee 1) = C1 = Cd \vee C1$ and $C(d \wedge 1) = Cd = 1 = Cd \wedge C1$.

3. Clearly, by the definition of I and C, we have $Ix \leq x$ and $x \leq Cx$.

4. $I1 = 1$ and $C0 = 0$.

5. $II0 = I0 = 0,\ IIa = I0 = 0 = Ia,\ IIb = Ib = b,\ IIc = I0 = 0 = Ic,\ IId = Ib = b = Id,\ II1 = I1 = 1$.
   $CC0 = C0 = 0,\ CCa = Cb = b = Cb,\ CCb = Cb = b,\ CCc = C1 = 1 = Cc,\ CCd = C1 = 1 = Cd,\ CC1 = C1 = 1$.

6. $CI0 = C0 = 0 = I0,\ CIa = C0 = 0 = Ia,\ CIb = Cb = b = Ib,\ CIc = C0 = 0 = Ic,\ CId = Cb = b = Id,\ CI1 = 1 = I1$.
   $IC0 = C0 = 0,\ ICa = Ib = b = Cb,\ ICb = Ib = b = Cb,\ ICc = I1 = 1 = Cc,\ ICd = I1 = 1 = Cd,\ IC1 = C1 = 1$.

7. Regularity: $Ix \leq Iy$ and $Cx \leq Cy$ imply $x \leq y$ is trivial.

8. $I(x \rightarrow y) = (Ix \rightarrow Iy) \wedge (Cx \rightarrow Cy)$:
   Let $y = 0$, then $I(0 \rightarrow x) = I1 = 1$. Also $I0 \rightarrow Ix = 1$ and $C0 \rightarrow Cx = 1$. Hence $I(0 \rightarrow x) = (I0 \rightarrow Ix) \wedge (C0 \rightarrow Cx)$.
   Let $a \leq x$. Then $I(a \rightarrow x) = 1$ and $(0 =)Ia \rightarrow Ix = 1$ and $(b =)Ca \rightarrow Cx = 1$ Hence $I(a \rightarrow x) = (Ia \rightarrow Ix) \wedge (Ca \rightarrow Cx)$. Moreover $I(a \rightarrow 0) = 0$ and $Ia \rightarrow I0 = 1$ and $Ca \rightarrow C0 = 0$. Hence $I(a \rightarrow 0) = (Ia \rightarrow I0) \wedge (Ca \rightarrow C0)$.
   Let $b \leq x$. Then $I(b \rightarrow x) = 1$ and $(b =)Ib \rightarrow Ix = 1$ and $(b =)Cb \rightarrow Cx = 1$ Hence $I(b \rightarrow x) = (Ib \rightarrow Ix) \wedge (Cb \rightarrow Cx)$. Moreover $I(b \rightarrow 0) = 0$ and

$Ib \to I0 = 0$ and $Cb \to C0 = 0$. Hence $I(b \to 0) = (Ib \to I0) \wedge (Cb \to C0)$. Also $I(b \to c) = Ic = 0$ and $Ib \to Ic = 0$ and $Cb \to Cc = 1$. Hence $I(b \to c) = (Ib \to Ic) \wedge (Cb \to Cc)$.

Let $c \leq x$. Then $I(c \to x) = 1$ and $(0 =)Ic \to Ix = 1$ and $(1 =)Cc \to Cx = 1$ Hence $I(c \to x) = (Ic \to Ix) \wedge (Cc \to Cx)$. Moreover $I(c \to 0) = 0$ and $Ic \to I0 = 1$ and $Cc \to C0 = 0$. Hence $I(c \to 0) = (Ic \to I0) \wedge (Cc \to C0)$. Also $I(c \to b) = Ib = b$ and $Ic \to Ib = 1$ and $Cc \to Cb = b$. Hence $I(c \to b) = (Ic \to Ib) \wedge (Cc \to Cb)$.

Now, $I(d \to 1) = 1$ and $(b =)Id \to I1 = 1$ and $(1 =)Cc \to C1 = 1$ Hence $I(d \to 1) = (Id \to I1) \wedge (Cd \to C1)$. Moreover $I(d \to 0) = 0$ and $(b =)Id \to I0 = 0$ and $Cd \to C0 = 0$. Hence $I(d \to 0) = (Id \to I0) \wedge (Cd \to C0)$. Also $I(d \to b) = Ib = b$ and $Id \to Ib = 1$ and $Cd \to Cb = b$. Hence $I(d \to b) = (Id \to Ib) \wedge (Cd \to Cb)$. $I(d \to c) = Ic = 0$ and $Id \to Ic = 0$ and $Cd \to Cc = 1$. Hence $I(d \to c) = (Id \to Ic) \wedge (Cd \to Cc)$.

Now let us consider $1 \to x$. But $1 \to x = x$, so $I(1 \to x) = Ix$. $I1 \to Ix = Ix$ and $C1 \to Cx = Cx$. But we already have $Ix \leq Cx$. Hence $I(1 \to x) = (I1 \to Ix) \wedge (C1 \to Cx)$.

9. $C(x \to y) = Cx \to Cy$:

Let $x = 0$, then $C(0 \to x) = C1 = 1$ and $C0 \to Cx = 1$. Hence $C(0 \to x) = C0 \to Cx$.

Let $a \leq x$. Then $C(a \to x) = 1$, $(b =)Ca \to Cx = 1$. Moreover, $C(a \to 0) = 0$ and $Ca \to C0 = 0$. Hence $C(a \to 0) = Ca \to C0$.

Let $b \leq x$. Then $C(b \to x) = 1$ and $(b =)Cb \to Cx = 1$. Hence $C(b \to x) = Cb \to Cx$. Moreover, $C(b \to 0) = 0$ and $Cb \to C0 = 0$. Hence $C(b \to 0) = Cb \to C0$. Also $C(b \to c) = Cc = 1$ and $Cb \to Cc = 1$. Hence $C(b \to c) = Cb \to Cc$.

Let $c \leq x$. Then $C(c \to x) = 1$ and $(1 =)Cc \to Cx = 1$. Hence $C(c \to x) = Cc \to Cx$. Moreover, $C(c \to 0) = 0$ and $Cc \to C0 = 0$. Hence $C(c \to 0) = Cc \to C0$. Also $C(c \to b) = Cb = b$ and $Cc \to Cb = b$. Hence $C(c \to c) = Cc \to Cb$.

$C(d \to 1) = 1$ and $(1 =)Cd \to C1 = 1$. Hence $C(d \to 1) = Cd \to C1$. Moreover, $C(d \to 0) = 0$ and $(1 =)Cd \to C0 = 0$. Hence $C(d \to 0) = Cd \to C0$. Also $C(d \to b) = Cb = b$ and $(1 =)Cd \to Cb = b$. Hence $C(d \to b) = Cd \to Cb$. $C(d \to c) = Cc = 1$ and $Cd \to Cc = 1$. Hence $C(d \to c) = Cd \to Cc$.

$C(d \to a) = Ca = b$ and $Cd \to Ca = 1 \to b = b$. Hence $C(d \to a) = Cd \to Ca$.

Now let us consider $1 \to x$. But $1 \to x = x$, so $C(1 \to x) = Cx$ and $C1 \to Cx = Cx$. Hence $C(1 \to x) = C1 \to Cx$.

Now consider the sets $D_H := \{a \in L : Ia = a\}$ and $IC_H = \{(Ia, Ca) : a \in H\}$ for any rough Heyting algebra H. Analogous to the observations in the previous section, we obtain.

Proposition 30. 1. D_H forms a Heyting subalgebra of $\mathcal{H}$.
2. IC_H is a Heyting subalgebra of $\mathcal{R}_{D_H}$.

Proof. 1. $a, b \in D_H$, $C(a \to b) = C(a) \to C(b) = a \to b$. Hence D_H also forms a Heyting algebra.

2. $(Ia, Ca) \to (Ib, Cb) := ((Ia \to Ib) \land (Ca \to Cb), Ca \to Cb)$, but by the definition of rough Heyting algebra, we have $((Ia \to Ib) \land (Ca \to Cb), Ca \to Cb) = (I(a \to b), C(a \to b)) (\in IC_H)$. Hence IC_H also forms a Heyting algebra.

Recall I' and C' as defined in the previous section.

$$I'(Ia, Ca) := (Ia, Ia); C'(Ia, Ca) := (Ca, Ca), \text{ for any } (Ia, Ca) \in IC_H.$$

Let us see how I' and C' interact with the Heyting algebra $\mathcal{IC}_H$.

$C'((Ia, Ca) \to (Ib, Cb)) = C'(((Ia \to Ib) \land (Ca \to Cb), Ca \to Cb)) = (Ca \to Cb, Ca \to Cb) = C'(Ia, Ca) \to C'(Ib, Cb)$.

$I'((Ia, Ca) \to (Ib, Cb)) = I'(((Ia \to Ib) \land (Ca \to Cb), Ca \to Cb)) = ((Ia \to Ib) \land (Ca \to Cb), (Ia \to Ib) \land (Ca \to Cb))$.

$(I'(Ia, Ca) \to I'(Ib, Cb)) \land (C'(Ia, Ca) \to C'(Ib, Cb)) = ((Ia, Ia) \to (Ib, Ib)) \land ((Ca, Ca) \to (Cb, Cb)) = (Ia \to Ib, Ia \to Ib) \land (Ca \to Cb, Ca \to Cb) = ((Ia \to Ib) \land (Ca \to Cb), (Ia \to Ib) \land (Ca \to Cb))$.

So for all $x, y \in IC_H$ we have $I'(x \to y) = (I'x \to I'y) \land (C'x \to C'y)$, and $C'(x \to y) = C'x \to C'y$.

Hence we have the following proposition.

Proposition 31. $\mathcal{IC}_H^{I'C'}$, *the lattice* $\mathcal{IC}_H$ *enhanced with the operators* I' *and* C', *is a rough Heyting algebra. Further,* $\mathcal{H}$ *is isomorphic to* $\mathcal{IC}_H^{I'C'}$ *via the map* $a \hookrightarrow (Ia, Ca)$.

Proof. Let us consider the map ϕ defined in Proposition 26, which is defined as $\phi(a) := (Ia, Ca)$. We have proved that ϕ is a rough lattice isomorphism, let us prove that ϕ also preserves $\to$.

$\phi(a \to b) = (I(a \to b), C(a \to b)) = ((Ia \to Ib) \land (Ca \to Cb), Ca \to Cb)) = (Ia, Ca) \to (Ib, Cb) = \phi(a) \to \phi(b)$.

Example 24. In Example 23, $D_H = \{a \in H : Ia = a\} = \{a \in H : Ca = a\} = \{0, b, 1\}$. Hence $D_H^{[2]} = \{(0, 0), (0, b), (0, 1), (b, b), (b, 1), (1, 1)\}$.

$IC_H = \{(Ix, Cx) : x \in H\} = \{(I0, C0), (Ia, Ca), (Ib, Cb), (Ic, Cc), (Id, Cd), (I1, C1)\} = \{(0, 0), (0, b), (b, b), (0, 1), (b, 1), (1, 1)\}$.

Hence $D_H^{[2]} = IC_H$, and $\mathcal{H}$ is isomorphic to $\mathcal{IC}^{I'C'}$ via the map ϕ given as:

$$\phi(0) = (I0, C0) = (0, 0)$$
$$\phi(a) = (Ia, Ca) = (0, b)$$
$$\phi(b) = (Ib, Cb) = (b, b)$$
$$\phi(c) = (Ic, Cc) = (0, 1)$$
$$\phi(d) = (Id, Cd) = (b, 1)$$
$$\phi(1) = (I1, C1) = (1, 1).$$

Theorem 75. *(Rough Set Representation)*

1. *Let* $\mathcal{H} := (H, \vee, \wedge, \rightarrow, I, C, 0, 1)$ *be a rough Heyting algebra. Then there exists an approximation space* (U, R) *such that* $\mathcal{H}$ *can be (Heyting) embedded in* $\mathcal{R}^{IC}$.
2. *Let* $\mathcal{H} := (H, \vee, \wedge, \rightarrow, I, C, 0, 1)$ *be a rough Heyting algebra. Then there exists an approximation space* (U, R) *such that* $\mathcal{H}$ *can be (Heyting) embedded in* $\mathcal{RS}^{IC}$.

Proof. Let $\mathcal{H}$ be a rough Heyting algebra. In particular, $\mathcal{H}$ is a Heyting algebra. Consider the approximation space as in Theorem 64. Then the map $(a, b) \hookrightarrow (h(a), h(b))$ provides an embedding of $\mathcal{R}_{D_H}$ into $\mathcal{R}$. $\phi : \mathcal{H} \rightarrow \mathcal{R}^{IC}$ given as $\phi(a) := (h(Ia), h(Ca))$, serves as the required embedding, as in the case of Theorem 74. Let us show that ϕ preserves the implication $\rightarrow$:
$\phi(a \rightarrow b) = (h(I(a \rightarrow b)), h(C(a \rightarrow b))) = (h((Ia \rightarrow Ib) \wedge (Ca \rightarrow Cb)), h(Ca \rightarrow Cb)) = ((h(Ia) \rightarrow h(Ib)) \wedge (h(Ca) \rightarrow h(Cb)), h(Ca) \rightarrow h(Cb)) = (h(Ia), h(Ca)) \rightarrow (h(Ib), h(Cb)) = \phi(a) \rightarrow \phi(b)$.

Example 25. Let us consider the rough Heyting algebra $\mathcal{H} := (H, \vee, \wedge, \rightarrow, I, C, 0, 1)$ as in Example 23, then $D_H = \{0, b, 1\}$. The prime filters of D_H are give as:
$$P_1 = \{b, 1\} \text{ and } P_2 = \{1\}.$$
Now let $U := \{P_1, P_2\}$. Then $h(0) = \emptyset$, $h(b) = \{P_1\}$, $h(1) = \{P_1, P_2\}$. Using the construction of Theorem 64, $R_C(P_1) = \{P_1\}$ and $R_C(P_2) = \{P_1, P_2\}$. Hence (U, R_C) is a QOCAS. The isomorphism of the lattices D_H and collection of definable sets $\mathcal{D}$ corresponding to the QOCAS (U, R_C) can be depicted as in Fig. 32. Now $\mathcal{D}$ and D_H are isomorphic as Heyting algebras. Hence using Lemma 11, the structures $\mathcal{R}_{\mathcal{D}}^{IC} = \mathcal{R}^{IC}$ and $\mathcal{R}_{D_H}^{IC}$ are isomorphic. Using Theorem 75 the rough Heyting algebra $\mathcal{H}$ can be embedded into the the rough Heyting algebra $\mathcal{R}_{D_H}^{IC}$. The illustration of this embedding ϕ is already given in Example 24. Hence the rough Heyting algebra $\mathcal{H}$ is embeddable into $\mathcal{R}^{IC}$. Let us explicitly mention this embedding Φ.

$$\Phi(0) = (h(I0), h(C0)) = (\emptyset, \emptyset)$$
$$\Phi(a) = (h(Ia), h(Ca)) = (\emptyset, \{P_1\})$$
$$\Phi(b) = (h(Ib), h(Cb)) = (\{P_1\}, \{P_1\})$$
$$\Phi(c) = (h(Ic), h(Cc)) = (\emptyset, U)$$
$$\Phi(c) = (h(Id), h(Cd)) = (\{P_1\}, U)$$
$$\Phi(c) = (h(I1), h(C1)) = (U, U)$$

The pictorial representation of this isomorphism is given in Fig. 33.

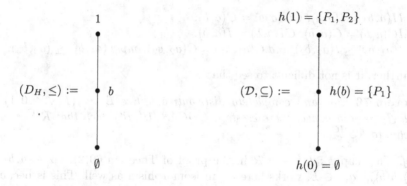

Fig. 32. $\mathcal{RS}$

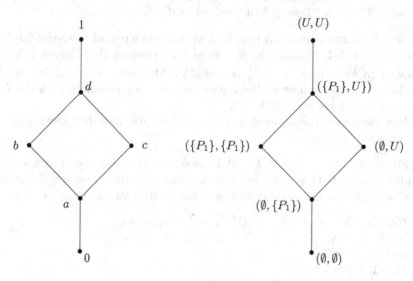

Fig. 33. $\mathcal{H} \cong \mathcal{R}^{IC}$

In Sect. 5, we have seen that the lattice $\mathcal{R}_L$, where L is a completely distributive lattice in which $\mathcal{J}_L$ is join dense, abstracts $\mathcal{R}$. In this section, we follow the same lines of study as in the previous section, and investigate the enhanced structure $\mathcal{R}_L^{IC}$. Let us first extend Proposition 24 to this case, i.e. when L is a completely distributive lattice in which $\mathcal{J}_L$ is join dense.

Proposition 32. *For* $(a, b), (a_i, b_i) \in \mathcal{R}_L^{IC}$, $i \in \Lambda$, Λ *being any index set,* I, C *satisfy the following properties.*

(i) $I(\wedge_{i \in \Lambda}(a_i, b_i)) = \wedge_{i \in \Lambda} I(a_i, b_i)$, $C(\vee_{i \in \Lambda}(a_i, b_i)) = \vee_{i \in \Lambda} C(a_i, b_i)$.
(ii) $I(\vee_{i \in \Lambda}(a_i, b_i)) = \vee_{i \in \Lambda} I(a_i, b_i)$, $C(\wedge_{i \in \Lambda}(a_i, b_i)) = \wedge_{i \in \Lambda} C(a_i, b_i)$.
(iii) $I(a, b) \le (a, b)$, $(a, b) \le C(a, b)$.
(iv) $I(1, 1) = (1, 1)$, $C(0, 0) = (0, 0)$.

(v) $II(a,b) = I(a,b)$, $CC(a,b) = C(a,b)$.
(vi) $IC(a,b) = C(a,b)$, $CI(a,b) = I(a,b)$.
(vii) $I(a_1,b_1) \leq I(a_2,b_2)$ and $C(a_1,b_1) \leq C(a_2,b_2)$ imply $(a_1,b_1) \leq (a_2,b_2)$.

Further, it is not difficult to see that

Theorem 76. *For any completely distributive lattice* $\mathcal{L} := (L,\vee,\wedge,0,1)$ *in which* $\mathcal{J}_L$ *is join dense, there exists a QOCAS* (U',R') *such that* $\mathcal{R}_L^{IC}$ *is isomorphic to* $\mathcal{RS}'^{IC}$.

Proof. The map $\phi : L^{[2]} \longrightarrow \mathcal{R}$ in the proof of Theorem 69(2), viz. $\phi(a,b) := (h(a),h(b))$, $a,b \in L$, works here as an isomorphism as well. This is because, for the additional operators I,C, we have, $\phi(I(a,b)) = \phi(a,a) = (h(a),h(a)) = I(h(a),h(b)) = I\phi(a,b)$ and similarly $\phi(C(a,b)) = C\phi(a,b)$.
Now using Theorem 73, we get the desired QOCAS.

It may be worth mentioning here that we can take a re-look at the lattice $\mathcal{R}_{L_2}$ defined in Sect. 5.4, through the operators I,C. Observe that the set $IC_{L_2} := \{(I(a,b),C(a,b)) : (a,b) \in L^{[2]}\}$ is just $L_2^{[2]}$. Moreover, Theorem 70 can be extended to the case where all the lattices involved are enhanced with the I,C operators. So, for instance, $\mathcal{R}_L^{IC} \cong \mathcal{R}_{L_2}^{IC}$.

Abstracting from the above observations, let us introduce the following lattice structure.

Definition 42. $\mathcal{L} := (L,\vee,\wedge,I,C,0,1)$ *is said to be a complete rough lattice, provided* $(L,\vee,\wedge,0,1)$ *is a completely distributive lattice in which* $\mathcal{J}_L$ *is join dense, and* I,C *are unary operators on* L *that satisfy the following properties.*

(i) $I(\wedge_{i\in\Lambda}a_i) = \wedge_{i\in\Lambda}I(a_i)$, $C(\vee_{i\in\Lambda}a_i) = \vee_{i\in\Lambda}C(a_i)$.
(ii) $I(\vee_{i\in\Lambda}a_i) = \vee_{i\in\Lambda}I(a_i)$, $C(\wedge_{i\in\Lambda}a_i) = \wedge_{i\in\Lambda}C(a_i)$.
(iii) $Ia \leq a$, $a \leq Ca$.
(iv) $I1 = 1$, $C0 = 0$.
(v) $IIa = Ia$ and $CCa = Ca$.
(vi) $ICa = Ca$ and $CIa = Ia$.
(vii) $Ia \leq Ib$ and $Ca \leq Cb$ imply $a \leq b$.

So $\mathcal{R}^{IC}$, for any QOCAS (U,R), and $\mathcal{R}_L^{IC}$ are both complete rough lattices.

Another example of a complete rough lattice is obtained by again considering the set $D_L := \{a \in L : Ia = a\}$. We see that $\mathcal{D}_L := (D_L,\vee,\wedge,0,1)$ is a complete sublattice of $\mathcal{L}$, where join and meet are inherited from $\mathcal{L}$. Further, we have

Proposition 33. $\mathcal{D}_L$ *is a completely distributive lattice in which the set* $\mathcal{J}_{D_L}$ *of completely join irreducible elements is join dense.*

Proof. $\mathcal{D}_L$ being a complete sublattice of $\mathcal{L}$, is also a completely distributive lattice. Now, let us characterize the completely join irreducible elements of $\mathcal{D}_L$. Let $a \in \mathcal{J}_L$, and suppose $Ca = \vee_{i\in\Lambda}a_i$, $a_i \in D_L$, for all $i \in \Lambda$. As $a \leq Ca$, $a = Ca \wedge a = \vee_{i\in\Lambda}a_i \wedge a = \vee_{i\in\Lambda}(a_i \wedge a)$. $a \in \mathcal{J}_L$ implies that $a = a \wedge a_i$, for some $i \in \Lambda$. So $a \leq a_i$, and $Ca \leq Ca_i = a_i$. But we already have $Ca_i \leq Ca$, whence $Ca = a_i$, for some $i \in \Lambda$. Hence Ca is a completely join irreducible element of D_L. Now let $a \in D_L$. As $a \in L$, there is an index set Λ such that $a = \vee_{i \in \Lambda} a_i$, $a_i \in \mathcal{J}_L$. So $Ca = a = \vee_{i \in \Lambda} Ca_i$, and for each $i \in \Lambda$, Ca_i is a completely join irreducible element of D_L. Hence $\mathcal{J}_{D_L} = \{Ca : a \in \mathcal{J}_L\}$ is join dense in D_L.

Using Proposition 32, we may then conclude that $\mathcal{R}_{D_L}^{IC}$ forms a complete rough lattice.

Yet another example comes from the set $IC_L := \{(Ia, Ca) : a \in L\}$, for a complete rough lattice $\mathcal{L}$. It is easy to see that IC_L forms a complete sublattice $\mathcal{IC}_L := (IC_L, \vee, \wedge, (0,0), (1,1))$ of the lattice $L \times L$. Further, we have

Proposition 34. $\mathcal{IC}_L$ *is a completely distributive lattice in which the set* $\mathcal{J}_{IC_L}$ *of completely join irreducible elements is join dense.*

Proof. It is easy to verify that $\mathcal{IC}_L$ is a complete sublattice of $L \times L$, and hence a completely distributive lattice. Now let us characterize the completely join irreducible elements of $\mathcal{IC}_L$. Let $a \in \mathcal{J}_L$ and suppose $(Ia, Ca) = \vee_{i \in \Lambda}(Ia_i, Ca_i)$. Then $(Ia, Ca) = (\vee_{i \in \Lambda} Ia_i, \vee_{i \in \Lambda} Ca_i) = (I(\vee_{i \in \Lambda} a_i), C(\vee_{i \in \Lambda} a_i))$, using (ii) of Definition 42. But (vii) of the definition then implies that $a = \vee a_i$. As $a \in \mathcal{J}_L$, $a = a_i$ for some $i \in \Lambda$, hence $Ia = Ia_i$ and $Ca = Ca_i$. So $(Ia, Ca) = (Ia_i, Ca_i)$. Hence, (Ia, Ca) is completely join irreducible in $\mathcal{IC}$. Similarly, using the fact that $\mathcal{J}_L$ is join dense in $\mathcal{L}$, one can show that $\mathcal{J}_{IC_L}$ is join dense in $\mathcal{IC}_L$.

Following the theme of this section, let us state the following proposition, whose proof is that of Proposition 26, extended to showing that the embedding ϕ preserves arbitrary joins and meets.

Proposition 35. $\mathcal{IC}_L^{I'C'}$, *the lattice* $\mathcal{IC}_L$ *enhanced with the operators* I' *and* C', *is a complete rough lattice. Further,* $\mathcal{L}$ *is isomorphic to* $\mathcal{IC}_L^{I'C'}$, *via the map* $a \hookrightarrow (Ia, Ca)$.

We have $IC_L \subseteq D_L^{[2]}$. The inclusion of IC_L in $D_L^{[2]}$ gives us a natural embedding of $\mathcal{IC}_L^{I'C'}$ into $\mathcal{R}_{D_L}^{IC}$. But then by Theorem 76, there exists a QOCAS (U', R') such that $\mathcal{R}_{D_L}^{IC}$ is isomorphic to $\mathcal{RS}'^{IC}$. Hence using Proposition 35, we have

Theorem 77. *For any complete rough lattice* $\mathcal{L}$, *there exists a QOCAS* (U, R) *such that* $\mathcal{L}$ *can be embedded into* $\mathcal{RS}^{IC}$.

There is a clear case when the embedding turns into an isomorphism. If for any $(a, b) \in D_L^{[2]}$ there exists a $c \in L$ such that $Ic = a, Cc = b$, then $(a, b) \in IC$. If this condition is true for all $(a, b) \in D_L^{[2]}$, we get $IC = D_L^{[2]}$. Hence we have

Theorem 78. *Let* $\mathcal{L} := (L, \vee, \wedge, I, C, 0, 1)$ *be a complete rough lattice such that for each* $(a, b) \in D_L^{[2]}$, *there exists a* $c \in L$ *with* $Ic = a, Cc = b$. *Then there exists a QOCAS* (U, R) *such that* $\mathcal{L}$ *is isomorphic to* $\mathcal{RS}^{IC}$.

Proof. If the condition of the theorem is satisfied, $\mathcal{L} \cong \mathcal{IC}^{I'C'} \cong \mathcal{R}_{D_L}^{IC}$. But for $\mathcal{R}_{D_L}^{IC}$, there exists an approximation space (U, R) (cf. Theorem 76) such that $\mathcal{R}_{D_L}^{IC}$ is isomorphic to $\mathcal{RS}^{IC}$.

6.2   Logics of Rough Sets from QOCAS

In this section, we study logics for rough set structures defined in the previous section, and, through the representation results proved therein, obtain rough set semantics for the logics.

Let us first define the logic $\mathcal{L}_R$ for rough lattices through the following.
The set of propositional variables, $\mathcal{P} := p, q, r, \dots$.
Propositional constants: $\top, \perp$.
The set of formulas, $\mathcal{F}$ is given by the scheme:$= \top \mid \perp \mid p \mid \phi \vee \psi \mid \phi \wedge \psi \mid \Box\phi \mid \Diamond\phi$.

Rules and Postulates of $\mathcal{L}_R$:

1. Rules and postulates of the logic $BDLL$ (cf. Sect. 1).
2. $\Box\alpha \vdash \alpha, \ \alpha \vdash \Diamond\alpha$.
3. $\Box\alpha \wedge \Box\beta \vdash \Box(\alpha \wedge \beta), \ \Diamond(\alpha \wedge \beta) \vdash \Diamond\alpha \wedge \Diamond\beta$.
4. $\Box(\alpha \wedge \beta) \vdash \Box\alpha \wedge \Box\beta, \ \Diamond\alpha \wedge \Diamond\beta \vdash \Diamond(\alpha \wedge \beta)$
5. $\top \vdash \Box\top, \ \Diamond\perp \vdash \perp$.
6. $\Box\alpha \vdash \Box\Box\alpha, \ \Diamond\Diamond\alpha \vdash \Diamond\alpha$.
7. $\Diamond\alpha \vdash \Box\Diamond\alpha, \ \Diamond\Box\alpha \vdash \Box\alpha$.
8. $\dfrac{\alpha \vdash \beta \qquad \alpha \vdash \beta}{\Box\alpha \vdash \Box\beta, \ \Diamond\alpha \vdash \Diamond\beta}$
9. $\dfrac{\Box\alpha \vdash \Box\beta, \ \Diamond\alpha \vdash \Diamond\beta}{\alpha \vdash \beta}$.

Let $\mathcal{L}$ be a rough lattice. Let $v : \mathcal{P} \rightarrow L$ be a mapping, then recursively, v can easily be extended to the set of all formulas, as,

$$v(\alpha \vee \beta) = v(\alpha) \vee v(\beta).$$
$$v(\alpha \wedge \beta) = v(\alpha) \wedge v(\beta).$$
$$v(\Box\alpha) = Iv(\alpha), v(\Diamond\alpha) = Cv(\alpha).$$

The pair $(\mathcal{L}, v)$ is a called a *model*. A sequent $\alpha \vdash_{\mathcal{L}_R} \beta$ is called *true* in a model $(\mathcal{L}, v)$ if $v(\alpha) \leq v(\beta)$. $\alpha \vdash_{\mathcal{L}_R} \beta$ is *valid in a class of models* if it is true in every model belonging to the class. Moreover, $\alpha \vdash_{\mathcal{L}_R} \beta$ is *valid* if it is true in each model.

Theorem 79. *(Soundness and Completeness)*

1. *The system $\mathcal{L}_R$ is sound with respect to all models.*
2. *Let $\alpha \vdash_{\mathcal{L}_R} \beta$ be true in all models $(\mathcal{L}, v)$. Then $\alpha \vdash_{\mathcal{L}_R} \beta$ is derivable in the logical system $\mathcal{L}_R$.*

Proof. The proof of soundness is direct, and that of completeness uses the standard algebraic technique employing the Lindenbaum-Tarski algebra of the logic.

Using representation Theorem 74 for rough lattices, we therefore obtain

Theorem 80. *(Rough Set Semantics)*

1. The sequent $\alpha \vdash_{\mathcal{L}_R} \beta$ is valid in the class of all rough lattices if and only if it is valid in the class of all rough lattices formed by $QOCAS$.
2. If $\alpha \vdash_{\mathcal{L}_R} \beta$ is valid in the class of all rough lattices formed by $QOCAS$ then it is derivable in $\mathcal{L}_R$.

Now, let us define the logic for rough Heyting algebras. It is well known that Heyting algebras give the algebraic semantics of intuitionistic logic. The Hilbert style axiomatic calculus of the intuitionistic system IPC (cf. e.g. [8]) is given by the following syntax and rules.

The set of propositional variables $\mathcal{P} := p, q, r, \ldots$
Propositional constants: $\top, \bot$.
Logical connectives:$= \vee, \wedge, \rightarrow$.
The set of formulas $\mathcal{F}$ is given by the scheme:$= \top \mid \bot \mid p \mid \alpha \vee \beta \mid \alpha \wedge \beta \mid \alpha \rightarrow \beta$.
The axioms and rule of inference of IPC are as follows.

Axioms:

1. $\alpha \rightarrow (\beta \rightarrow \alpha)$.
2. $(\alpha \rightarrow (\beta \rightarrow \gamma)) \rightarrow ((\alpha \rightarrow \beta) \rightarrow (\alpha \rightarrow \gamma))$.
3. $(\alpha \wedge \beta) \rightarrow \alpha$.
4. $(\alpha \wedge \beta) \rightarrow \beta$.
5. $\alpha \rightarrow (\beta \rightarrow (\alpha \wedge \beta))$.
6. $\alpha \rightarrow \alpha \vee \beta$.
7. $\beta \rightarrow \alpha \vee \beta$.
8. $(\alpha \rightarrow \beta) \rightarrow ((\beta \rightarrow \gamma) \rightarrow (\alpha \vee \gamma \rightarrow \beta))$.
9. $\bot \rightarrow \alpha$.

The only rule of inference is Modus Ponens (MP): From α and $\alpha \rightarrow \beta$ infer β. In the following, we recall in brief, the algebraic semantics of IPC. Let $\mathcal{H} := (H, \vee, \wedge, \rightarrow, 0, 1)$ be a Heyting algebra. Let $v : \mathcal{P} \rightarrow H$. Extend this map v to $\tilde{v}$ to the set $\mathcal{F}$ of all formulas as usual. A formula α is *valid in a Heyting algebra* if and only if $v(\alpha) = 1$, for all valuations v. A formula α is *valid in a class of Heyting algebras* if it is valid in all Heyting algebras belonging to that class. Let $\mathbb{H}$ be the class of all Heyting algebras. α is valid in $\mathbb{H}$, is denoted by $\vDash_{\mathbb{H}} \alpha$. The following soundness and completeness theorem is obtained in a standard manner.

Theorem 81. *(Soundness and Completeness)* $\vdash_{IPC} \alpha$ if and only if $\vDash_{\mathbb{H}} \alpha$.

Adding modalities to IPC, started from Prior's philosophical work in 1957. He studied this by the name of $MIPC$. We refer to [7] for its presentation. Two connectives $\Box$ and $\Diamond$ are added to the syntax of IPC, and $MIPC$ is defined as the logic which contains IPC, the following axioms:

$$\Box\alpha \rightarrow \alpha \qquad\qquad \alpha \rightarrow \Diamond\alpha$$
$$\Box\alpha \wedge \Box\beta \rightarrow \Box(\alpha \wedge \beta) \qquad\qquad \Diamond(\alpha \vee \beta) \rightarrow \Diamond\alpha \vee \Diamond\beta$$
$$\Diamond\alpha \rightarrow \Box\Diamond\alpha \qquad\qquad \Diamond\Box\alpha \rightarrow \Box\alpha$$
$$\Box(\alpha \rightarrow \beta) \rightarrow (\Diamond\alpha \rightarrow \Diamond\beta)$$

and the rule of Necessitation: $\alpha/\Box\alpha$.

Monadic Heyting algebras, first studied by Monteiro and Varsavsky (cf. [7]), are the algebraic models of Prior's $MIPC$. Let us provide the definition of monadic Heyting algebra; in the following we use the same notation as in [7].

Definition 43. *An abstract algebra* $(H, \vee, \wedge, \rightarrow, \forall, \exists, 0, 1)$, *in brief* $(H, \forall, \exists)$, *is a monadic Heyting algebra if* $(H, \vee, \wedge, \rightarrow, 0, 1)$ *is a Heyting algebra and* $\forall, \exists$ *are unary operators on* H *which satisfy the following properties:* $\forall a, b \in H$,

1. $\forall a \leq a$ $a \leq \exists a$
2. $\forall(a \wedge b) = \forall a \wedge \forall b$ $\exists(a \vee b) = \exists a \vee \exists b$
3. $\forall 1 = 1$ $\exists 0 = 0$
4. $\forall \exists a = \exists a$ $\exists \forall a = \forall a$
5. $\exists(\exists a \wedge b) = \exists(a \wedge b)$

Note that monadic Heyting algebras have also been studied in the name of bitopological pseudo-Boolean algebras by Ono [54] and Suzuki [72].

Observation 5. *Any rough Heyting algebra is, in particular, a monadic Heyting algebra.*

Now, let us define the logic, $\mathcal{L}_{roughInt}$ for rough Heyting algebras. Similar to $MIPC$, we add two modalities $\Box, \Diamond$ to the syntax of IPC. Rules and axioms of $\mathcal{L}_{roughInt}$ are as follows.

1. Rules and axioms of IPC.
2. $\Box\alpha \rightarrow \alpha$.
3. $\alpha \rightarrow \Diamond\alpha$.
4. $\top \rightarrow \Box\top$.
5. $\Diamond\bot \rightarrow \bot$.
6. $\Box(\alpha \vee \beta) \rightarrow \Box\alpha \vee \Box\beta$.
7. $\Box\alpha \wedge \Box\beta \rightarrow \Box(\alpha \wedge \beta)$.
8. $\Diamond(\alpha \vee \beta) \rightarrow \Diamond\alpha \vee \Diamond\beta$.
9. $\Diamond\alpha \wedge \Diamond\beta \rightarrow \Diamond(\alpha \wedge \beta)$.
10. $\Diamond(\alpha \rightarrow \beta) \leftrightarrow (\Diamond\alpha \rightarrow \Diamond\beta)$.
11. $\Box(\alpha \rightarrow \beta) \leftrightarrow ((\Box\alpha \rightarrow \Box\beta) \wedge (\Diamond\alpha \rightarrow \Diamond\beta))$.
12. $\Box\alpha \rightarrow \Box\Box\alpha, \Diamond\Diamond\alpha \rightarrow \Diamond\alpha$.
13. $\Diamond\alpha \rightarrow \Box\Diamond\alpha, \Diamond\Box\alpha \rightarrow \Box\alpha$.
14. $\dfrac{\alpha \rightarrow \beta \quad\quad \alpha \rightarrow \beta}{\Box\alpha \rightarrow \Box\beta, \Diamond\alpha \rightarrow \Diamond\beta}$
15. $\dfrac{\Box\alpha \rightarrow \Box\beta, \Diamond\alpha \rightarrow \Diamond\beta}{\alpha \rightarrow \beta}$.

In view of Observation 5, we have the following theorem.

Theorem 82. *If* $\vdash_{MIPC} \alpha$ *then* $\vdash_{\mathcal{L}_{roughInt}} \alpha$.

Let $\mathcal{H} := (H, \vee, \wedge, \rightarrow, I, C, 0, 1)$ be a rough Heyting algebra. Let $v : \mathcal{P} \rightarrow H$. Extend this map v to $\tilde{v}$ on all formulas as usual.

1. $\widetilde{v}(\alpha \vee \beta) = \widetilde{v}(\alpha) \vee \widetilde{v}(\beta)$.
2. $\widetilde{v}(\alpha \wedge \beta) = \widetilde{v}(\alpha) \wedge \widetilde{v}(\beta)$.
3. $\widetilde{v}(\alpha \to \beta) = \widetilde{v}(\alpha) \to \widetilde{v}(\beta)$.
4. $\widetilde{v}(\Box\alpha) = I\widetilde{v}(\alpha)$.
5. $\widetilde{v}(\Diamond\alpha) = C\widetilde{v}(\alpha)$.
6. $\widetilde{v}(\bot) = 0$.
7. $\widetilde{v}(\top) = 1$.

As before, we have the following definitions. A formula α is *valid in a rough Heyting algebra* if and only if $v(\alpha) = 1$, for all valuations v. A formula α is *valid in a class of rough Heyting algebras* if it is valid in all Heyting algebras belonging to that class. Let $\mathbb{RH}$ denote the class of all rough Heyting algebras. α is valid in $\mathbb{RH}$, is denoted by $\models_{\mathbb{RH}} \alpha$. The following soundness and completeness theorems are then obtained in the standard manner.

Theorem 83. *(Soundness and Completeness)* $\vdash_{\mathcal{L}_{roughInt}} \alpha$ *if and only if* $\models_{\mathbb{RH}} \alpha$.

Now, let us recall Theorem 75, which states that given any rough Heyting algebra $\mathcal{H}$, there exists an approximation space (U, R) such that $\mathcal{H}$ can be embedded into $\mathcal{R}^{IC}$. So if α is valid in the class of all rough Heyting algebras, it is also valid in the class of all rough Heyting algebras formed by QOCAS. Moreover, we have

Theorem 84. *(Rough Set Semantics)*

1. $\models_{\mathbb{RH}} \alpha$ *if and only if α is valid in the class of all rough Heyting algebras formed by QOCAS.*
2. *If α is valid in the class of all rough Heyting algebras formed by QOCAS then* $\vdash_{\mathcal{L}_{roughInt}} \alpha$.

6.3   New Negations from Pseudo Negation

Let B be a Boolean algebra. The set $B^{[2]}$ forms various structures by inducing various negations by the Boolean negation. In literature, there is a work of Vakarelov, where he constructed strong negation with the help of pseudo negation, while constructing the algebraic model of constructive logic with strong negation. In this section we follow the same lines of work. We define unary operators with the help of pseudo complement and dual pseudo complement and establish correspondence results in classes of compatibility and exhaustive frames.

We have seen in Sect. 5 that for a given QOCAS (U, R), the collection of definable sets $\mathcal{D}$ is a pseudo complemented lattice. In the classical case definable sets form a Boolean algebra, which induces many non-classical negations. It is natural to ask here 'what kind of unary operators are induced by the pseudo complement?'. We have already noted that the pseudo complement $\sim$ on $\mathcal{D}$ induces the pseudo complement $(\sim)$ on $\mathcal{R}$ as

$\sim (D_1, D_2) := (\sim D_2, \sim D_2)$, $D_1, D_2 \in \mathcal{D}$. Motivated by the Vakarelov construction, we consider the following unary operations on $\mathcal{R}_L$ for a given pseudo complemented lattice $\mathcal{L} := (L, \vee, \wedge, \sim, 0, 1)$: $\sim_1, \sim_2, \sim_3 \colon \mathcal{R}_L \to \mathcal{R}_L$, $(a, b) \in \mathcal{R}_L$,

$$\sim_1 (a, b) := (\sim b, \sim a).$$

$$\sim_2 (a, b) := (\sim b, \sim b).$$

$$\sim_3 (a, b) := (\sim a, \sim a).$$

Can we call these unary operations, negations? Again, we follow Dunn's model of negations, and to answer the question show that these unary operations can be characterized by compatibility frames or can be looked upon as perp. Let us discuss some properties of $\sim_1, \sim_2, \sim_3$ in the following proposition.

Proposition 36. *Let $\mathcal{L} := (L, \vee, \wedge, \sim, 0, 1)$ be a pseudo complemented lattice. Then for all $x, y \in L^{[2]}$,*

1. $\sim_1$ *satisfies the following properties.*
    (a) $x \leq y \Rightarrow \sim_1 y \leq \sim_1 x$.
    (b) $x \leq \sim_1 \sim_1 x$.
    (c) $\sim_1 \sim_1 \sim_1 x = \sim_1 x$
    (d) $x \wedge y = 0 \Rightarrow x \leq \sim_1 y$.
2. $\sim_2$ *is a pseudo complement on $\mathcal{R}_L$.*
3. $\sim_3$ *satisfies the following properties.*
    (a) $x \leq y \Rightarrow \sim_3 y \leq \sim_3 x$.
    (b) $\sim_3 \sim_3 \sim_3 x = \sim_3 x$.
    (c) $x \wedge y = 0 \Rightarrow x \leq \sim_3 y$.

Proof. 1. a) Let $(a, b) \leq (c, d)$. We have $a \leq c$ and $b \leq d$. But then $\sim c \leq \sim a$ and $\sim d \leq \sim b$. Hence we have $\sim_1 (c, d) = (\sim d, \sim c) \leq (\sim b, \sim a) = \sim_1 (a, b)$.
b) As, $a \leq \sim\sim a$ and $b \leq \sim\sim b$, hence we have $(a, b) \leq (\sim\sim a, \sim\sim b) = \sim_1 \sim_1 (a, b)$.
c) $\sim a = \sim\sim\sim a$ and $\sim b = \sim\sim\sim b$, we have $(\sim b, \sim a) = (\sim\sim\sim b, \sim\sim\sim a)$. So, we have $\sim_1 (a, b) = \sim_1 \sim_1 \sim_1 (a, b)$.
d) Let $(a, b) \wedge (c, d) = (0, 0)$. This implies $a \wedge c = 0$ and $b \wedge d = 0$. Hence $a \leq \sim c$ and $b \leq \sim d$. But $a \leq b$ gives $a \leq \sim d$. $c \leq d$ implies $\sim d \leq \sim c$, hence $b \leq \sim c$. So, $(a, b) \leq (\sim d, \sim c) = \sim_1 (c, d)$.
2. Follows from Proposition 19.
3. a) Let $(a, b) \leq (c, d)$. This implies $a \leq c$ and $b \leq d$. But then $\sim c \leq \sim a$, Hence we have $\sim_3 (c, d) \leq \sim_3 (a, b)$.
b) $\sim a = \sim\sim\sim a$. So, we have $\sim_3 (a, b) = (\sim a, \sim a) = (\sim\sim\sim a, \sim\sim\sim a) = \sim_3 \sim_3 \sim_3 (a, b)$.
c) Let $(a, b) \wedge (c, d) = (0, 0)$. This implies $a \wedge c = 0$ and $b \wedge d = 0$. Hence $a \leq \sim c$ and $b \leq \sim d \leq \sim c$. So, $(a, b) \leq (\sim c, \sim c) = \sim_3 (c, d)$

Let us show through some examples that these unary operations do not satisfy some typical properties of negations.
1. $\sim_3$ does not satisfy $x \leq \sim_3 \sim_3 x$.

Example 26. Let us recall Example 20, $\sim_3 (\emptyset, U) = (\sim \emptyset, \sim \emptyset) = (U, U)$. Moreover, $\sim_3\sim_3 ((\emptyset, U) =\sim_3 (U, U) = (\emptyset, \emptyset)$. But $(\emptyset, U) \not\le (\emptyset, \emptyset)$.

2. It is possible that $\sim_i\sim_i x \not\le x$, i = 1,3.

Example 27. In Example 20, $\sim_1 (\emptyset, b) = (\sim b, U) = (\emptyset, U)$ and $\sim_1\sim_1 (\emptyset, b) =\sim_1$ $(\emptyset, U) = (\emptyset, U)$. But $\sim_1\sim_1 (\emptyset, b) = (\emptyset, U) \not\le (\emptyset, b)$.
$\sim_3 (b, b) = (\sim b, \sim b) = (\emptyset, \emptyset)$ and $\sim_3\sim_3 (b, b) =\sim_3 (\sim b, \sim b) =\sim_3 (\emptyset, \emptyset) = (U, U)$. But $\sim_3\sim_3 (b, b) = (U, U) \not\le (b, b)$.

3. $x \wedge y \le z \not\Rightarrow x \wedge \sim_i z \le\sim_i y$, i = 1,3.

Example 28. From Example 20, we have $(\emptyset, U) \wedge (b, b) \le (\emptyset, b)$, $\sim_1 (\emptyset, b) = (\emptyset, U)$, and $\sim_1 (b, b) = (\emptyset, \emptyset)$. Hence $(\emptyset, U) \wedge \sim_1 (\emptyset, b) = (\emptyset, U) \not\le\sim_1 (b, b) = (\emptyset, \emptyset)$.
Moreover, we also have $\sim_3 (\emptyset, b) = (U, U)$ and $\sim_3 (b, b) = (\emptyset, \emptyset)$. Hence $(\emptyset, U) \wedge \sim_3 (\emptyset, b) = (\emptyset, U) \not\le\sim_3 (b, b) = (\emptyset, \emptyset)$.

It is well known that the logic K_i is sound with respect to all compatibility frames (cf. Sect. 3). The axiom $\alpha \vdash\sim\sim \alpha$ is characterized by Dunn in [29]. Let us prove correspondence results for $\sim\sim\sim \alpha \vdash\sim \alpha$ and $\sim \alpha \vdash\sim\sim\sim \alpha$.

Theorem 85. $\sim \alpha \vdash\sim\sim\sim \alpha$ *is valid in a compatibility frame if and only if the frame satisfies the following first order condition.*

$$\forall x \forall y (xCy \to \exists w (yCw \wedge \forall w' (wCw' \to xCw'))). \qquad (*)$$

Proof. Let $(U, C, \le)$ be a compatibility frame and let (*) hold. Let $x \vDash\sim \alpha$, $x \in U$, then our claim is $x \vDash\sim\sim\sim \alpha$.
So, let xCy. As (*) holds, there exists w such that $(yCw \wedge \forall w' (wCw' \to xCw'))$. So, if wCw' then xCw'. But as $x \vDash\sim \alpha$, hence $w' \nvDash \alpha$. We have $w \vDash\sim \alpha$. As yCw, hence $y \nvDash\sim\sim \alpha$. y was chosen arbitrary, so, $x \vDash\sim\sim\sim \alpha$.
Now, let (*) not hold. This implies

$$\exists x \exists y (xCy \wedge \forall w (yCw \to \exists w' (wCw' \wedge not(xCw')))).$$

Define, $z \vDash p$ if and only if $not(xCz)$. Let $z \vDash p$ and $z \le z'$. If xCz', then $x \le x$, xCz' and $z \le z' \Rightarrow xCz$, which is a contradiction. Hence $not(xCz')$ (i.e., $z' \vDash p$) and $\vDash$ is indeed an evaluation.
It is clear that $x \vDash\sim p$. Let yCw. This implies there exists w' such that wCw' and $not(xCw')$. Hence $w' \vDash p$. wCw' implies that $w \nvDash\sim p$. As, w was chosen arbitrary such that yCw, hence $y \vDash\sim\sim p$. Finally, we have $x \nvDash\sim\sim\sim p$. Hence (*) holds.

Theorem 86. $\sim\sim\sim \alpha \vdash\sim \alpha$ *is valid in a compatibility frame if and only if the following frame condition holds.*

$$\forall x \forall y (xCy \to \exists z (xCz \wedge \forall z' (zCz' \to \exists z'' (z'Cz'' \wedge y \le z'')))). \qquad (*)$$

Proof. Let (*) hold in a given compatibility frame $(U, C, \leq)$. Let $x \vDash \sim\sim\sim \alpha$, $x \in U$. We want to show that $x \vDash \sim \alpha$.

So, let xCy. Our claim is $y \nvDash \alpha$.

But as (*) holds, hence we have:

$$\exists z(xCz \wedge \forall z'(zCz' \to \exists z''(z'Cz'' \wedge y \leq z''))). \tag{1}$$

We have the following. As xCz and $x \vDash \sim\sim\sim \alpha$, $z \nvDash \sim\sim \alpha$. This implies there exists w such that zCw and $w \vDash \sim \alpha$.

As (1) holds, there exists w' such that wCw' and $y \leq w'$. But then $w \vDash \sim \alpha$ and wCw' imply $w' \nvDash \alpha$. So $y \nvDash \alpha$, and $\sim\sim\sim \alpha \vdash \sim \alpha$ is valid.

Now let (*) not hold. This means,

$$\exists x \exists y(xCy \wedge \forall z(xCz \to \exists z'(zCz' \wedge \forall z''(z'Cz'' \to y \nleq z'')))).$$

Define, $w \vDash p$ if and only if $y \leq w$.

Let us show that $x \vDash \sim\sim\sim p$. So, let xCz.

Hence there exists z' such that zCz' and $\forall z''(z'Cz'' \to y \nleq z'')$. So we have, by definition of $\vDash$, $z'' \nvDash p$ for all such z''.

Hence $z' \vDash \sim p$ and $z \nvDash \sim\sim p$. So we have $x \vDash \sim\sim\sim p$.

Clearly, we have $x \nvDash \sim p$, as xCy and $y \vDash p$.

$$\frac{\alpha \wedge \beta \vdash \gamma}{}$$

The characterization of the rule $\overline{\alpha \wedge \sim \gamma \vdash \sim \beta}$ is provided by Dunn (in [29], Theorem 2.13), and characterized by the first order condition $\forall x \forall y(xCy \to$

$$\frac{\alpha \wedge \beta \vdash \bot}{}$$

$\exists z(x \leq z \wedge y \leq z \wedge xCz))$. The rule $\overline{\alpha \vdash \sim \beta}$ is the special case of the rule $\frac{\alpha \wedge \beta \vdash \gamma}{\alpha \wedge \sim \gamma \vdash \sim \beta}$, when we replace γ by $\bot$, as mentioned in [29]. We note the char-

acterization of the rule $\dfrac{\alpha \wedge \beta \vdash \bot}{\alpha \vdash \sim \beta}$ in the following.

Theorem 87. *The rule* $\dfrac{\alpha \wedge \beta \vdash \bot}{\alpha \vdash \sim \beta}$ *is valid in a compatibility frame if and only if the frame satisfies*

$$\forall x \forall y(xCy \to \exists z(x \leq z \wedge y \leq z)). \tag{*}$$

Proof. Let (*) hold in a compatibility frame $(W, C, \leq)$. Let the consequent $\alpha \wedge \beta \vdash \bot$ be valid, and let $x \vDash \alpha$, $x \in W$. Then we have $x \nvDash \beta$. Now let xCy. As (*) holds, we have, $\exists z(x \leq z \wedge y \leq z)$.

$x \leq z$ implies $z \vDash \alpha$. This implies $z \nvDash \beta$, as $\alpha \wedge \beta \vdash \bot$.

As $y \leq z$, we have $y \nvDash \beta$.

Hence $x \vDash \sim \beta$. So, the given rule is valid.

Let (*) not hold. Then we have:

$$\exists x \exists y(xCy \wedge \forall z(x \nleq z \vee y \nleq z)).$$

Let us define $w \vDash p$ if and only if $x \leq w$, and $w' \vDash q$ if and only if $y \leq w'$. Clearly for any w, we have $w \nvDash p \wedge q$. But $x \leq x$, so $x \vDash p$. Also, we have xCy and $y \vDash q$. Hence $x \nvDash \sim q$.

The enhanced Kite of negations is given in Fig. 34.

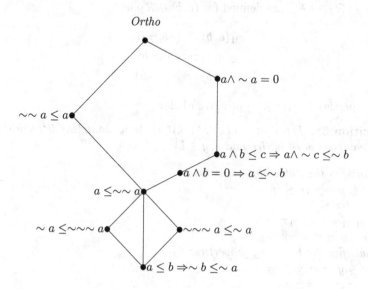

Fig. 34. Enhanced kite of negations

Let (U, R) be a QOCAS. Then the collection of definable sets $\mathcal{D}$ forms a dual pseudo complemented lattice, where dual pseudo complement ($\neg$) is defined as:

$$\neg D := min\{D' : D \cup D' = U\}, \; D \in \mathcal{D}.$$

This dual pseudo complement induces a dual pseudo complement on $\mathcal{R}$ as:

$$\neg_{dpseudo}(D_1, D_2) := (\neg D_1, \neg D_1), \; D_1, D_2 \in \mathcal{D}.$$

In fact, the above is true for any dual pseudo complemented lattice, as we see in the following.

Proposition 37. *Let $\mathcal{L} := (L, \vee, \wedge, \neg, 0, 1)$ be a dual pseudo complemented lattice. Then $\mathcal{R}_L$ is a dual pseudo complemented lattice, where dual pseudo complement ($\neg$) is defined as:*

$$\neg(a, b) := (\neg a, \neg a).$$

Proof. Our claim is $\neg(a, b) = min\{(c, d) : (a, b) \vee (c, d) = (1, 1)\} = (\neg a, \neg a)$.
    Indeed, we have $(a, b) \vee \neg(a, b) = (a, b) \vee (\neg a, \neg a) = (a \vee \neg a, b \vee \neg a) = (1, 1)$.

Now, let $(a, b) \vee (c, d) = (1, 1)$. This implies $(a \vee c, b \vee d) = (1, 1)$. Hence $a \vee c = 1$ and $b \vee d = 1$, but this gives us $\neg a \leq c$ and $\neg b \leq d$. Hence $(\neg a, \neg a) \leq (c, d)$. Hence the claim.

Following the theme of this section, we consider the following unary operations on $\mathcal{R}_L$ for a given dual pseudo complemented lattice $\mathcal{L} := (L, \vee, \wedge, \neg, 0, 1)$: $\neg_1, \neg_2, \neg_3 : \mathcal{R}_L \to \mathcal{R}_L$ are defined for $(a, b) \in \mathcal{R}_L$ as

$$\neg_1(a, b) := (\neg b, \neg a).$$
$$\neg_2(a, b) := (\neg b, \neg b).$$
$$\neg_3(a, b) := (\neg a, \neg a).$$

Some properties of $\neg_1, \neg_2, \neg_3$ are given below.

Proposition 38. *Let* $\mathcal{L} := (L, \vee, \wedge, \neg, 0, 1)$ *be a bounded distributive dual pseudo complemented lattice and* $x, y \in L^{[2]}$.

1. $\neg_1$ *satisfies the following properties.*
   (a) $x \leq y \Rightarrow \neg_1 y \leq \neg_1 x$.
   (b) $\neg_1 \neg_1 x \leq x$.
   (c) $\neg_1 \neg_1 \neg_1 x = \neg_1 x$.
   (d) $x \vee y = 1 \Rightarrow \neg x \leq y$.
2. $\neg_2$ *satisfies the following properties:*
   (a) $x \leq y \Rightarrow \neg_3 y \leq \neg_3 x$.
   (b) $\neg_3 \neg_3 \neg_3 x = \neg_3 x$.
   (c) $x \vee y = 1 \Rightarrow \neg x \leq y$.
3. $\neg_3$ *is a dual pseudo negation on* $\mathcal{R}_H$.

Proofs are dual to those presented for Proposition 36.
   We now prove correspondence results for $\neg\neg\neg\alpha \vdash \neg\alpha$ and $\neg\alpha \vdash \neg\neg\neg\alpha$. The same first order conditions for compatibility frames are used for characterizations.

Theorem 88. $\neg\neg\neg\alpha \vdash \neg\alpha$ *is valid in an exhaustive frame if and only if the frame satisfies the following first order condition.*

$$\forall x \forall y (xRy \to \exists w(yRw \wedge \forall w'(wRw' \to xRw'))). \tag{*}$$

Proof. Let $(U, R, \leq)$ be an exhaustive frame and (*) hold. Let $x \vDash \neg\neg\neg\alpha$, $x \in U$.

$$\Rightarrow \exists y (xRy \wedge y \nvDash \neg\neg\alpha).$$
$$\Rightarrow \forall w (yRw \Rightarrow w \vDash \neg\alpha). \tag{B}$$
$$\Rightarrow \exists w' (wRw' \wedge w' \nvDash \alpha), \text{ for all such } w. \tag{C}$$

Claim: $x \vDash \neg\alpha$.
As (*) holds and xRy, we have $\exists w(yRw \wedge \forall w'(wRw' \to xRw'))$.
Using (B) and (C) we have $w \vDash \neg\alpha$, and there is w' such that $w' \nvDash \alpha$ and wRw'. But wRw' imply xRw'. Hence $x \vDash \neg\alpha$.

Now, let (*) not hold. This implies:

$$\exists x \exists y (xRy \wedge \forall w(yRw \to \exists w'(wRw' \wedge not(xRw')))).$$

Define, $z \models p$ if and only if xRz. Let $z \models p$ and $z' \leq z$. We have $x \leq x$ and xRz and $z \geq z'$; this implies xRz'. Hence $\models$ is well defined.

It is clear that $x \not\models \neg p$. Let us prove that $x \models \neg\neg\neg p$. Note that $y \not\models \neg\neg p$, as if yRw then $\exists$ a w' such that wRw' and $not(xRw')$, i.e., $w' \not\models p$. Hence $w \models \neg p$ and $y \models \neg\neg p$. Finally, we have $x \models \neg\neg\neg p$. Hence (*) holds.

Theorem 89. $\neg\alpha \vdash \neg\neg\neg\alpha$ is valid in an exhaustive frame if and only if the following frame condition holds.

$$\forall x \forall y (xRy \to \exists z(xRz \wedge \forall z'(zCz' \to \exists z''(z'Rz'' \wedge y \leq z')))). \qquad (*)$$

Proof. Let (*) hold in an exhaustive frame $(W, R, \leq)$. Let $x \models \neg\alpha$, $x \in W$.

$$\Rightarrow \exists y(xRy \wedge y \not\models \alpha).$$

Claim: $x \models \neg\neg\neg\alpha$, i.e., there exists x' such that xRx' and $x' \not\models \neg\neg\alpha$.
As xRy and (*) holds, hence we have:

$$\exists z(xRz \wedge \forall z'(zRz' \to \exists z''(z'Rz'' \wedge y \leq z''))).$$

We show $z \not\models \neg\neg\alpha$. Let zRz'.
This shows that there is z'' such that $z'Rz''$ and $y \leq z''$. Thus $z'' \not\models \alpha$, as $y \not\models \alpha$.
So $z' \models \neg\alpha$, whence $z \not\models \neg\neg\alpha$.
Now let (*) not hold. So,

$$\exists x \exists y (xRy \wedge \forall z(xRz \to \exists z'(zRz' \wedge \forall z''(z'Rz'' \to y \not\leq z'')))).$$

Define, $w \models p$ if and only if $y \not\leq w$. $\models$ is well defined: if $w \models p$ and $w' \leq w$, then $y \not\leq w'$. Clearly, we have $x \models \neg p$ as xRy and $y \not\models p$.
We show $x \not\models \neg\neg\neg p$. So let xRz and we show $z \models \neg\neg p$. Now by our assumption, there exists z' such that zRz', and for all z'' such that $z'Rz''$ we have $z'' \models p$. Hence $z' \not\models \neg p$. So, $z \models \neg\neg p$ and $x \not\models \neg\neg\neg p$.

Theorem 90. The rule $\dfrac{\top \vdash \alpha \vee \beta}{\neg\beta \vdash \alpha}$ is valid in an exhaustive frame if and only if the frame satisfies

$$\forall x \forall y (xRy \to \exists z(x \leq z \wedge y \leq z)). \qquad (*)$$

Proof. Let (*) hold in an exhaustive frame $(W, R, \leq)$. Let the consequent $\top \vdash \alpha \vee \beta$ be valid, i.e., for any x in W, we have $x \models \alpha \vee \beta$. Let $x \models \neg\beta$, hence there exists y in W such that xRy and $y \not\models \beta$. As (*) holds, there exists z in W such that $x \leq z$ and $y \leq z$. We have, using backward hereditary property, $z \not\models \beta$. Hence $z \models \alpha$ and $x \models \alpha$.

Let (*) not hold. So we have:

$$\exists x \exists y (xRy \wedge \forall z (x \nleq z \vee y \nleq z)).$$

Let us define $w \vDash p$ if and only if $x \nleq w$, and $w' \vDash q$ if and only if $y \nleq w'$. Clearly for any w in W, we have $w \vDash p \vee q$. Also, we have xRy and $y \nvDash q$, hence $x \nvDash \neg q$. But $x \leq x$, so $x \nvDash p$.

The enhanced dual kite of negations is given in Fig. 35.

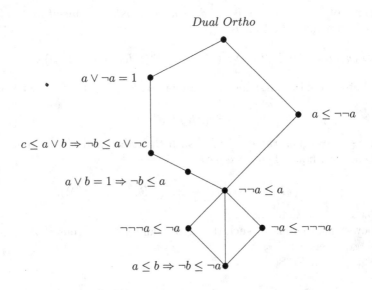

Fig. 35. Enhanced dual lopsided kite of negations

7   Conclusions and Future Work

Our work in this paper involves three aspects related to rough set theory.

1. Algebraic and logical aspects of classical rough set theory.
2. Algebraic and logical aspects of a generalization of rough set theory.
3. Semantic analysis of some negations appearing in classical and generalized rough set structures.

7.1   Summary and Conclusions

Let us present the summary and conclusions, sectionwise.

Section 2

1. Variety of Kleene algebras is generated by the 3 element Kleene algebra. Using this fact, a structural theorem for Kleene algebras has been proved, which asserts that an element of a Kleene algebra can be looked upon as a monotone ordered pair of sets. It is shown that the Kleene negations in Kleene algebras are definable using the Boolean negation.
2. A logic $\mathcal{L}_K$ is presented, which is sound and complete with respect to the class of all Kleene algebras. We consider a 3-valued semantic consequence relation $\vDash_{t,f}$, with respect to which $\mathcal{L}_K$ is proved to be sound and complete.
3. Rough set representation of Kleene algebras is obtained, through which we get a rough set semantics for the logic $\mathcal{L}_K$ as well.
4. Algebraic semantics, 3-valued semantics and rough set semantics for the logic $\mathcal{L}_K$ are equivalent.
5. The Kleene algebra **3** plays the same fundamental role among the class of Kleene algebras and the logic $\mathcal{L}_K$, as that played by the Boolean algebra **2** among the class of all Boolean algebras and classical propositional logic.

Section 3

1. Perp semantics for the logic $\mathcal{L}_K$ for Kleene algebras and the logic $\mathcal{L}_S$ for Stone algebras is presented. As a result, the Kleene and Stone negations can be treated as modal operators, and moreover, yield new positions in Dunn's lopsided kite of negations.
2. Characterization and completeness results for the logic $\mathcal{L}_K$ and logic of dual Stone algebras is given in exhaustive frames. Hence, Dunn's dual lopsided kite of negations is enhanced with new positions occupied by dual Kleene and dual Stone negations.
3. Semantics of the logic $\mathcal{L}_{RDSA}$ of regular double Stone algebras is given with respect to K_- frames. Negations in $\mathcal{L}_{RDSA}$ can thus be looked upon as modal operators. Moreover, the rough set semantics and semantics in K_- frames for $\mathcal{L}_{RDSA}$ become equivalent.

Section 4

Discrete duality between a number of classes of algebras and frames is provided. More precisely, the duality results are obtained between the classes of Kleene, dual Kleene, Stone, double Stone, double Kleene, regular double Stone algebras and the corresponding classes of frames defined in the section. As a consequence, we obtain representations for all the above classes of algebras in terms of algebras based on set lattices.

Section 5

1. A a granule based definition of lower and upper approximation operators in a generalized approximation space (QOCAS) is given, which appears to capture the essence of 'approximation by granule' in an approximation space, as in classical rough set theory.

2. Topologically, definable sets for a QOCAS are open sets in an Alexandrov topological space (generated by a quasi order).
3. Algebraically, the collection of definable sets for a QOCAS forms a distributive lattice, Heyting algebra as well as a completely distributive lattice in which the set of join irreducible elements is join dense.
4. For a given QOCAS, the collection $\mathcal{RS}$ of rough sets may not form a lattice, but the collection $\mathcal{R}$ of generalized rough sets always forms a distributive lattice, in fact a Heyting algebra, and a completely distributive lattice in which the set of join irreducible elements is join dense.
5. We observe that the Heyting algebras and completely distributive lattices in which the set of join irreducible elements is join dense, abstract the algebraic structures formed by the collections $\mathcal{R}$ and $\mathcal{RS}$ of QOCAS.
6. We obtain rough set representations of the classes of Heyting algebras, and Heyting algebras based on completely distributive lattices in which the set of join irreducible elements is join dense.

Section 6

1. The algebraic structures formed by $\mathcal{R}$ and $\mathcal{RS}$, as presented in Sect. 5, are enhanced by introducing two unary operators I and C. Abstraction of the resulting structures leads to the introduction of rough lattices, rough Heyting algebras and complete rough lattices.
2. Rough set representations of these new classes of algebras are provided.
3. A logic $\mathcal{L}_{\mathcal{R}}$, which is an extension of the system $BDLL$, is shown to be sound and complete with respect to the class of all rough lattices. Due to the rough set representation result for this class of algebras, we obtain a rough set semantics for the logic $\mathcal{L}_{\mathcal{R}}$ also.
4. We extend Priors' $MIPC$ to the logic $\mathcal{L}_{roughInt}$, to observe that it is sound and complete with respect to the class of all rough Heyting algebras. Rough set semantics for $\mathcal{L}_{roughInt}$ is also provided.
5. A number of unary operators induced on rough set structures by pseudo complement and dual pseudo complement are studied. It is shown that these can be characterized in the framework of either compatibility frames or exhaustive frames. Hence these can be considered as negations, interpreted as impossibility or unnecessity operators. New positions are achieved in Dunn's lopsided kite and dual kite of negations.

7.2   Future Work

In Sect. 2, we provided a structural theorem for the class of Kleene algebras. For any approximation space (U, R), the collection $\mathcal{R}$ of generalized rough sets, when considered as a Kleene algebra, is isomorphic to $\mathbf{3}^I$ for some index set I. This also leads to an embedding of a given Kleene algebra into the collection $\mathcal{RS}$ of rough sets for some approximation space. It is natural to enquire, what is the class $\mathcal{K}$ of Kleene algebras such that each $K \in \mathcal{K}$ is *isomorphic* to $\mathcal{RS}$ for some approximation space?

We have introduced the notion of a double Kleene algebra in Sect. 3. Properties of the structure need to be explored further. Immediate questions could be regarding representations of this class, possibly with some known class of algebras, or by some class of rough set structures. Rough lattices, rough Heyting algebras and their logics are introduced in Sect. 6. These should be subject to further study, in particular regarding relationships with other known systems.

Various negations arising from pseudo complement and dual pseudo complement have been characterized in compatibility or exhaustive frames in Sect. 6. However, the following questions on canonicity remain open. (a) The canonicity of the logics $K_i + \sim \alpha \vdash \sim\sim\sim \alpha$ and $K_i + \sim\sim\sim \alpha \vdash \sim \alpha$ with respect to classes of compatibility frames. (b) The canonicity of the logics $K_u + \neg\alpha \vdash \neg\neg\neg\alpha$ and $K_u + \neg\neg\neg\alpha \vdash \neg\alpha$ with respect to classes of exhaustive frames.

Acknowledgements. I would like to express my sincere gratitude to my thesis advisor Prof. Mohua Banerjee for her warm encouragement and thoughtful guidance. She was always there to listen and to give advice. The discussion hours between us was exceptionally helpful for me, without her help I could not have finished my thesis work successfully.

References

1. Avron, A., Konikowska, B.: Rough sets and 3-valued logics. Studia Logica **90**, 69–92 (2008)
2. Balbes, R., Dwinger, P.: Distributive Lattices. University of Missouri Press, Columbia (1974)
3. Banerjee, M.: Rough sets and 3-valued Lukasiewicz logic. Fundamenta Informaticae **31**, 213–220 (1997)
4. Banerjee, M., Chakraborty, M.K.: Rough sets through algebraic logic. Fundamenta Informaticae **28**(3–4), 211–221 (1996)
5. Banerjee, M., Chakraborty, M.K.: Algebras from rough sets. In: Pal, S.K., Polkowski, L., Skowron, A. (eds.) Rough-neuro Computing: Techniques for Computing with Words, pp. 157–184. Springer, Berlin (2004). https://doi.org/10.1007/978-3-642-18859-6_7
6. Banerjee, M., Khan, M.: Propositional logics from rough set theory. In: Peters, J.F., Skowron, A., Duntsch, I., Grzymała-Busse, J., Orłowska, E., Polkowski, L. (eds.) Transactions on Rough Sets VI. LNCS, vol. 4374, pp. 1–25. Springer, Heidelberg (2007). https://doi.org/10.1007/978-3-540-71200-8_1
7. Bezhanishvili, G.: Varieties of monadic Heyting algebras. part I. Studia Logica **61**(3), 367–402 (1998)
8. Bezhanishvili, N., de Jongh, D.: Intuitionistic logic. Institute for Logic, Language and Computation (ILLC), University of Amsterdam PP-2006-25 (2006)
9. Birkhoff, G.: Lattice Theory. Colloquium Publications, vol. XXV, 3rd edn., American Mathematical Society, Providence (1995)
10. Blackburn, P., de Rijke, M., Venema, Y.: Modal Logic. Cambridge University Press (1991)
11. Boicescu, V., Filipoiu, A., Georgescu, G., Rudeanu, S.: Lukasiewicz-Moisil Algebras. North-Holland, Amsterdam (1991)

12. Bonikowski, Z.: Algebraic structures of rough sets in representative approximation spaces. Electronic Notes Theoret. Comput. Sci. **82**, 1–12 (2003)

13. Bonikowski, Z., Bryniariski, E., Skardowska, V.W.: Extension and intension in the rough set theory. Inf. Sci. **107**, 149–167 (1998)

14. Chakraborty, M.K., Banerjee, M.: Rough sets: some foundational issues. Fundamenta Informaticae **127**, 1–15 (2013)

15. Cignoli, R.: Boolean elements in Łukasiewicz algebras. I. Proc. Japan Acad. **41**, 670–675 (1965)

16. Cignoli, R.: The class of Kleene algebras satisfying an interpolation property and Nelson algebras. Algebra Universalis **23**(3), 262–292 (1986)

17. Cignoli, R.: The algebras of Łukasiewicz many-valued logic: a historical overview. In: Aguzzoli, S., Ciabattoni, A., Gerla, B., Manara, C., Marra, V. (eds.) Algebraic and Proof-theoretic Aspects of Non-classical Logics. LNAI, vol. 4460, pp. 69–83. Springer, Heidelberg (2007). https://doi.org/10.1007/978-3-540-75939-3_5

18. Ciucci, D., Dubois, D.: Three-valued logics, uncertainty management and rough sets. In: Peters, J.F., Skowron, A. (eds.) Transactions on Rough Sets XVII. LNCS, vol. 8375, pp. 1–32. Springer, Heidelberg (2014). https://doi.org/10.1007/978-3-642-54756-0_1

19. Comer, S.: Perfect extensions of regular double Stone algebras. Algebra Universalis **34**(1), 96–109 (1995)

20. Dai, J.-H.: Logic for rough sets with rough double stone algebraic semantics. In: Slezak, D., Wang, G., Szczuka, M., Düntsch, I., Yao, Y. (eds.) RSFDGrC 2005. LNCS (LNAI), vol. 3641, pp. 141–148. Springer, Heidelberg (2005). https://doi.org/10.1007/11548669_15

21. Davey, B.A., Priestley, H.A.: Introduction to Lattices and Order. Cambridge University Press (2002)

22. Degang, C., Wenxiu, Z., Yeung, D., Tsang, E.: Rough approximations on a complete completely distributive lattice with applications to generalized rough sets. Inf. Sci. **176**, 1829–1848 (2006)

23. Dunn, J.: The algebra of intensional logic. Doctoral dissertation, University of Pittsburgh (1966)

24. Dunn, J.: Star and perp: two treatments of negation. In: Tomberlin, J. (ed.) Philosophical Perspectives, vol. 7, pp. 331–357. Ridgeview Publishing Company, Atascadero, California (1994)

25. Dunn, J.: Positive modal logic. Studia Logica **55**, 301–317 (1995)

26. Dunn, J.: Generalised ortho negation. In: Wansing, H. (ed.) Negation: A Notion in Focus, pp. 3–26. Walter de Gruyter, Berlin (1996)

27. Dunn, J.: A comparative study of various model-theoretic treatments of negation: a history of formal negations. In: Gabbay, D., Wansing, H. (eds.) What is Negation?, pp. 23–51. Kluwer Academic Publishers, Netherlands (1999)

28. Dunn, J.: Partiality and its dual. Studia Logica **66**, 5–40 (2000)

29. Dunn, J.: Negation in the context of gaggle theory. Studia Logica **80**, 235–264 (2005)

30. Düntsch, I.: A logic for rough sets. Theoret. Comput. Sci. **179**, 427–436 (1997)

31. Düntsch, I., Orłowska, E.: Discrete dualities for double Stone algebras. Studia Logica **99**(1), 127–142 (2011)

32. Fidel, M.: An algebraic study of a propositional system of Nelson. In: Arruda, A.I., da Costa, N.C.A., Chuaqui, R. (eds.) Mathematical Logic: Proceedings of First Brazilian Conference, pp. 99–117. Lecture Notes in Pure and Applied Mathematics, vol. 39, M.Dekker Inc., New York (1978)

33. Gehrke, M., van Gool, S.J.: Distributive envelopes and topological duality for lattices via canonical extensions. Order **31**(3), 435–461 (2014)
34. Gehrke, M., Walker, E.: On the structure of rough sets. Bull. Polish Acad. Sci. Math. **40**(3), 235–255 (1992)
35. Greco, S., Matarazzo, B., Słowinski, R.: Rough sets theory for multi-criteria decision analysis. Eur. J. Oper. Res. **129**, 1–47 (2001)
36. Greco, S., Matarazzo, B., Slowinski, R.: Multicriteria classification by dominance based rough set approach. In: Kloesgen, W., Zytkow, J. (eds.) Handbook of Data Mining and Knowledge discovery. Oxford University Press, New York (2002)
37. Greco, S., Matarazzo, B., Słowinski, R.: Algebra and topology for dominance-based rough set approach. In: Ras, Z., Tsay, L.S. (eds.) Advances in Intelligent Information Systems, pp. 43–78. Springer, Heidelberg (2010). https://doi.org/10.1007/978-3-642-05183-8_3
38. Iturrioz, L.: Rough sets and three-valued structures. In: Orłowska, E. (ed.) Logic at Work: Essays Dedicated to the Memory of Helena Rasiowa. Studies in Fuzziness and Soft Computing, vol. 24, pp. 596–603. Springer, Heidelberg (1999)
39. Järvinen, J., Pagliani, P., Radeleczki, S.: Information completeness in Nelson algebras of rough sets induced by quasiorders. Studia Logica **101**(5), 1073–1092 (2013)
40. Järvinen, J., Radeleczki, S.: Representation of Nelson algebra by rough sets determined by quasiorder. Algebra Universalis **66**, 163–179 (2011)
41. Järvinen, J., Radeleczki, S.: Rough sets determined by tolerances. Int. J. Approximate Reason. **55**(6), 1419–1438 (2014)
42. Järvinen, J., Radeleczki, S., Veres, L.: Rough sets determined by quasiorders. Order **26**, 337–355 (2009)
43. Kalman, J.: Lattices with involution. Trans. Am. Math. Soc. **87**, 485–491 (1958)
44. Katriňák, T.: Construction of regular double p-algebras. Bull. Soc. Roy. Sci. Liege **43**, 238–246 (1974)
45. Khan, M.A., Banerjee, M.: Logics for information systems and their dynamic extensions. ACM Trans. Comput. Logic **12**(4), art. no. 29 (2011)
46. Khan, M.A., Banerjee, M., Rieke, R.: An update logic for information systems. Int. J. Approximate Reasoning **55**(1), 436–456 (2014)
47. Kozen, D.: On kleene algebras and closed semirings. In: Rovan, B. (ed.) MFCS 1990. LNCS, vol. 452, pp. 26–47. Springer, Heidelberg (1990). https://doi.org/10.1007/BFb0029594
48. Kumar, A., Banerjee, M.: Definable and rough sets in covering-based approximation spaces. In: Li, T., Nguyen, H., Wang, G., Grzymała-Busse, J., Janicki, R., Hassanien, A., Yu, H. (eds.) Rough Sets and Knowledge Technology, pp. 488–495. Springer, Heidelberg (2012). https://doi.org/10.1007/978-3-642-31900-6_60
49. Kumar, A., Banerjee, M.: Algebras of definable and rough sets in quasi order-based approximation spaces. Fundamenta Informaticae **141**, 37–55 (2015)
50. Kumar, A., Banerjee, M.: Kleene algebras and logic: boolean and rough set representations, 3-valued, rough set and perp semantics. Studia Logica **105**, 439–469 (2017)
51. Kumar, A., Banerjee, M.: A semantic analysis of Stone and dual Stone negations with regularity. In: Ghosh, S., Prasad, S. (eds.) ICLA 2017, pp. 139–153. Springer, Heidelberg (2017). https://doi.org/10.1007/978-3-662-54069-5_11
52. Li, T.J.: Rough approximation operators in covering approximation spaces. In: Greco, S., et al. (eds.) RSCTC 2006, pp. 174–182. Springer, Heidelberg (2006). https://doi.org/10.1007/11908029_20
53. Nagarajan, E.K.R., Umadevi, D.: A method of representing rough sets system determined by quasi orders. Order **30**, 313–337 (2013)

54. Ono, H.: On some intuitionistic modal logic. Publ. Inst. Math. Sci. Kyoto Univ. **13**, 687–722 (1977)
55. Orłowska, E., Rewitzky, I.: Duality via truth: semantic frameworks for lattice-based logics. Logic J. IGPL **13**(4), 467–490 (2005)
56. Orłowska, E., Rewitzky, I.: Discrete duality for Heyting algebras with operators. Fundamenta Informaticae **81**(1–3), 275–295 (2007)
57. Orłowska, E., Rewitzky, I., Düntsch, I.: Relational semantics through duality. In: MacCaull, W., Winter, M., Düntsch, I. (eds.) Relational Methods in Computer Science, pp. 17–32. Springer, Heidelberg (2006). https://doi.org/10.1007/11734673_2
58. Pagliani, P.: Rough sets and Nelson algebras. Fundamenta Informaticae **27**(2–3), 205–219 (1996)
59. Pagliani, P.: Rough set theory and logic-algebraic structures. In: Orłowska, E. (ed.) Incomplete Information: Rough Set Analysis. Studies in Fuzziness and Soft Computing, vol. 13, pp. 109–190. Springer, Heidelberg (1998). https://doi.org/10.1007/978-3-7908-1888-8_6
60. Pawlak, Z.: Rough sets. Int. J. Comput. Inf. Sci. **11**, 341–356 (1982)
61. Pawlak, Z.: Rough Sets: Theoretical Aspects of Reasoning About Data. Kluwer Academic Publishers (1991)
62. Pomykała, J., Pomykała, J.A.: The Stone algebra of rough sets. Bull. Polish Acad. Sci. Math. **36**, 495–508 (1988)
63. Pomykała, J.A.: Approximation, similarity and rough construction. ILLC prepublication series CT-93-07, University of Amsterdam (1993)
64. Qin, K., Gao, Y., Pei, Z.: On covering rough sets. In: Yao, J.T., Lingras, P., Wu, W.-Z., Szczuka, M., Cercone, N.J., Ślęzak, D. (eds.) RSKT 2007. LNCS (LNAI), vol. 4481, pp. 34–41. Springer, Heidelberg (2007). https://doi.org/10.1007/978-3-540-72458-2_4
65. Rasiowa, H.: An Algebraic Approach to Non-classical Logics. North-Holland (1974)
66. Restall, G.: Defining double negation elimination. L. J. IGPL **8**(6), 853–860 (2000)
67. Saha, A., Sen, J., Chakraborty, M.K.: Algebraic structures in the vicinity of prerough algebra and their logics. Inf. Sci. **282**, 296–320 (2014)
68. Saha, A., Sen, J., Chakraborty, M.K.: Algebraic structures in the vicinity of prerough algebra and their logics II. Inf. Sci. **333**, 44–60 (2016)
69. Samanta, P., Chakraborty, M.K.: Generalized rough sets and implication lattices. In: Peters, J.F., et al. (eds.) Transactions on Rough Sets XIV. LNCS, vol. 6600, pp. 183–201. Springer, Heidelberg (2011). https://doi.org/10.1007/978-3-642-21563-6_10
70. Samanta, P., Chakraborty, M.K.: Interface of rough set systems and modal logics: a survey. In: Peters, J.F., Skowron, A., Ślęzak, D., Nguyen, H.S., Bazan, J.G. (eds.) Transactions on Rough Sets XIX. LNCS, vol. 8988, pp. 114–137. Springer, Heidelberg (2015). https://doi.org/10.1007/978-3-662-47815-8_8
71. Sanjuan, E.: Heyting algebras with Boolean operators of rough sets and information retrieval applications. Discrete Appl. Math. **156**, 967–983 (2008)
72. Suzuki, N.: An algebraic approach to intuitionistic modal logics in connection with intermediate predicate logics. Studia Logica **48**(2), 141–155 (1989)
73. Urquhart, A.: Basic many-valued logic. In: Gabbay, D., Guenthner, F. (eds.) Handbook of Philosophical Logic, vol. 2, pp. 249–295. Springer, Heidelberg (2001). https://doi.org/10.1007/978-94-017-0452-6_4
74. Vakarelov, D.: Notes on N-lattices and constructive logic with strong negation. Studia Logica **36**, 109–125 (1977)
75. Varlet, J.: A regular variety of type (2,2,1,1,0,0). Algebra Universalis **2**, 218–223 (1972)

76. Yao, Y., Chen, Y.: Rough set approximations in formal concept analysis. In: Peters, J.F., Skowron, A. (eds.) Transactions on Rough Sets V. LNCS, vol. 4100, pp. 285–305. Springer, Heidelberg (2006). https://doi.org/10.1007/11847465_14
77. Yao, Y.: Relational interpretations of neighborhood operators and rough set approximation operators. Inf. Sci. **111**, 239–259 (1998)
78. Yao, Y.: Granular computing using neighborhood system. In: Roy, R., Furuhashi, T., Chawdhry, P.K. (eds.) Advances in Soft Computing: Engineering Design and Manufacturing, pp. 539–553. Springer, New York (1999). https://doi.org/10.1007/978-1-4471-0819-1_40
79. Yao, Y.: Rough sets, neighborhood system and granular computing. In: Proceedings of IEEE Canadian Conference on Electrical and Computer Engineering, pp. 1553–1558. IEEE Press (1999)
80. Yao, Y.: Information granulation and rough set approximation. Int. J. Intell. Syst. **16**(1), 87–104 (2001)
81. Yao, Y.: Covering based rough set approximations. Inf. Sci. **200**, 91–107 (2012)
82. Yao, Y., Lin, T.: Generalization of rough sets using modal logic. Intell. Autom. Soft Comput. Int. J. **2**(2), 103–120 (1996)
83. Zhang, Y., Luo, M.: Relationships between covering-based rough sets and relation based rough sets. Inf. Sci. **225**, 55–71 (2013)
84. Zhoua, N., Hua, B.: Rough sets based on complete completely distributive lattice. Inf. Sci. **269**, 378–387 (2014)
85. Zhu, W.: Topological approaches to covering rough sets. Inf. Sci. **177**, 1499–1508 (2007)
86. Zhu, W., Wang, F.Y.: Reduction and axiomatization of covering generalized rough sets. Inf. Sci. **152**, 217–230 (2003)
87. Zhu, W., Wang, F.Y.: Relationship among three types of covering rough sets. In: Proceedings of IEEE International Conference on Granular Computing, pp. 43–48 (2006)

Similarity-based Rough Sets
and Its Applications in Data Mining

Dávid Nagy$^{(\boxtimes)}$

Department of Computer Science, Faculty of Informatics, University of Debrecen,
Kassai út 26, Debrecen 4028, Hungary
nagy.david@inf.unideb.hu

Abstract. Pawlakian spaces rely on an equivalence relation which represent indiscernibility. As a generalization of these spaces, some approximation spaces have appeared that are not based on an equivalence relation but on a tolerance relation that represents similarity. These spaces preserve the property of the Pawlakian space that the union of the base sets gives out the universe. However, they give up the requirement that the base sets are pairwise disjoint. The base sets are generated in a way where for each object, the objects that are similar to the given object, are taken. This means that the similarity to a given object is considered. In the worst case, it can happen that the number of base sets equals those of objects in the universe. This significantly increases the computational resource need of the set approximation process and limits the efficient use of them in large databases. To overcome this problem, a possible solution is presented in this dissertation. The space is called similarity-based rough sets where the system of base sets is generated by the correlation clustering. Therefore, the real similarity is taken into consideration not the similarity to a distinguished object. The space generated this way, on the one hand, represents the interpreted similarity properly and on the other hand, reduces the number of base sets to a manageable size. This work deals with the properties and applicability of this space, presenting all the advantages that can be gained from the correlation clustering.

Keywords: Rough set theory · Correlation clustering · Set approximation · Representatives · Similarity

1 Introduction

Nowadays the amount of data is growing exponentially. However, data are often incomplete or inconsistent. There can be many reasons if a value is missing. For example, it can be unknown, unassigned or even inapplicable. Inconsistency occurs when the data are contradictory. These issues can cause some undesirable events (bad prediction, inappropriate decision making, etc). In computer science, there are numerous ways to handle these kinds of inaccuracies.

Rough set theory can be considered as a rather new field in computer science. Its fundamentals were proposed by professor Pawlak in the 80's [42,43,45]. The

© Springer-Verlag GmbH Germany, part of Springer Nature 2020
J. F. Peters and A. Skowron (Eds.): TRS XXII, LNCS 12485, pp. 252–323, 2020.
https://doi.org/10.1007/978-3-662-62798-3_5

pawlakian spaces handle the uncertainty among the data with a relation that is based on the indiscernibility of objects. In many cases, based on the available knowledge, two objects cannot be distinguished from each other. Two arbitrary objects can be treated as indiscernible if all of their considered, relevant properties are the same. This indiscernibility can be modeled by an equivalence relation which represents our background knowledge or its limits. It can affect the membership relation by making the judgment on this relation uncertain. It makes a set vague because a decision about a certain object has an effect on the decisions about all the objects that are indiscernible from the given object. This uncertainty can be represented by set-approximation tools. If one wants to extract as much useful information as possible from large-scale information systems, then it is inevitable to handle the indiscernibility. Rough set theory tries to answer how certain sets can be characterized or if a given object belongs to a set generated by some property.

In the last 40 years, many generalizations of the original pawlakian spaces saw the light of day. In some cases, the equivalence relation, which can usually be too strict for practical applications, is replaced with a tolerance relation representing similarity [24,46,49]. Many rough set models exist that are based on the probability theory [26,44,50,56,57]. Last but not least, the hybridization with fuzzy set theory needs to be mentioned [18,19,55].

Data mining became a very important and growing field in computer science due to the incredible increase in data. Data mining is a technique by which useful information can be extracted automatically from a large amount of data. Its goal is to search for new and useful patterns, which could otherwise remain unknown, in data repositories. Data mining methods can be applied in many areas of life. With its help, one can answer questions like: is it true that if a customer buys diapers then they will also buy beer. Naturally, not every information retrieval task can be considered as data mining. For instance, searching for records by a database system or finding certain web pages by a web search engine are tasks of the information retrieval field.

There are 3 main steps of data mining: pre-processing, knowledge discovery (data mining) and post-processing. The goal of pre-processing is to convert the raw data into an appropriate format. Its basic steps contain the following: uniting the data coming from different sources, cleaning the data from noise and redundancy and choosing the records and variables that are essentials in the given task. After the first main step, the data mining algorithm gets the pre-processed data as its input. The result is a pattern, a model or sometimes one can just say "knowledge". However, these patterns can be uninterpretable or useless in their format. That is why the so-called post-processing is needed which helps the decision-making with various visualization and evaluation techniques.

Rough set theory can be crucial in data sciences [10,30,47,51,52] because handling the uncertainty is necessary in case of a large amount of data. In the field of data pre-processing, there are many methods based on rough set theory.

Discretization is a process where a continuous variable is converted to a nominal one by applying a set of cuts to the domain of the original attribute and treating each interval as a discrete value. Important rough set-based discretization techniques can be seen in [31,41].

The so-called feature selection is a process where the irrelevant features (attributes) are discarded from the system. With an increasing number of attributes, the execution time and the resource requirements of an algorithm also increase. The rough set-based feature selection methods rest on the concept reduct. Essentially, a reduct is a minimal subset of features that generates the same granulation of the universe as that induced by all features [29,53,54].

Another very important pre-processing task is the so-called instance selection. In data mining, the supervised learning methods (e.q., classification) divide the input dataset into two parts: training and test data. The models can be taught by the training set, while with the help of the test dataset it can be evaluated. The aim of the instance selection is to reduce the number of examples in order to bring down the size of the training set. As a result, a new training set can be obtained by which the efficiency of the system can be improved. A rough set-based technique is described in [12].

Naturally, rough set theory can be applied not only in pre-processing. In many data mining techniques, it proved to be very useful. For example decision rule induction [25,27], association rule mining [17,28], clustering [48] etc.

Pawlakian spaces rely on an equivalence relation which represent indiscernibility. As a generalization of these spaces, some approximation spaces have appeared that are not based on an equivalence relation but on a tolerance relation that represents similarity. These spaces preserve the property of the Pawlakian space that the union of the base sets gives out the universe. However, they give up the requirement that the base sets are pairwise disjoint. The base sets are generated in a way where for each object, the objects that are similar to the given object, are taken. This means that the similarity to a given object is considered. In the worst case, it can happen that the number of base sets equals those of objects in the universe. This significantly increases the computational resource need of the set approximation process and limits the efficient use of them in large databases. To overcome this problem, a possible solution is presented in this dissertation. The space is called similarity-based rough sets [36] where the system of base sets is generated by the correlation clustering. Therefore, the real similarity is taken into consideration not the similarity to a distinguished object. The space generated this way, on the one hand, represents the interpreted similarity properly and on the other hand, reduces the number of base sets to a manageable size. This work deals with the properties and applicability of this space, presenting all the advantages that can be gained from the correlation clustering. The structure of the dissertation is the following. In Sect. 2, the fundamentals of rough set theory and some of the main types of approximation spaces are presented. In the second chapter, the similarity-based rough sets approximation space is introduced. In this chapter, some of its tools and improvements are also presented [5–7,35,37]. Similarity-based rough sets was applied to graphs so with its help graphs can be approximated as well. This work is presented in Sect. 11. This method can be used in the field of feature selection.

2   Theoretical Background

In this chapter, the basic notations and techniques are introduced which are the basis of this dissertation. In the first subsection, the fundamentals of rough set theory is presented and then some of its possible generalizations. The collected definitions and methods are not part of the work of the author of this dissertation. They are merely required to understand the ideas presented in this dissertation.

In practice, a set is a collection of objects and it is uniquely identified by its members. It means that if one would like to decide, whether an object belongs to this set, then a precise answer can be given which is yes or no. A good example is the set of numbers which are divisible by 3 because it can be decided if an arbitrary number is divisible by 3 or not. Of course, it is required that one knows how to use the modulo operation. This fact can be considered as a background knowledge and it allows someone to decide if a number belongs to the given set. Naturally, not everybody knows how to use the modulo operation for each number. Some second graders may not be able to divide numbers greater than 100. They would not be able to decide if 142 is divisible by 3. For them, 142 is neither divisible nor indivisible by 3. So there is some uncertainty (vagueness) based on their background knowledge. Rough set theory was proposed by professor Pawlak in 1982 [42]. The theory offers a way to handle vagueness determined by some background knowledge. Each object of a universe can be described by a set of attribute values. If two objects have the same known attribute values, then these objects cannot be distinguished. The indiscernibility generated in this way is the mathematical basis of rough set theory.

Definition 1. *A general approximation space is an ordered 5-tuple* $\langle U, \mathfrak{B}, \mathfrak{D}, \mathsf{l}, \mathsf{u} \rangle$ *where:*

- $U \neq \emptyset$ *is the universe of objects*
- $\mathfrak{B}$ *is the set of base sets for which the following properties hold:*
  - $\mathfrak{B} \neq \emptyset$
  - *if* $B \in \mathfrak{B}$ *then* $B \subseteq U$ *and* $B \neq \emptyset$
- $\mathfrak{D}$ *is the set of definable sets which can be given by an inductive definition whose base is* $\{\emptyset\} \cup \mathfrak{B}$
- $\mathsf{l}, \mathsf{u} : 2^U \rightarrow \mathfrak{D}$ *form an approximation pair*

Definition 2. *The set of definable sets* $\mathfrak{D}$ *can be given by the following inductive definition.*

1. $\mathfrak{B} \subseteq \mathfrak{D}$;
2. $\emptyset \in \mathfrak{D}$;
3. *if* $D_1, D_2 \in \mathfrak{D}$, *then* $D_1 \cup D_2 \in \mathfrak{D}$.

The members of the Boole algebra generated by $\mathfrak{B}$ *appear as definable sets.*

The set $\mathfrak{D}$ defines how the background knowledge represented by the base sets can be used.

Many interesting properties can be checked on approximation pairs. Here, the following properties are examined (full description can be seen in [16, 20]):

Definition 3. – *Monotonicity:* l *and* u *are said to be monotone if* $S \subseteq S'$ *then*
$\mathsf{l}(S) \subseteq \mathsf{l}(S')$ *and* $\mathsf{u}(S) \subseteq \mathsf{u}(S')$
- *Weak approximation property:* $\forall S \in 2^U : \mathsf{l}(S) \subseteq \mathsf{u}(S)$
- *Strong approximation property:* $\forall S \in 2^U : \mathsf{l}(S) \subseteq S \subseteq \mathsf{u}(S)$
- *Normality of* l: $\mathsf{l}(\emptyset) = \emptyset$
- *Normality of* u: $\mathsf{u}(\emptyset) = \emptyset$

Definition 4. *The functions* l, u *form a Pawlakian approximation pair* $\langle \mathsf{l}, \mathsf{u} \rangle$ *if the followings are true for an arbitrary set* S:

1. $\mathsf{l}(S) = \bigcup \{B \mid B \in \mathfrak{B} \text{ and } B \subseteq S\}$;
2. $\mathsf{u}(S) = \bigcup \{B \mid B \in \mathfrak{B} \text{ and } B \cap S \neq \emptyset\}$.

In this dissertation, only Pawlakian approximation pairs are used.

Theorem 1. *All of the properties described in Definition 3 are true for the Pawlakian approximation pair*

Definition 5. *A general approximation space is a Pawlakian approximation space if the set of base sets is the following:* $\mathfrak{B} = \{B \mid B \subseteq U, \text{ and } x, y \in B \text{ if } x\mathcal{R}y\}$, *where* $\mathcal{R}$ *is an equivalence relation on* U. *The approximation pair is a Pawlakian approximation pair.*

A Pawlakian approximation space [42, 43, 45] can be characterized as an ordered pair $\langle U, \mathcal{R} \rangle$ where U is the same as in the case of a general approximation space. $\mathcal{R}$ is an equivalence relation on U. $\mathcal{R}$ is an equivalence relation based on the indiscernibility of objects. The system of base sets represents the background knowledge or its limit. The functions l and u give the lower and upper approximation of a set. The lower approximation contains objects that surely belong to the set, and the upper approximation contains objects that possibly belong to the set.

A rough set can be defined in several ways. In a general sense, it can be treated as an orthopair [13] which means it is a pair of sets such that the two sets are disjoint. The other possible way is when it is considered as a pair of sets such that one of them is a subset of the other. In the case of approximation spaces, both definitions are commonly used in the literature. If the rough set corresponding to the set S is considered as an orthopair, then it is defined as $\langle \mathsf{l}(S), \mathsf{l}(\overline{S}) \rangle$, where $\overline{S}$ is the complement of S. In the other case, it can be given as $\langle \mathsf{l}(S), \mathsf{u}(S) \rangle$.

Definition 6. *The set* $NEG(S) = \mathsf{l}(\overline{S})$ *is called the negative region of the set* S.

A rough set can also be characterized numerically by the accuracy of the approximation.

Definition 7. *Let* U *be finite, then* α_S *is called the accuracy of the approximation* (// *denotes the cardinality*),

$$\alpha_S = \frac{|\mathsf{l}(S)|}{|\mathsf{u}(S)|}$$

Naturally $0 \leq \alpha \leq 1$.

Definition 8. *A set S is crisp if $\alpha_S = 1$.*

In many real-world applications, information on objects are stored in datasets or databases. Datasets can be given by an information system, which is a pair $IS = (U, A)$, where U is a set of objects called the universe and A is a set of attributes. Let $a : U \rightarrow V_a$ be a function, where V_a denotes the domain of attribute a. An information system can be represented by a table, where each row contains data on an entity of the universe, and the columns represent the attributes. Any pair (x, a), where $x \in U$ and $a \in A$ in the table is a cell whose value is $a(x)$.

Table 1 shows a very simple information system containing 8 rows. Each of them represents a patient and each has 3 attributes: headache, body temperature, and muscle pain. The base sets contain patients with the same symptoms (which means they are indiscernible from each other) and it is the following:

$$\mathfrak{B} = \{\{x_1\}, \{x_2\}, \{x_3\}, \{x_4\}, \{x_5, x_7\}, \{x_6, x_8\}\}$$

Based on some background knowledge, let the patients x_1, x_2 and x_5 have the flu. Let S be the following set of these patients: $\{x_1, x_2, x_5\}$. The approximation of the set S is the following:

- $\mathsf{l}(S) = \{\{x_1\}, \{x_2\}\}$
- $\mathsf{u}(S) = \{\{x_1\}, \{x_2\}, \{x_5, x_7\}\}$

Here, $\mathsf{l}(S)$ relates to patients that surely have the flu. Patient x_5 is not in the lower approximation because there is one other patient, x_7, who is indiscernible from x_5, and there is no information about, whether x_7 has the flu or not. So the base set $\{x_5, x_7\}$ can only be in the upper approximation.

Table 1. Information system

Object	Headache	Body temp	Muscle pain
x_1	YES	Normal	NO
x_2	YES	High	YES
x_3	YES	Very high	YES
x_4	NO	Normal	NO
x_5	NO	High	YES
x_6	NO	Very high	YES
x_7	NO	High	YES
x_8	NO	Very High	YES

The indiscernibility modeled by an equivalence relation represents a sort of limit of our knowledge embedded in an information system (or background knowledge). Indiscernibility has an effect on the membership relation. In some situation, it makes our judgment of the membership relation uncertain – making the set vague – because a decision about a given object affects the decision about all the other objects which are indiscernible from the given object.

In practical applications not only the indiscernible objects must be handled in the same way but also those that are similar to each other based on some property. (Irrelevant differences for the purpose of the given applications should not be taken into account.) Over the years, many new approximation spaces have been created as the generalization of the original Pawlakian space [33]. The main difference between these kinds of approximation spaces (with a Pawlakian approximation pair) lies in the base sets (members of $\mathfrak{B}$). Only four main kinds of approximation spaces are mentioned in this dissertation: the original Pawlakian; covering generated by a tolerance relation; general covering; general (partial).

Pawlakian approximation spaces have been generalized using tolerance relations that are based on similarity. These relations are symmetric and reflexive but not necessarily transitive.

Covering-based approximation spaces generated by tolerance relations [49] generalize Pawlakian approximation spaces in the following points.

Definition 9. *A general approximation space is a covering-based approximation space generated by a tolerance relation if there is a tolerance relation $\mathcal{R}$ such that $\mathfrak{B} = \{[x] \mid x \in U\}$, where $[x] = \{y \mid y \in U, x\mathcal{R}y\}$.*

In these covering spaces, a base set contains objects that are similar to a distinguished member. This means that the similarity to a given element generates the system of base sets.

General covering approximation spaces [58, 59] are not necessarily based a tolerance.

Definition 10. *A general approximation space is a general covering approximation space if $\bigcup \mathfrak{B} = U$.*

In these covering spaces a property generates the system of base sets.

In the case of general (partial) approximation spaces [15] the last requirement is also given up: any family $\mathfrak{B}$ of nonempty subsets of U can be a set of base sets. In these spaces also some property generates the system of base sets, but there is also a lack of information due to partiality.

Why is the system of base sets important from the theoretical point of view? Because it represents a sort of limit of our knowledge embedded in an information system. Sometimes it makes the judgment of the membership relation uncertain – thus making the set vague – because a decision about a given object affects the decision about all the other objects that are in the same base set.

The main source of uncertainty is in our background knowledge. Let S be a subset of U, and $x, y \in U$. What can be said about y with respect to x?

1. In an original Pawlakian space:
   - if $x \in I(S)$ (i.e. x is surely a member of S), then $y \in S$ for all $y, x\mathcal{R}y$;
   - if $x \in u(S) \setminus I(S)$ (i.e. x is possibly a member of S), then y may be a member of S for all $y, x\mathcal{R}y$ (it means that there are y_1, y_2 such that $x\mathcal{R}y_1, y_1 \in S$, and $x\mathcal{R}y_2, y_2 \notin S$);

– if $x \in \mathsf{I}(\overline{S})(= U \setminus \mathsf{u}(S))$ (i.e. x is surely not a member of S), then $y \notin S$ for all $y, x\mathcal{R}y$.

This means that if an object x is in the lower approximation of some set S, then all the objects that are indiscernible from x are in S as well. If an object x is in the boundary region of S, then there is at least one object which is indiscernible from x and a member of S and also there is at least one object which is indiscernible from x and not a member of S. If an object x is in the negative region of S, then none the objects that are indiscernible from x are in S.

2. In a covering space generated by a tolerance relation $\mathcal{R}$:
   – if $x \in \mathsf{I}(S)$ (i.e. x is surely a member of S), then $y \in S$ for all $y, y \in [x']$ where $x' \in [x]$ and $[x'] \in \{B \mid B \in \mathfrak{B}$ and $B \subseteq S\}$;
   – if $x \in \bigcup(\{B \mid B \in \mathfrak{B}$ and $B \cap S \neq \emptyset\} \setminus \{B \mid B \in \mathfrak{B}$ and $B \subseteq S\})$ (i.e. x is possibly a member of S), then there is an x' and a base set $[x']$ such that $x \in [x']$, $[x'] \cap S \neq \emptyset$, $[x'] \not\subseteq S$ and y may be a member of S for all $y \in [x']$;
   – if $x \in \mathsf{I}(\overline{S})(= U \setminus \mathsf{u}(S))$ (i.e. x is surely not a member of S), then $y \notin S$ for all $y, x\mathcal{R}y$.

3. In a general covering space:
   – if $x \in \mathsf{I}(S)$ (i.e. x is surely a member of S), then there is a base set B, such that $x \in B$ and $B \in \{B \mid B \in \mathfrak{B}$ and $B \subseteq S\}$)therefore $y \in S$ for all $y \in B$;
   – if $x \in \bigcup(\{B \mid B \in \mathfrak{B}$ and $B \cap S \neq \emptyset\} \setminus \{B \mid B \in \mathfrak{B}$ and $B \subseteq S\})$ (i.e. x is possibly a member of S), then there is a base set B such that $x \in B$, $B \cap S \neq \emptyset$ and $B \not\subseteq S$ therefore y may be a member of the set S for all $y \in B$;
   – if $x \in \mathsf{I}(\overline{S})(= U \setminus \mathsf{u}(S))$ (i.e. x is surely not a member of S), then there is a base set B such that $B \cap S = \emptyset$ therefore $y \notin S$ for all $y \in B$.

4. In a general partial space:
   – if $x \in \mathsf{I}(S)$ (i.e. x is surely a member of S), then there is a base set B, such that $x \in B$ and $B \in \{B \mid B \in \mathfrak{B}$ and $B \subseteq S\}$)therefore $y \in S$ for all $y \in B$;
   – if $x \in \bigcup(\{B \mid B \in \mathfrak{B}$ and $B \cap S \neq \emptyset\} \setminus \{B \mid B \in \mathfrak{B}$ and $B \subseteq S\})$ (i.e. x is possibly a member of S), then there is a base set B such that $x \in B$, $B \cap S \neq \emptyset$ and $B \not\subseteq S$ therefore y may be a member of the set S for all $y \in B$;
   – if $x \in \mathsf{I}(\overline{S})$ (i.e. x is surely not a member of S), then there is a base set B such that $B \cap S = \emptyset$ therefore $y \notin S$ for all $y \in B$;
   – otherwise nothing is known about x (i.e. there is no base set B such that $x \in B$), therefore nothing can be said about y with respect to x.

In the dissertation, there is a strict distinction between the general covering spaces and the covering spaces generated by a tolerance relation. This distinction is based on the observation that, although spaces generated by tolerance relations are covering ones, not all covering space can be generated by some tolerance relation. The main reason for this is the symmetric property of the tolerance relation and that in spaces generated by the tolerance relation each base set has at least one generator object. If $\mathcal{R}$ is a tolerance relation on U, then the system of base sets is $\mathfrak{B} = \{[x] \mid x \in U\}$, where $[x] = \{y \mid y \in U, x\mathcal{R}y\}$ (x is the generator object of $[x]$). In support of this statement, the followings highlight the difference between the general and the tolerance-based covering spaces.

1. At first, it is shown that in some cases, a tolerance relation generated by a covering does not necessarily generate the same base sets. Let $\mathfrak{B}$ a general covering. The tolerance relation $R^\star$ generated by the $\mathfrak{B}$ is as follows: $xR^\star y$ if there is $B \in \mathfrak{B}$ such that $x, y \in B$.

   (a) *Base set-loss*: Suppose that $\mathfrak{B}$ has B_1, B_2 members ($B_1, B_2 \subseteq U$) such that $B_2 \subseteq B_1$. In this case, the base set B_2 is not created by the tolerance relation $R^\star$, since every member of B_2 generates at least B_1. This makes the approximation of certain sets less efficient. Generally speaking, a tolerance relation $R^\star$ (generated by a covering) cannot generate base sets, that are part of the general covering, where $B_2 \subseteq B_1$. On the left side of Fig. 1 some base sets of a general covering space can be seen. On the right, some members of the base sets generated by the covering based on $R^\star$ can be seen. (For the sake of simplicity, only a few base sets are shown in both figures, so the union of the base set does not give out U). In the case of the Pawlakian definition of the lower approximation of the set S, the space generated by the tolerance relation $R^\star$ contains members of B_3 but does not contain members of B_2, and the general covering space contains both the members of B_2 and B_3. This happens because B_2 does not appear at all among the base sets generated by the tolerance relation $R^\star$. So in this case, the general covering space gave a finer approximation.

   (b) *Base set-gain*: Let B_1 and B_2 be some members of the general covering $\mathfrak{B}$ such that their intersection is not empty. Such a pair does not exist only if the base sets are pairwise disjoint. In this case, however, the $R^\star$ relation (generated by the general covering space) becomes an equivalence relation and it generates the classical Pawlakian space. Let us suppose that $B_1 \cup B_2 \notin \mathfrak{B}$ and that $B_1 \cup B_2$ does not have a common member with the other members of $\mathfrak{B}$. Let $x \in B_1 \cap B_2$. Based on the $R^\star$ tolerance relation, the set of objects similar to x is $B_1 \cup B_2$. Therefore, $B_1 \cup B_2$ is a member of the system of base set generated by $R^\star$. The appearance of this base set (in the case of the Pawlakian approximation pair) significantly modifies the upper approximation, consequently, it degrades the efficiency of the approximation. On the left side of Fig. 2 some base sets of a general covering space can be seen. On the right, some members of the base sets generated by the covering based on $R^\star$ can be seen. (For the sake of simplicity, only a few base sets are shown in both figures, so the union of the base set does not give out U). If one wants to define the upper approximation of the set S (using a Pawlakian type approximation), then B_2 will not be a subset of the upper approximation in the case of the general covering. However, it will be a subset of the system of base sets generated by the $R^\star$ tolerance relation because $B_1 \cup B_2$ appears as a base set. So, the upper approximation became larger in the case of the space generated by R^* which increases the uncertainty relative to the set S.

2. Secondly, it is shown that there is a general covering space for which there is no tolerance relation that generates the same base sets that are the members of the general covering. Indirectly, suppose that $\mathfrak{B}$ is a general covering space and R is a tolerance relation that generates exactly the member of $\mathfrak{B}$. Let U be the universe of the general covering, $x \in U$, $\{x\} \in \mathfrak{B}$, $B_1, B_2 \in \mathfrak{B}$ be two base sets which are different from the base set $\{x\}$ such that $\{x\} = B_1 \cap B_2$. If B_1, B_2 are generated by the tolerance relation R, then there must be some $y, z \in U$ such that $B_1 = \{u \mid u \in U, yRu\}$, and $B_2 = \{u \mid u \in U, zRu\}$ (y, z are the generator objects of B_1 and B_2 respectively). As B_1 and B_2 are different from the base set $\{x\}$, $x \neq y$ and $x \neq z$. The base set $\{x\}$ is a singleton, therefore its generator object must be x, so $[x] = \{u \mid u \in U, xRu\} = \{x\}$. yRx and zRx hold because $x \in B_1$ and $x \in B_2$. Due to the similarity of the tolerance relation R, xRy and xRz also hold. This means that $y, z \in [x] (= \{x\})$ which is a contradiction.

Boundary regions are also essential in the representation of uncertainty. In [13] the authors showed that theoretically different boundary regions can be introduced into a general partial approximation space $\langle U, \mathfrak{B}, \mathfrak{D}_\mathfrak{B}, \mathsf{l}, \mathsf{u} \rangle$:

1. $\mathsf{b}_1(S) = \mathsf{u}(S) \setminus \mathsf{l}(S)$;
2. $\mathsf{b}_2(S) = \bigcup(\{B \mid B \in \mathfrak{B} \text{ and } B \cap S \neq \emptyset\} \setminus \{B \mid B \in \mathfrak{B} \text{ and } B \subseteq S\})$;
3. $\mathsf{b}_3(S) = \bigcup \mathcal{C}^b(S)$, where $\mathcal{C}^b(S) = \{B \mid B \in \mathfrak{B}, B \cap S \neq \emptyset, \text{ and } B \not\subseteq S\}$.

In original Pawlakian spaces there is no difference between the aforementioned boundary regions, i.e. if $\langle U, \mathfrak{B}, \mathfrak{D}_\mathfrak{B}, \mathsf{l}, \mathsf{u} \rangle$ is an original Pawlakian space characterized by an ordered pair $\langle U, \mathcal{R} \rangle$, then $\mathsf{b}_1(S) = \mathsf{b}_2(S) = \mathsf{b}_3(S)$ for all $S \subseteq U$. In general case, the boundary regions defined according to the first point are not definable sets necessarily, therefore this definition cannot be used in general approximations spaces where one wants to rely on only definable sets. If there are only finite number of base sets (i.e. $\mathfrak{B}$ is finite), then the sets $\mathsf{b}_2(S), \mathsf{b}_3(S)$ are definable for all $S \subseteq U$. Some important connections between different types of boundary regions were showed in [13, 14]:

- $\mathsf{b}_1(S) \subseteq \mathsf{b}_2(S) \subseteq \mathsf{u}(S)$;
- $\mathsf{b}_1(S) = \mathsf{b}_2(S)$ if and only if $\mathsf{b}_2(S) \cap \mathsf{l}(S) = \emptyset$;
- if $\mathfrak{B}$ is one-layered (i.e. the base sets are pairwise disjoint), then there is no difference between different types of boundary regions, i.e.
  - $\mathsf{b}_1(S) = \mathsf{b}_2(S) = \mathsf{b}_3(S)$;
  - $\mathsf{b}_1(S)$ is definable;
  - $\mathsf{b}_i(S) \cap \mathsf{l}(S) = \emptyset$, where $i = 1, 2, 3$;
  - $\mathsf{u}(S) = \mathsf{l}(S) \cup \mathsf{b}_i(S)$, where $i = 1, 2, 3$.

Notice that only lower and upper approximations (and so only background and embedded knowledge represented by base sets) are used, and in a finite one-layered case there is no real difference between different types of boundary regions.

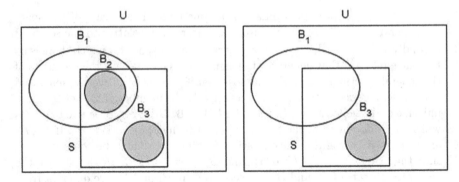

Fig. 1. Base set-loss

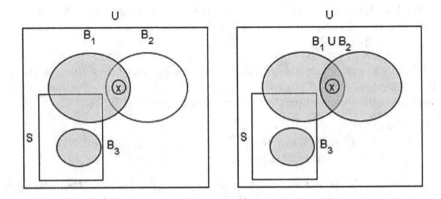

Fig. 2. Base set-gain

The next step is to make clear the 'nature', the usage and the influences of background (and embedded) knowledge.

1. In the original Pawlakian case, the limit of our knowledge appears explicitly: base sets consist of indiscernible objects, there is no way to distinguish them from each other.

2. In covering structures generated by tolerance relations, a base set contains objects which are similar to a given object, and therefore they are treated in the same way. Being similar to a given object is a property, but it is a very special (not a general) one, it is generated by the tolerance relation.

3. In general covering spaces, base sets can be considered as the representations of real properties, and it is supposed that all objects have at least one (known, represented) property. Objects with the same property (members of a base set) are handled in the same way. (The system of base sets cannot be generated by tolerance relations in some cases.)

4. General partial spaces are similar to general covering ones, but it is not supposed that all objects have at least one property represented by a base set. In

practical cases information systems are not total, there is no relevant information about an object: it may be in our database but some information is missing, and so it does not have any property represented by a base set.

Some problems appear in different cases. In practical applications indiscernibility (as an equivalence relation) may be too strong. In the case of a huge number of objects if there is a reflexive and symmetric relation, then it may be difficult to decide whether it is transitive. Covering spaces generated by tolerance relations give possibilities to use only reflexive and symmetric relations, but to many base sets appear, (each object generate a base set). These base sets are not about similarity (in general), but only about the similarity to given objects (to their generators). In general covering and partial spaces, there is no room for similarity, these spaces rely on only common properties of objects. A pairwise disjoint system of base sets generated from a covering space (relying on a tolerance relation or a family of properties) or a general partial space is not a real solution: it is difficult to give any meaning represented by received base sets and too many small base sets appear, therefore the system may become very close to classical set theory. If every base set is a singleton (contains only one member), then the lower and upper approximation of any set will coincide. Therefore, every set will be crisp in this case.

As mentioned earlier, in practical applications indiscernibility can be too strong. Sometimes the similarity of objects is enough. From a mathematical point of view, it can be modeled by a tolerance relation. In computer science, correlation clustering a typical method that uses a tolerance relation. So a natural question arises: can the clusters (gained by the correlation clustering) represent the system of base sets? If so, then how do they modify the approximation space? In the next chapter, a new approximation space is presented that tries to answer the previous question.

3   Similarity-based Rough Sets[1]

Some covering approximation spaces use tolerance relations, which represent similarity, (described in Sect. 2) instead of equivalence relations, but the usage of these relations is very special. It emphasizes the similarity to a given object and not the similarity of objects 'in general'. One can recognize this feature when they try to understand the precise meaning of the answer coming from an approximation relying on a covering approximation space (based on a tolerance relation). If one is interested in whether $x \in S$ (where S is the set to be approximated), then there are three possible answers[2] (see Fig. 3):

- if $x \in I(S)$ (i.e. x is surely a member of S), then $y \in S$ for all $y, y \in [x']$ where $x' \in [x]$ and $[x'] \in \{B \mid B \in \mathfrak{B} \text{ and } B \subseteq S\}$;

---

[1] The work described in this chapter was based on [5–7,21,35–38].

[2] The formal definition of these answers can also be seen in the first chapter but for better readability, they are also listed here again focusing on covering approximation spaces based on a tolerance relation.

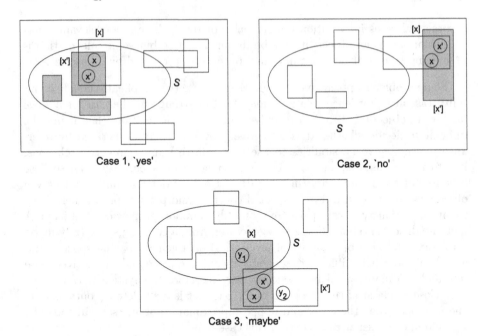

Case 1, 'yes'   Case 2, 'no'

Case 3, 'maybe'

Fig. 3. Some base sets in covering (based on a tolerance relation) cases

– if $x \in \bigcup(\{B \mid B \in \mathfrak{B}$ and $B \cap S \neq \emptyset\} \setminus \{B \mid B \in \mathfrak{B}$ and $B \subseteq S\})$ (i.e. x is possibly a member of S), then there is an x' and a base set $[x']$ such that $x \in [x']$, $[x'] \cap S \neq \emptyset$, $[x'] \nsubseteq S$ and y may be a member of S for all $y \in [x']$;
– if $x \in \mathsf{I}(\overline{S})(= U \setminus \mathsf{u}(S))$ (i.e. x is surely not a member of S), then $y \notin S$ for all $y, x\mathcal{R}y$.

Some practical problems of covering approximation spaces based on a tolerance relation:

1. The former answers show, that generally the lower and upper approximations are not close in the following sense (see Fig. 4):
   (a) If $x \in \mathsf{I}(S)$, then it cannot be said that $[x] \subseteq S$.
   (b) If $x \in \mathsf{u}(S)$, then it cannot be said that $[y] \cap S \neq \emptyset$ for all $y \in [x]$.
2. The number of base sets is not more than the number of members of U, so there are too many base sets for practical applications.

If one wants to avoid these problems, then a Pawlakian approximation space can be generated by constructing a system of disjoint base sets [13] (see Fig. 5). If there are two base sets B_1, B_2, such that $B_1 \cap B_2 \neq \emptyset$, then they can be substituted with the following three sets: $B_1 \setminus B_2$, $B_2 \setminus B_1$, $B_1 \cap B_2$. Applying this iteratively the reduction can be got. Although, it is not a real solution from the practical point of view. The base sets can become too small for practical applications. The smaller the base sets are, the closer the system gets to the classical set theory. (If all base sets are singleton, then there is no difference between the approximation space and classical set theory)

In rough set theory, the members of a given base set share some common properties.

- In Pawlak's original space all members of a given base set have the same attributes (i.e. they have the same properties with respect to the represented knowledge).
- In covering approximation spaces based on a tolerance relation, all members of a given base set are similar to a distinguished object (which is used to generate the given base set).

A further generalization is possible: general (partial) Pawlakian approximation spaces can be obtained by the generalization of the set of base sets:

- let $\mathfrak{B}$ be an arbitrary nonempty set of nonempty subsets of U.

These spaces are Pawlakian in the sense that they use Pawlakian definition of definable sets and approximation pairs. This generalization is very useful because a base set can be taken as a collection of objects with a given property, and one can use very different properties to define different base sets. The members of the base set can be handled in the same way relying on their common property. In this case there is no way to give a corresponding relation which can generate base sets (similarly to covering approximation spaces), so a general (partial) Pawlakian approximation space can be characterized only by the pair $\langle U, \mathfrak{B} \rangle$, since the lower and upper approximations of a subset of U are determined by the members of $\mathfrak{B}$. However, any system of base sets induces a tolerance relation $\mathcal{R}$ on U: $x\mathcal{R}y$ if there is a base set $B \in \mathfrak{B}$ such that $x, y \in B$. If this relation is used to get the system of base sets, the result can be totally different from our original base system (see Fig. 6).

In Fig. 6 x is in the intersection of B_1 and B_2 (B_1 and B_2 are defined by some properties). It means that it has common properties with all y_i and z_i, where $i = 1, 2, 3$. So if some $x \in B_1 \cap B_2$ it means that:

- $x\mathcal{R}y$ for all $y \in B_1$
- $x\mathcal{R}z$ for all $z \in B_2$

Therefore the base set generated by x is the following: $[x] = B_1 \cup B_2$ (In this example only two base sets were used, but it is the same when there are more.)

When one would like to define the base sets, then the background knowledge embedded in a given information system is used. In the case of a Pawlakian space, two objects are called indiscernible if all of their known attribute values are the same. In many cases covering spaces rely on a tolerance relation based similarity. As it was mentioned earlier, some problems can come up using these covering spaces. A base set contains members that are similar to a distinguished member. This means that these spaces do not consider the similarity itself but the similarity with respect to a distinguished object. If correlation clustering

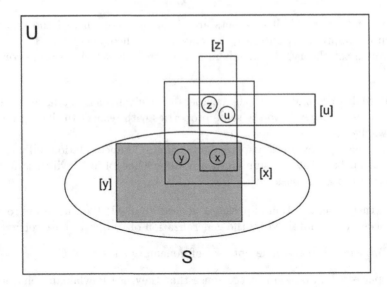

Fig. 4. In covering (based on a tolerance relation) the lower and upper approximations are not closed

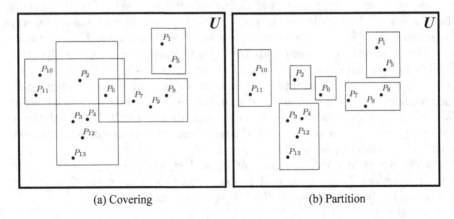

(a) Covering

(b) Partition

Fig. 5. Covering (based on a tolerance relation) and its reduction to a partition

(described in Sect. 4) is used, based on the tolerance relation, a partition of the universe is obtained. The clusters contain objects which are mostly similar to each other (not just to a distinguished member). So the partition can be understood as a system of base sets. As a result, a new approximation space appears which uses the same tolerance relation as the aforementioned covering spaces and has the following features:

- the similarity of objects relying on their properties (and not the similarity to a distinguished object) plays a crucial role in the definition of base sets;

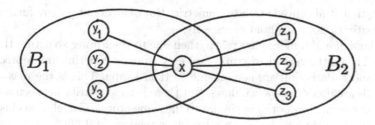

Fig. 6. Base sets by properties of objects

- the system of base sets consists of disjoint sets, so the lower and upper approximation are closed;
- only the necessary number of base sets appears (in applications, an acceptable number of base sets must be used);
- the size of base sets is not too small, or too big.

The formal definition can be seen here:

Definition 11.

$$\mathfrak{B} = \{B \mid B \subseteq U, \ and \ x, y \in B \ if \ p(x) = p(y)\},$$

where p is the partition gained from the correlation clustering.

Singleton clusters sets represent very little information (its member is only similar to itself). Without increasing the number of conflicts, its member cannot be considered similar to any object. So, they always require an individual decision. By deleting the singletons, a partial system of base sets can be defined. In Sect. 7 a method is proposed

4 Correlation Clustering

Cluster analysis is a widely used technique in data mining. Our goal is to create groups in which objects are more similar to each other than to those in other groups. Usually, the similarity and dissimilarity are based on the attribute values describing the objects. Although there are some cases when the objects cannot be described by numbers, but something about their similarity or dissimilarity can still be stated. Think of humans for example. It is hard to detail someone's looks by a number, but one still makes statements whether two persons are similar to each other or not. Of course, these opinions are dependent on the persons. Some can treat two random persons as similar, while others treat them dissimilar. If one wants to formulate the similarity and dissimilarity by using mathematics, then a tolerance relation is needed. If this relation holds for two objects, then they are similar. If this relation does not hold, then they are dissimilar. Of course, each object is similar to itself, so the relation needs to be reflexive, and it is easy to show, that it also needs to be symmetric. But one cannot go any further, e.g. the transitivity does not hold necessarily.

If a human and a mouse are taken, then due to their inner structure they are similar. This is the reason why mice are used in drug experiments. Moreover, a human and a Paris doll are similar due to their shape. This is the reason why these dolls are used in show-windows. But there is no similarity between a mouse and a doll except that both are similar to the same object. Correlation clustering is a clustering technique based on a tolerance relation [8,9,60].

Our task is to find an R $\subseteq U \times U$ equivalence relation *closest* to the tolerance relation. A (partial) tolerance relation $\mathcal{R}$ [32,49] can be represented by a matrix M. Let matrix $M = (m_{ij})$ be the matrix of the partial relation $\mathcal{R}$ of similarity: $m_{ij} = 1$ whenever objects i and j are similar, $m_{ij} = -1$ whenever objects i and j are dissimilar, and $m_{ij} = 0$ otherwise.

A relation is partial if there exist two elements (i, j) such that $m_{ij} = 0$. It means that if there is an arbitrary relation R $\subseteq U \times U$, then there are two sets of pairs. Let R$_{\text{true}}$ be the set of those pairs of elements for which the R holds, and R$_{\text{false}}$ be the one for which R does not hold. If R is partial then R$_{\text{true}} \cup$ R$_{\text{false}} \subseteq U \times U$. If R is total then R$_{\text{true}} \cup$ R$_{\text{false}} = U \times U$.

A partition of a set S is a function $p : S \to \mathbb{N}$. Objects $x, y \in S$ are in the same cluster at partitioning p, if $p(x) = p(y)$.

The cost function counts the negative cases i.e. it gives the number of cases whenever two dissimilar objects are in the same cluster, or two similar objects are in different clusters. The cost function of a partition p and a relation R_M with matrix M is

$$f(p, M) = \frac{1}{2} \sum_{i<j} (m_{ij} + abs(m_{ij})) - \sum_{i<j} \delta_{p(i)p(j)} m_{ij},$$

where δ is the Knockecker delta symbol [39]. For a fixed relation, the partition with the minimal cost function value is called optimal. Solving a correlation clustering problem is equivalent to minimizing its cost function, for the fixed relation. If the value of this optimal cost function is 0, the partition is called perfect. Given the $\mathcal{R}$ and R the value f is called the distance of the two relations. The partition given this way generates an equivalence relation. This relation can be considered as the closest to the tolerance relation.

It is easy to check that the solution cannot be generally perfect for a tolerance relation (based on similarity). Consider the simplest such case, given three objects x, y and z, and x is similar to both y and z, but y and z are dissimilar. In this situation, the following 5 partitions can be given:

$$\{\{x, y, z\}, \{\{x, y\}, \{z\}\}, \{\{x, z\}, \{y\}\}, \{\{y, z\}, \{x\}\}, \{\{x\}, \{y\}, \{z\}\}\}.$$

In every of one them there is at least 1 conflict.

The number of partitions can be given by the Bell number [1], which grows exponentially. Hence, in general—even in the case of some dozens of objects—the optimal partition cannot be determined in a reasonable time, thus a search algorithm that produces a quasi-optimal partition would be more useful in practical cases. However, in practical examples, it gives us the right to handle objects, which are in the same class, the same way.

5 Comparison Between Covering Based on a Tolerance Relation and Similarity-based Rough Sets

As mentioned in the previous subsection, in these covering based spaces there can be too many base sets. Naturally, this can be very problematic if the dataset is huge because the approximation process can take a lot of time. Therefore, in practice, these spaces can be a little bit slow. So a natural question arises. How fast can the approximation be with similarity-based rough sets? In [36] a comparison between covering (based on a tolerance relation) spaces (and its disjoint variant) and similarity-based rough sets can be seen.

The execution time of the approximation process can be seen in Fig. 7. The axis x represents the number of points, and the axis y represents the execution time in milliseconds. The points here were chosen randomly.

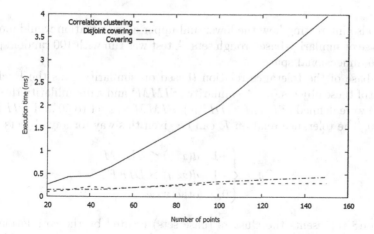

Fig. 7. Execution time

If one takes a look at the figure, then it can be seen that the approximation by covering (based on a tolerance relation) is the slowest. This was expected because there are a lot of base sets to work with. Between the disjoint covering and the correlation clustering, there is no significant difference. Nevertheless, as the number of points increases, the correlation clustering gives the fastest way to approximate. It is an interesting fact that there is such a great difference between the covering (based on a tolerance relation) and its disjoint variant. Despite the fact that a disjoint covering has the largest number of base sets, their cardinality is much less (most of them are singleton) than that of covering based on a tolerance relation.

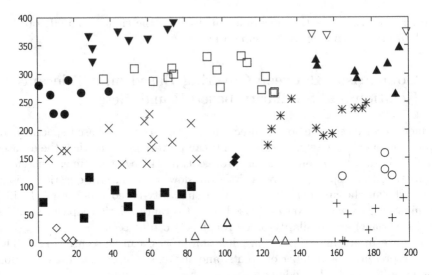

Fig. 8. The clusters

It is also interesting how the lower and upper approximation would look like in the case of similarity-based rough sets. A test was run with 100 random points on a two-dimensional space.

The base of the tolerance relation (based on similarity) was the Euclidean distance of these objects (d). A similarity ($SIMM$) and a dissimilarity threshold ($DIFF$) were defined ($SIMM < DIFF$). $SIMM$ was set to 50 and $DIFF$ was set to 90. The tolerance relation $\mathcal{R}$ can be given this way for any objects x, y:

$$x\mathcal{R}y = \begin{cases} +1 & d(x,y) \leq SIMM \\ -1 & d(x,y) > DIFF \\ 0 & \text{otherwise} \end{cases} \tag{1}$$

Figure 8 represents the clusters (base sets) created by the correlation clustering. The set to be approximated is shown in Fig. 9. The members of this set are denoted by the × symbols, and the other members are denoted by the cross symbol. The members were chosen randomly.

The approximation generated by the correlation clustering is displayed in Fig. 10/A. The cardinality of the base sets is relatively great so the lower approximation consists of only a few members. (Only the members denoted by the empty circle and filled diamond are in the set.)

The approximation generated by the covering (based on a tolerance relation) is shown in Fig. 10/B. Like in correlation clustering the lower approximation consists of only a few members. The two lower approximations have some difference, but they only differ in a set which has two members.

Between the upper approximations, a significant difference can be seen. The upper approximation defined by covering (based on a tolerance relation) contains many more objects, almost twice as much as the one defined by correlation clustering.

The approximation generated by the disjoint covering is shown in Fig. 10/C. It can be seen that among the methods this generated the finest approximation (lower and upper approximation coincide). The reason is that almost all base sets are singletons. As mentioned before, if there were only singleton clusters, one would get the common set theory back.

6   Search Algorithms[3]

In a reasonable time, correlation clustering can only be solved by using search/optimization algorithms due to its NP-hard property. However, each algorithm can provide different clusters. So the system of base sets can also be different. It is a natural question to ask how the search algorithms can affect the structure of the base sets. As the approximation based on correlation clustering is a completely new way of approximating sets, so it is crucial to use the best possible search algorithm. In this section, some widely-known search algorithms are shown and also a comparison of how they work in the case of similarity-based rough sets [38].

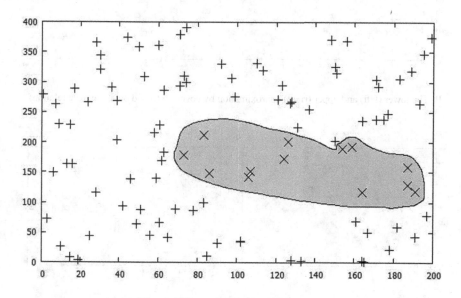

Fig. 9. The set to be approximated

The following list shows the used algorithms in the experiments. Each of them can be downloaded from [3] and their description can be seen in Appendix A:

---

[3] The work described in this section was based on [38].

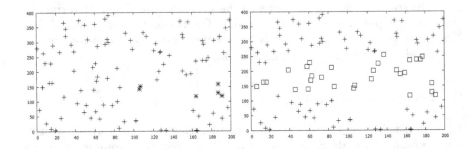

A. The lower (left) and upper (right) approximation by correlation clustering

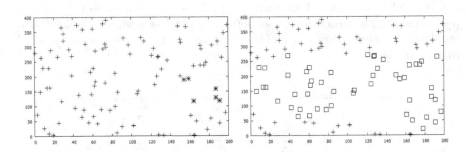

B. The lower (left) and upper (right) approximation by covering (based on a tolerance relation)

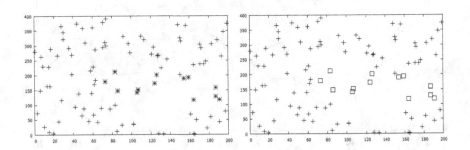

C. The lower (left) and upper (right) approximation by disjoint covering

Fig. 10. The outputs of the approximations by the software

- Hill Climbing Algorithm
- Stochastic Hill Climbing Algorithm
- Tabu Search
- Simulated Annealing
- Parallel Tempering
- Genetic Algorithm
- Bees Algorithm
- Particle Swarm Optimization
- Firefly Algorithm

To compare the algorithms, the following values were computed:

- Number of singleton clusters
- Standard deviation of the base set sizes
- The range of the base set sizes
- Execution time of the algorithm

Here, the cardinality of a base set is referred to as its size. As previously mentioned, singleton clusters mean little information. The greater their number is, the more unclear the information based on our knowledge becomes. For a search algorithm, the most optimal is, if it provides the least number of these clusters to have a precise result of the system.

The sizes of the base sets are also worth to be checked. For set approximations, it is more suitable if the sizes do not vary much. So the standard deviation of the base set sizes should be minimized as well as the range of the base set sizes.

An important parameter is the execution time of the search algorithms. It is especially crucial when there are a huge number of objects.

6.1 Results

Similarity-based rough sets always requires a tolerance relation which represents similarity. In the experiments, random graphs were only used to define these tolerance relations.

Most of the algorithms have many parameters, and changing them can result in different outputs. Many possible combinations were tried during the research. Dozens of tables were generated and these tables are not present in this dissertation, but they can be downloaded from the following link: https://bit.ly/2sO4UoD.

For comparing the parameters, the same graph (with 100 points, p = 0.6) were used for each algorithm. Each algorithm was run three times to exclude the randomness. In each case, the optimal parameter combination was the one that minimized the above mentioned 4 values. If the differences between the standard deviations, the ranges and the numbers of singletons were not significant, then the judgment was made by the execution time.

In this part of the experiments, Erdős-Rényi graphs were used [22,23]. This random graph generating method is very simple. Its pseudo-code can be seen in

---

Algorithm 1. Erdős-Rényi random graph generating method

1: **procedure** GENERATE(p)
2:     **for** $i = 1, \ldots, N$ **do**
3:         **for** $j = 1, \ldots, N$ **do**
4:             Generate a random x value between 0 and 1
5:             **if** $x < p$ **then**
6:                 There is an edge between objects i and j
7:             **end if**
8:         **end for**
9:     **end for**
10: **end procedure**

Algorithm 1. In the experiments, the following values were used $p = 0.5$, $p = 0.6$ and $p = 0.7$. Half of the generated edges denoted the similarity between the two objects and half of them the difference. Naturally, any other kind of graph can be used. 100, 200, 300 and 400 points were generated. For each test case, each algorithm was run 3 times on the same graphs, then the averages of the values, described in Sect. 6, were determined. The results can be seen in the following tables and can be downloaded from the following link: https://bit.ly/2sOqMA6.

From Table 2 the average standard deviations of the base set sizes can be read. Even for a small number of points, the differences are apparent. The simulated annealing provided the best result in most cases. Its parallel version has almost the same output. The bees algorithm also returned a rather acceptable result.

Table 2. Average standard deviations of the base set sizes for Erdős-Rényi graphs

Points	HC	SHC	TABU	SA	PT	GE	BEE	PSO	FF
200 q = 0.5	37	31	32	13	12	40	15	34	27
400 q = 0.5	63	61	51	15	24	82	21	66	49
200 q = 0.6	37	35	37	18	19	39	15	35	31
400 q = 0.6	69	63	61	21	23	83	26	65	49
200 q = 0.7	37	31	26	18	20	38	15	39	30
400 q = 0.7	68	61	66	14	19	78	18	63	55

In Table 3 the distance of the sizes of the biggest and the smallest base sets can be seen. The values show almost the same tendency as in the last table. The simulated annealing, the parallel tempering, and the bees algorithm proved to be the most acceptable. For 400 points, the other algorithms provided twice or three times as large values as the other 3 which is not suitable.

In Table 4 the average numbers of singletons are listed. Here, the differences are not so significant as before. The number of points does not affect it very much.

Table 3. Average ranges of the base set sizes for Erdős-Rényi graphs

Points	HC	SHC	TABU	SA	PT	GE	BEE	PSO	FF
200 q = 0.5	105	92	80	48	45	98	49	92	82
400 q = 0.5	176	189	144	57	78	213	65	186	193
200 q = 0.6	105	94	97	46	46	98	47	89	92
400 q = 0.6	192	173	178	87	62	209	81	193	197
200 q = 0.7	104	94	74	52	64	104	49	104	88
400 q = 0.7	198	179	191	50	53	228	81	186	210

Table 4. Average numbers of singletons for Erdős-Rényi graphs

Points	HC	SHC	TABU	SA	PT	GE	BEE	PSO	FF
200 q = 0.5	0	0	0	1	0	1	0	1	1
400 q = 0.5	0	0	0	2	2	2	0	2	7
200 q = 0.6	1	1	1	1	1	1	0	1	2
400 q = 0.6	0	0	0	3	3	1	1	1	3
200 q = 0.7	1	0	0	1	1	1	0	1	2
400 q = 0.7	0	1	0	2	3	1	1	1	4

Table 5. Average execution time for Erdős-Rényi graphs

Points	HC	SHC	TABU	SA	PT	GE	BEE	PSO	FF
200 q = 0.5	5	7	163	3	16	82	53	569	274
400 q = 0.5	142	181	1549	6	39	360	247	7971	1167
200 q = 0.6	4	6	170	3	17	91	66	734	317
400 q = 0.6	156	248	1665	6	40	457	274	8611	1287
200 q = 0.7	3	8	156	3	17	87	59	818	283
400 q = 0.7	138	256	1652	7	43	404	284	9878	1336

In Table 5 the average run-time of the algorithms can be seen in seconds. The values here vary the most. The simulated annealing was the least affected by the increasing number of points. For 200 points, the hill climbing algorithm and its stochastic version provided the fastest output. Although, as soon as the number of points was increased, they could not compete with the simulated annealing. For 400 points, the simulated annealing was more than 20–35 times faster than the other two. The particle swarm optimization was the slowest of all the algorithms. For a huge number of points, it is pointless to be used.

In this part of our experiments, random two-dimensional points were generated. The tolerance relation (based on similarity) is defined in equation 1. 100, 150, 200, 300, 500 points were generated and each algorithm was run 3 times for each point set and calculated the averages of the values described in Sect. 6. In the following tables, the results can be seen.

In Table 6 the average standard deviations of the base set sizes can be seen. In the case of a small number of points, the difference was not so considerable, but it became larger as the number of points was increased. In every case, the simulated annealing provided the most acceptable result. Interestingly, the parallel tempering fell short against the simulated annealing for a small number of points. However, in the 500 points test case the difference was negligible. Local search algorithms (hill climbing, its variant, tabu search) were rather good for a small number of points. In almost every situation, the firefly algorithm, genetic algorithm and particle swarm optimization provided the worst result.

Table 6. Average standard deviations of the base set sizes

Points	HC	SHC	TABU	SA	PT	GE	BEE	PSO	FF
100	7.6	8.1	7.5	7.1	7.6	8.9	7.6	7.7	7.6
150	11.8	10.2	10.4	7	11	14.5	8.7	12.2	11.5
200	13.1	16.2	12.8	10.3	13.2	17.8	12.7	13.8	13.3
300	25.7	28.9	27.2	17.1	18.4	31.4	23	29.1	28.9
500	46.6	34.4	39.4	26.5	27	51.4	34.2	47.8	48.4

Table 7 shows the ranges of the base set sizes. The outcome was quite similar as in the previous table. For a small number of points, the difference was not so high. For 500 points, it can be more noticeable. Like before, the firefly algorithm, genetic algorithm and particle swarm optimization ended up in the last places, and simulated annealing proved to be the most optimal.

Table 7. Average ranges of the base set sizes

Points	HC	SHC	TABU	SA	PT	GE	BEE	PSO	FF
100	23	24	22	21	26	23	23	26	23
150	27	25	35	20	31	37	21	37	33
200	38	38	38	30	41	47	37	40	35
300	67	68	70	48	61	74	64	71	68
500	111	94	99	73	81	119	103	117	116

Table 8 shows how many singleton clusters appeared. The results were quite the same in all cases. It is interesting that most of the algorithms were not affected by the increasing number of points. In the 500 points test case, some differences can be observed. In this case, the simulated annealing and its parallel version provided a rather inadequate result compared to the others. However, this difference was not so high.

Table 8. Average numbers of singletons

Points	HC	SHC	TABU	SA	PT	GE	BEE	PSO	FF
100	1	1	0	0	0	0	0	0	1
150	1	1	2	0	1	2	0	2	2
200	1	1	0	0	0	1	0	1	2
300	0	1	1	2	1	0	0	1	3
500	2	1	0	5	4	1	3	1	5

In Table 9 the average execution time is listed in seconds. As expected, these values were the most dependent on the number of points. For less than 200 points, the hill climbing algorithm and its stochastic variant provided the fastest run-time. However, after 200 points they could not compete with the simulated annealing which could find the quasi-optimal partition in less than 5 seconds for each test case. The parallel tempering also proved to be quite fast, but not as fast as its single-threaded variant. The other algorithms executed in an unreasonable time which is unacceptable for a great number of points. Especially the particle swarm optimization proved to be very slow, it finished running after 3.5 h for 500 points.

The experiments showed that the simulated annealing proved to be the best choice. In almost every test case, it provided the most suitable result. However, its most important property is that it was the least affected by the increasing number of points, so it can also finish in a reasonable time even for large amounts of points.

Table 9. Average execution time

Points	HC	SHC	TABU	SA	PT	GE	BEE	PSO	FF
100	0.2	0.3	15.6	1.1	5.4	15.7	15.5	32.5	54.1
150	0.5	0.4	43.5	0.7	7.3	14.5	11.9	130.8	58.3
200	6.4	9.5	172.4	3	16.8	92.7	66.3	564	281.1
300	40.1	43.5	647.4	4.7	25.2	207	161.1	1937.8	579.6
500	69.4	108.5	3550	3	50.8	207.4	135.9	13251	612.3

7 Similarity-based Rough Sets with Annotation[4]

Singleton clusters represent very little information because the system could not consider their members similar to any other objects without increasing the value of the cost function (see in Sect. 4). As they mean little information, they can be

---

[4] The work described in this section was based on [37].

left out. If the singleton clusters are not considered, then a partial system of base
sets can be generated from the partition. Sometimes it can happen that an object
does not belong to any cluster because the system could not consider it similar to
any other objects based on the background information. This does not mean that
this object is only similar to itself, but without proper information, the system
could not insert it into any cluster in order not to increase the number of conflicts.
In medical applications, it can occur that a patient has a similar disease as some
other patients but has different data in the information system. In this case, the
search algorithm would consider this patient different from the others and so the
patient would not belong to any non-singleton cluster. Although, a doctor or an
expert could recognize that the patient could belong to a non-singleton cluster.
The original partial space was defined by the correlation clustering. However, the
user has some background knowledge. They can use this knowledge to help the
system by inserting the members of some singletons into base sets (non-singleton
clusters). With the help of the annotation process, the user can put their own
knowledge into the system. It also decreases the partiality by decreasing the
number of singletons. After the annotation, a new approximation space appears.

Let S be the set to be approximated, $\{x\}$ a singleton gained from the correlation clustering and B a base set. The following cases can happen with the
base set B after the annotation if $B \subseteq l(S)$:

- If $x \in S$, then $B' = \{x\} \cup B$ and $B' \subseteq l(S)$ This way the approximation of
  the set S becomes more precise.
- If $x \notin S$, then $B' = \{x\} \cup B$ and $B' \subseteq u(S)$ but $B' \not\subseteq l(S)$ This increases the
  uncertainty relative to the set S.

The following cases can happen with the base set B after the annotation if
$B \subseteq u(S)$:

- If $x \in S$, then $B' = \{x\} \cup B$ and $B' \subseteq u(S)$
- If $x \notin S$, then $B' = \{x\} \cup B$ and $B' \subseteq u(S)$

The following cases can happen with the base set B after the annotation if
$B \subseteq u(S) \setminus l(S)$:

- If $x \in S$, then $B' = \{x\} \cup B$ and $B' \subseteq u(S) \setminus l(S)$
- If $x \notin S$, then $B' = \{x\} \cup B$ and $B' \subseteq u(S) \setminus l(S)$

In both cases, the upper approximation and the boundary region become
larger. It can be said that the annotation depends on the set to be approximated.
It could be useful if:

- $x \in S$, then the user could only choose from those B base sets which are in
  $l(S)$.
- $x \notin S$, then the user could only choose from those B base sets which are in
  $l(\overline{u(S)})$, where $\overline{u(S)}$ denotes the complement of the upper approximation.

This relative annotation looks very promising.

The order of the annotation is also worth to be checked. If the members x_1, x_2 of 2 different singletons were to be inserted into the same base set B, then the following question needs to be answered. Is it still relevant to insert x_2 into B after putting x_1 into B?

- If the answer is yes, then the two members are interchangeable. This means that x_1, x_2 has some sort of similarity that was hidden in the tolerance relation (based on similarity).
- If the answer is no, then the two members are not interchangeable. This means that annotating x_1 makes it irrelevant to insert x_2 into B.

In a real-world application, it can happen that an attribute value of an object is missing. This means that it can be unknown, unassigned or inapplicable (i.e. maiden name of a male). Coping with these data is usually a hard task. In many cases, these values are often substituted. It is common to replace a missing value with the mean or the most frequent value. Typically this gives a rather good result in many situations. In early-stage diabetes, it is not unusual that only the blood sugar level is higher than the normal level. If this value is missing for a patient, then it should not be replaced by the mean because the mean can be the normal blood sugar level. After the substitution, this patient can be treated as a healthy one. This type of substitution does not consider the information of an object itself but the information of a collection of objects, therefore it can lead to a false conclusion. The following method is proposed to handle missing data. If an object has a missing attribute value, then it cannot be treated as similar to any other object, so this entity forms a cluster alone. As mentioned earlier, these clusters cannot be treated as base sets. However, with the annotation, the user has the possibility to decide whether an object with missing data is similar to other objects or not. The user has some background knowledge that can be used this way to cope with the missing values. In this case, the information of an object itself is considered.

7.1 Annotation with Random Points

In this subsection, a possible example of the annotation process is shown. In the following figures, 20 points can be seen. The tolerance relation (based on similarity) is defined in Eq. 1. The similarity threshold $SIMM$ was set to 50, and $DIFF$ was set to 90. On the left side of Fig. 11/A the clusters generated by the correlation clustering can be seen. The singleton clusters contain the objects denoted by the ◇ symbol, the ◯ symbol and the ▼ symbol. Some points were selected for approximation. The members of this set are denoted by the × symbols, and the other members are denoted by the star symbol. The members were chosen randomly. This set can be seen on the right side of Fig. 11/A. In Fig. 11/B the reader can see the lower and upper approximation defined by the base sets gained from clustering after leaving the singletons out. The members of two singletons were inserted into two different base sets. The singleton denoted

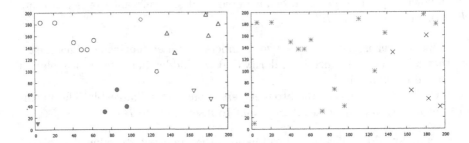

A. Clusters (left) and the set to be approximated (right)

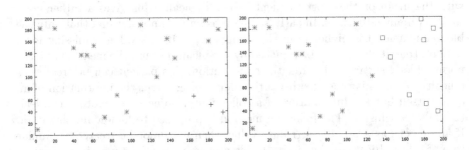

B. The lower (left) and upper (right) approximation by clustering

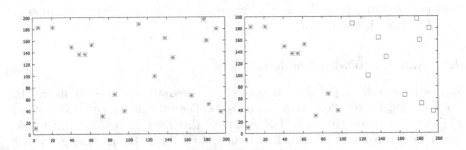

C. The lower (left) and upper (right) approximation by clustering with annotation

Fig. 11. Annotation with random points

by the ◇ symbol was merged with the base set denoted by the △ symbol. The base set denoted by the ▽ symbol was extended with the singleton denoted by the ◌ symbol. The result of the annotation can be seen in Fig. 11/C. None of the members of the chosen singletons were members of the set to be approximated. This is the reason why the lower approximation became the empty set, and the upper approximation had more members.

8 Tools of the Annotation[5]

Annotation is a very useful improvement of the similarity-based rough sets space because it creates a possible interaction between the system and the user. Naturally, this process is based on the user's expertise as they have to override the decision made by the system. Of course, this does not mean that there should not be any help or suggestions provided by the system itself. Two main techniques are introduced in the next part of the dissertation. The first one is a graphical method which tries to give a visual representation of the tolerance relation based on similarity. If two objects are close in this representation, then this indicates that those two objects should be treated as similar. If a member of a singleton is close to a base set, then maybe they should be merged. The second method is a mathematical way that aims to find those members in each cluster that can represent the entire cluster. During the computations, only the representative members should be considered. This way a lot of time and resources can be saved.

It is sometimes possible that there are more than one suitable base sets into which the user should insert the member of a singleton. In this case, the recommended base set should be the one whose representative member is the most similar to the member of the given singleton. In this way, there is no need to compare it to each member of each base set.

The annotation process can also be qualified as relevant or irrelevant regarding how it changes the representatives.

1. Relevant: After inserting a member of a singleton into a base set B, the representative member of the new base set B' is changed. In this case, some real information is implemented into the system. Let us assume that the objects are members of political parties and the representative members are the leaders of these parties. The annotation process is when a new member is elected to a party. If the annotation is relevant, then it means that the balance of the party has changed, and a new leader rises.
2. Irrelevant: After inserting a member of a singleton into a base set B, the representative member of the new base set B' is unchanged. In this case, the implemented information is not relevant because it does not alter the base sets gained from the correlation clustering.

---

[5] The work described in this section was based on [5–7, 35].

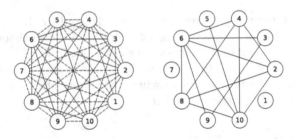

Fig. 12. Tolerance relation based on GCD.

In either case, the annotation can modify the set of possible representatives. As conclusion, one can say that if after the annotation something was changed, then the user had some useful information that was not embedded in the tolerance relation (based on similarity).

8.1 Visualization of Tolerance Relations[6]

Correlation clustering is based on similarity which can be represented by a tolerance relation. Visualizing these tolerance relations is sometimes problematic. In this chapter, an algorithm is presented [5] which can display similarity (based on a tolerance relation) in such a way that the user can easily interpret it. In the next paragraph, it can be seen why it is difficult to visualize these kinds of relations and why it is important to have a proper algorithm that can solve this issue. Let us suppose that two natural numbers are similar if their greatest common divisor is greater than 1. 1 is assumed to be similar to itself. And they are treated dissimilar if their greatest common divisor is 1. The analysis of this problem and its surprising solution can be found in [4].

If one takes this relation on numbers 1, 2, ..., 10, then the picture on the left in Fig. 12 can be constructed. Here the circles representing the numbers are positioned on a circle, to make it easy to connect the numbers. In this picture, the similarity is denoted by solid lines and the dissimilarity by dashed ones. Even though each number is similar to itself (except for 1), it is not displayed in the figure. This picture is transparent, but imagine a similar picture denoting the tolerance relation of numbers 1, 2, ..., 100! Why not leave out the dashed lines? The picture on right in Fig. 12 shows only the similarities. It is slightly more understandable, but it is hard to see the nexus. If the numbers are repositioned, everything becomes clearer. The numbers on the left of the circle have no similarities in the picture on the left of Fig. 13, and one can easily discover the groups 2-4-6-8-10, 3-6-9, 5-10. If the circle is disposed, the structure can become even clearer. The software yEd produced the picture in Fig. 13 on the right.

Numbers are easy to compare. However, it is not certain that 2 random objects are comparable. Or they are comparable, but nobody compared them yet. Hence, the relation can be partial. This means that there are three cases:

---

[6] The work described in this section was based on [5].

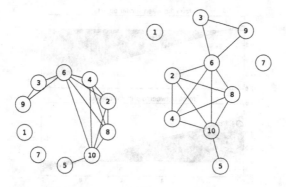

Fig. 13. Repositioned tolerance relation of GCD

two objects can be similar, dissimilar, or neutral. This can be visualized with three colors or three types of lines. If the lines for neutrality are not drawn, then the solid and dashed lines are enough.

In some graph visualization methods (force-directed graph) edges are modeled by springs, and the nodes are electrically charged particles. In these methods, the similarity (the edge between two nodes) is handled with springs, and the dissimilarity (the absence of the edge between two nodes) is handled with electricity. The graphs visualized with these methods are sparse graphs, i.e. the number of edges is a linear function of the number of nodes.

In this case, three different kinds of springs are used according to the three different values of the partial tolerance relation. With these three values, the graph of this relation is a dense one (all pairs of nodes are somehow connected), i.e. the number of edges is a quadratic function of the number of nodes.

A physical metaphor is used to give a quasi-optimal solution of correlation clustering, where similar objects attract each other and dissimilar objects repulse each other. The same is used here, but the objects are not grouped but arranged on the plane (or in space). The requirements are:

– similar objects get close, and
– dissimilar objects get far from each other.

Néda et al. presented a model using electric particles to solve the problem of correlation clustering in [40]. In this model, the particles could move on a circle based on the superposition of the forces acting on the particle.

Here, imaginary springs are used. Each node of the graph moves by the superposition of the forces of its springs. To simplify the problem in this section total tolerance relations are used, i.e. any pair of objects are comparable (similar or dissimilar). To get a suitable location for each of the objects, the following constraints are defined:

– it is not appropriate, if some object hides the others, so an optimal (minimal) distance ($SIMM$) is fixed for similar objects,

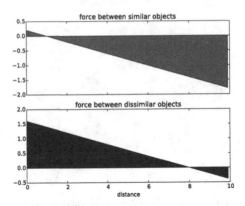

Fig. 14. Spring functions

– to get a finite picture, an optimal distance ($DIFF$) is determined for dissimilar objects.

It can be translated this for springs: there are short springs for similar objects and long springs for dissimilar objects. By Hooke's law, the force needed to extend or compress a spring by some distance d is proportional to that distance: $F = kd$.

Some placement of the objects can be treated as a network of these short and long springs. If two similar object (connected by a short spring) are closer than $SIMM$, then they repulse, and if they are farther than $SIMM$, then they attract each other. For the long springs, the same hold, but with $DIFF$ instead of $SIMM$, as Fig. 14 shows. Therefore, two functions are introduced: $f(\mathbf{d}) = (SIMM - |\mathbf{d}|)$ and $f'(\mathbf{d}) = (DIFF - |\mathbf{d}|)$, where $\mathbf{d}$ is the (distance) vector between two objects.

The periodic (sinusoidal) motion of a mass is used on a spring. One wants to get a location of objects and not a motion, so some attenuation is needed, a negative feedback. After some trial, the cube of the distance was found to be similar to Coulomb's rule in some sense. Finally, the resultant is the superposition of the forces:

$$\mathbf{F_i} = \sum_j \frac{f(\mathbf{d_{ij}})\mathbf{d_{ij}}}{|\mathbf{d_{ij}}|^3} + \sum_l \frac{f'(\mathbf{d_{il}})\mathbf{d_{il}}}{|\mathbf{d_{il}}|^3} \tag{2}$$

Here, the first part is summing for the objects similar, and the second part is for the objects dissimilar to object i.

Some relations are partial, where there are no constraints on the unrelated objects. It means that they can be at any distance from each other, so they can be positioned at the same place and partially or totally hide each other, or they can be very far from each other, so the picture can become very big. To solve these problems, a new spring function $\widehat{f}$ is introduced for the third type of springs (used for undefined values). This function is similar to the modified f', but the optimal distance is the interval $[SIMM, DIFF]$. If $d < SIMM$ then

$\widehat{f}(d) > 0$, and if $d > DIFF$, then $\widehat{f}(d) < 0$, i.e. for small distances it repulses and for big distances it attracts.

After presenting the algorithm and its background, it is time to show it in practice. The algorithm (by László Aszalós) can be downloaded from https://github.com/aszalosl/visualize_tolerance

The resulting image for some simple, but typical tolerance relation is shown. The first example is the snake, where adjacent objects are similar, and the others are dissimilar. One could think that the result is a straight line. Although the left side of Fig. 15 shows that this theory does not hold. If the dissimilar objects get too far from each other, then they attract, so an arch appears. In case of a non-defined relation for non-adjacent objects, a different image is obtained, because these objects do not repulse each other, so the snake can move in any direction, as right side of Fig. 15 shows.

Fig. 15. Total and partial snake.

Figure 16 shows some total trees. Here each non-leaf node has exactly three successors. In the first two pictures, the non-adjacent nodes are dissimilar, and in the last one, the non-adjacent nodes are unrelated. The tree on the left is totally symmetric (if the difference of the size of nodes is omitted according to one and two digits). But not every run gives such a nice picture. The middle one was generated with the same parameters as the left one, but the nodes 8 and 13 got to the wrong place at the beginning, and since this layout is stable, these nodes cannot escape. In the last picture the non-adjacent nodes, since unrelated, do not need to get far from the other nodes, e.g. nodes 3 and 4 comply with the minimal distance constraint, so they are in a stable state.

Of course, the shapes of these pictures can be formed by changing the values of parameters $SIMM, DIFF$. Several combinations were tested, to get specious pictures, e.g. in the case of Fig. 17. Here, the GCD relation was presented but did not connect the similar nodes. From the numbers, the reader can reconstruct these lines. The result of the correlation clustering for a few nodes gives a partition where each prime number (plus the 1) has a cluster, and each number gets into its smallest prime divisor's cluster. Although, the collision of nodes is inappropriate (when they partly hide each other), but in this case, it clarifies the picture.

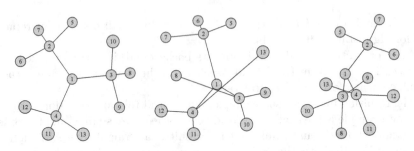

Fig. 16. Total trees.

In the middle of the picture are the even numbers. On the right are the multiplies of three. 5 is just below 25 (it can be known from the data of the image), as 7 is below 49. As the multiplicity of the divisors is not counted, the power of primes get to almost the same position (as 5 and 7 have shown). On the edge of the picture are the large prime numbers, because they differ from every other number, and as they are dissimilar to each other, they are positioned uniformly. It is worth to examine the subtleties of the picture: 35 is positioned between the numbers of clusters of 5 and 7, but it is slightly moved towards the even numbers because there are eight similar number (five numbers are divisible by five and three are divisible by seven), and three similar numbers in the cluster of 3. Similarly, the number 50 is slightly moved towards the numbers of the cluster of 5 and the number 48 towards the numbers of the cluster of 3. By zooming the picture, other subtleties could be explored, too.

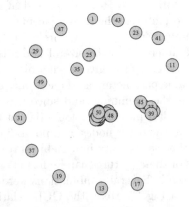

Fig. 17. Visualization of the relation GCD.

Finally let us consider a picture that does not use abstract concepts of number theory, but comes from real life. Figure 18 denotes the members of two departments. Two researchers are treated as similar, if they are co-authors, and dissimilar if there is no such third person who is co-author to both. Of course, this is a partial tolerance relation, because if x-y and y-z are pairwise similar, but x and z are not similar, then one cannot say that they are dissimilar according to the definition. Numbers 1-10 and 11-19 denote the members of the departments, respectively. The center of the picture is empty, hence the research areas are orthogonal. The numbers of the second department have higher densities, so their publication is stricter in the same themes mostly by the same co-authors. The numbers of the second department can be grouped into three clusters, and there are not many relations between these clusters.

Researchers 5, 6, 9 and 13 usually publish alone or with external colleagues, hence it is no wonder that they are alone in the picture. Researchers 1, 2, 7 and 8 wrote a common article, so they construct a strong core. As they publish with other authors too; they are positioned more widely on the picture according to the repulsion of other co-authors. From this group, 1 and 2 are the only co-authors of 4. Moreover 4 is the regular co-author of 10 and has no other co-author. Hence 7 and 8 repulse 10, who gets far from 4, and the attraction of 10 moves 4 away from its other co-authors. In the case of the other department, there is an attraction between 12 and 19 (they moved toward each other), but the chain of co-authors generates a repulsion, so they cannot get any closer.

This simple method could visualize complex systems. In this model, neither the quality and quantity of the common publications nor the date of these publications were taken into account.

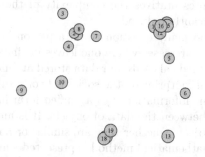

Fig. 18. Research activities of two departments.

8.2   Representative Members[7]

Instead of examining the entire population, polls usually only survey a small sample. This can be done because the results obtained are very close to what one would get by examining the entire population. However, the sample should be selected carefully. Many people think that the larger the sample, the better which is not true. The sample is representative in some respects, i.e., the specific properties are as similar in the sample as in the entire population. The sample

---

[7] The work described in this section was based on [6,7,21,35].

can be representative in one aspect, while not representative in another. There are various standard methods for determining a sample.

If the population is significantly inhomogeneous, i.e. it has high variability according to the survey, then the stratified (random) sampling can be used. In this case, the population is divided into several sub-populations (strata), where these sub-populations are homogeneous according to the examined criteria. From a homogeneous strata, the individuals can be randomly selected to be sampled (i.e., the representative of the group), typically in proportion to the size of the group.

If an object can be represented with a vector of numbers, the difference of vectors belonging to each object can be considered, where this difference/distance usually meets the requirements of metrics. Using this distance function, many clustering methods have been developed over the last sixty years. The most well-known is the k-means method in which a cluster is represented by its centre (one representative). The k-medoids algorithm is a version of this k-means method, and it replaces the cluster with the sample element closest to the cluster centre. The CURE method (clustering using representatives) goes one step further, replacing non-ellipsoid clusters with a maximum c sample elements. The k-means algorithm can be used in the k-nearest neighbours (k-NN) classification algorithm, where newly added objects are categorized into an existing cluster/class. Since comparing the new elements with all stored elements in a large database is a time costly task, by replacing the elements of the clusters with some of their representatives the complexity of the classification of new elements can be significantly reduced.

Polls cannot ask too many questions from a person because their patience is finite. However, there are cases where one leaves behind a lot of information. Think for example our medical cards, our data stored at different kinds of service providers, or our digital footprint on the social network. In these cases, it is not worth transforming this information into a unified form in order to be able to define the differences between the data of objects. It is much easier to directly decide for two given objects whether they are similar or not.

In this section, a mathematical method is presented which—having an existing partition and tolerance relation (based on similarity)—determines which is the most typical object in a cluster, i.e. which one can be considered representative. A real number, a rank is assigned to each of these objects which determines the "representativeness" of the objects.

Here, two possible ways of choosing the representatives are proposed.

8.3   First Method[8]

In this method, an object is called a representative if it is similar to most and different from the least of the members in its group.

For any member x four values are stored:

– α - the number of elements that are similar to x and are in the same cluster.

---

[8] The work described in this section was based on [21].

- β - the number of elements that are different from x and are the same cluster.
- γ - the number of elements that are similar to x and are in different clusters.
- δ - the number of elements that are different from x and are in different clusters.

Figure 19 shows a very simple example to the method. In this example, the tolerance relation (based on similarity) is based on the Euclidean distance of the objects. The smaller circle denotes the similarity threshold and the greater one denotes the difference threshold. For member A the four values are:

- $\alpha = 3$. Because there are four members (A,B,C) that are similar to A.
- $\beta = 2$. Because there are two members $(E$ and $F)$ that are different from A.
- $\gamma = 2$. Because there are two members $(H$ and $G)$ that are similar to A and are in a different cluster;
- $\delta = 3$ Because there are three members $(J, K$ and $L)$ that are different from A and are in a different cluster.

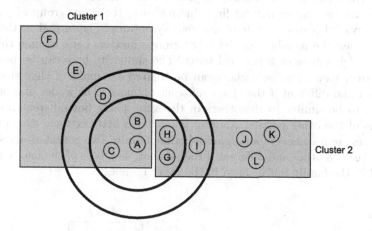

Fig. 19. α and β values for the member A

In the first case, a member can be considered a possible representative if the following fraction (rank) is maximal:

$$r_1 = \frac{\alpha^w - \beta^v}{\alpha + \beta + 1} \quad v, w \in \mathbb{R}, v, w \geq 1, w \geq v \tag{3}$$

In the second case, a member can be considered a possible representative if the following fraction (rank) is maximal:

$$r_2 = \frac{\alpha^w - \beta^v}{\alpha + \beta + u * \gamma + 1} \quad u, v, w \in \mathbb{R}, v, w \geq 1, w \geq v \tag{4}$$

If two arbitrary objects have the same r_2 value, then the δ value decides.

The values r_1 and r_2 show how much an object represents a given cluster. In the case of r_1, only the similarities and differences in the given cluster are considered. Its value is high if the object is similar to the others in the cluster which means that there are only a few elements that are different from the given object. r_2 takes into account every cluster. The r_2 value of an object is high if it is similar to most of the elements in the same cluster and there are very few objects that are similar to the given object but are in different clusters. Those cases, when the object is similar to an element that is in a different cluster, are punished. That is why the γ value is only in the denominator. In some cases, it can happen that the similarity and the difference cannot be taken with the same weights. In the formulas, the u, v, w denote these weights.

The first method can be used when the members of the other groups do not matter. For example, let us assume that the objects are patients. Here the similarity is based on sharing some common symptoms. If the patient, who is the most similar to the others, needs to be found, then the patients from the other groups are irrelevant. For instance, if the task is to find a new possible way to cure a certain disease that a group of patients has, then it can be useful to test it on the representative patient first. In this case, the other group of patients is not relevant because they have different symptoms. The second method can be used when the members of the other groups matter. Let's assume that the objects are members of a political party. The similarity here can be based on the political view. Two politicians can be treated as similar if they share the same idea and different if they have different opinions. The leader of a party is expected to be similar to the others in the same party but different from the members of the other parties. Another good but a little extreme example is if the objects are members of an organized crime family. Two gangsters are similar if they like each other and different if they do not. The boss of the family should be liked in the family but disliked in the other families.

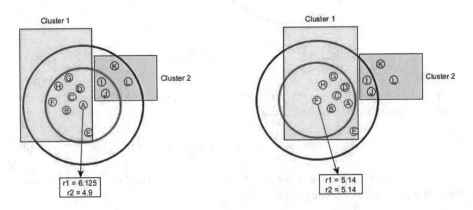

Fig. 20. Difference between r_1 and r_2 maximization

Section 8.3 shows the difference between the two methods. In the left side of the figure, the first method was used. Member A was the representative of $cluster1$, because there are seven objects that are similar to A and no such ones that are different from A ($\alpha = 7, \beta = 0, \gamma = 2, \delta = 2, v = 2, w = 2, u = 1$). So its r_1 value is maximal. In the right side of the figure, the second method was used. Here the member F was the representative of $cluster1$. Its r_1 is less than that of member A, because it has only 6 similar objects. However, the r_2 value is higher than that of member A, because it has no such objects that are similar to it and are in a different cluster, while the member A has 2 objects (I, J) that are similar to it and are in $cluster2$ ($\alpha = 6, \beta = 0, \gamma = 0, \delta = 2, v = 2, w = 2, u = 1$).

The method was introduced because it is very simple and it can be easily used. However, every member has the same vote. It means that each object decreases or increases the rank of the other members exactly the same. Let us assume that object A and B, C are in different clusters and A is different from B and C. Let us also assume that the rank of B is much higher than that of C. In this method, B and C decrease the rank of A and this decrease is the same in both cases, even though that the rank of B is much higher than that of C (B should decrease it much more). To solve this issue, another method is proposed.

8.4   Second method – Ranking Algorithm[9]

In this section, a method is shown on how to describe relations using signed graphs. Each element in a relation is a vertex of the graph and two vertices are connected with an edge if and only if their two corresponding elements are in relation.

In the case of graphs, one can speak of the distance between two vertices (as the shortest path between the two vertices), but it carries much less information than the difference between two large vectors. Therefore, the similarity information should already be included in the graph, so the graph will correspond to a tolerance relation (based on similarity). As there are usually partial tolerance relations (based on similarity) in practice, there will be edges in the graph that denote the similarity and there will be ones that denote the dissimilarity. The partiality is represented by missing edges.

For example, links between individual websites or citations between scientific articles define a directed graph, i.e. a partial tolerance relation (based on similarity), but there is no representation of the dissimilarity.

Google's PageRank algorithm [2] is a great example of a ranking system on directed graphs. Considering the web pages (vertices) and the links between them (edges) as a directed graph, the boundary distribution of the random walk on the graph gives the rank of each page. For example, if the web page p is more likely to be accessed than the web page q, then the rank of p will be higher than that of q and will be ranked higher in the hit list. Here, if a page with a low rank refers to a page with a high rank, or a novice author refers to a well-known author in his article, it raises the rank of the page/author to a higher rank,

---

[9] The work described in this section was based on [6,7].

but—through a non-symmetrical relation—this reference has no effect on the rank of the page/author with the lower rank.

However, if the graph is not directed, the edge between the two vertices affect the rank of both vertices. Since similarity is represented by a tolerance relation (reflexive, symmetric, but not necessarily transitive), the associated graph is not directed.

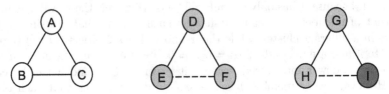

Fig. 21. Simple ranking problems

Let's see our (naive) expectations of a ranking method. In Fig. 21 on the left, there are three vertices (A, B and C) in a common cluster. In this figure, a cluster is represented by vertices of the same colour, while the similarity of vertices is denoted by a solid, and the difference by a dashed line. Because in this subgraph each element is similar to each other, the same rankings are expected due to symmetry. In the middle graph of Fig. 21—where D, E and F are in a common cluster—a difference appears. This graph is called a minimal frustrated graph, because there is no such partition of vertices where similar elements are common, and different elements are clustered separately. In this graph, vertex D has only similar vertices, while vertices E and F both have similar and also different vertices. The fact that an object differs from an object in its own cluster reduces the rank of the object/vertex and thus the chance of being a representative of the cluster. Conversely, if an object is similar to an object in its cluster, then this increases its rank. Based on these, this cluster will be represented by vertex D because it has the highest rank. Moreover—according to the symmetry—the rank of vertices E and F should be the same.

Finally, take the graph on the right side of Fig. 21. Here, the vertices were divided into two clusters: $\{G, H\}$ and $\{I\}$. The fact that the vertices of G and I are similar, but are found in different clusters also reduces the rank of both vertices, because similarity to vertices in other clusters means deviation from the idealized characteristics of the group. The vertex h is similar to vertex G which belongs to H's cluster, and H is different from the vertex I belongs to another cluster. This latter also raises the rank of the vertex H and hence H becomes the representative of its own cluster. In the other cluster, the only vertex will be the representative.

Based on the examples above, the similar objects of the same cluster and dissimilar objects of other clusters can be called the fosterer of the object, while the similar objects of different clusters and dissimilar objects of the same cluster can be called the adversary of the object. The fosterer objects help an object become a representative, while the adversary objects prevent it from happening.

Here, a list can be seen of what is expected from the rankings.

- Be symmetric, that is, if two vertices have the same number of vertices of the same rank in the same type of relation (fosterer or adversary), then their rank is the same.
- The rank of a particular item is immediately raised if:
  - one of its fosterer object increases in rank, or
  - one of its adversary object falls in rank, or
  - a new fosterer object appears.
- The rank of a particular item is immediately reduced if:
  - one of its adversary object increases in rank, or
  - one of its fosterer object falls in rank, or
  - a new adversary object appears.
- It does not directly change the rank of a particular item if:
  - another object that is not compared to it or is incomparable, appears in any cluster, or
  - the rank of such an object changes.

If a new object that is dissimilar to the current object, but similar to another object of the cluster, is added to this cluster, then it raises the rank of objects that are similar to it. This will have a ripple effect on objects that are similar to objects that are similar to the new object, and so on. Therefore, if this should be represented by an algorithm, then an iterative method, that would escalate these effects step by step, would be needed. On the other hand, since almost every object is related to every object, the rank of all objects should be treated altogether.

The Ranking Method. Let U denote the set of objects/vertices, and for simplicity, denote the objects with numbers: $U = \{1, 2, \ldots, n\}$. The set of clusters means a partition. This partition is interpreted as a function—denoted by p— that assigns a number to each object, so $p : U \to \mathbb{N}$. The objects x and y are in the same cluster, if $p(x) = p(y)$; and they are in different clusters, if $p(x) \neq p(y)$. Our tolerance relation (based on similarity) is a (possibly partial) tolerance relation $\mathcal{R}$—that is reflexive, symmetric, but not necessarily transitive.

Social Ranking. A similar approach as PageRank or as various evaluation sites is used (accommodations, restaurants, marketplaces), where the rankings of individual websites, hotels, restaurants are summed up by aggregating individual ratings.

The rank of each object could be taken as the difference between the number of fosterer and adversary objects, but a more sophisticated method would be more appropriate in practice. There is a significant difference between the cases where an adversary object is the representative of another cluster, or it just a marginal object there. In the former case, it decreases the rank of the current object to a greater extent. A value proportional to the rank of the adversary or fosterer objects could define a good amount by which the rank of the current object can be decreased or increased.

The rank of object i is determined by its relation to all objects. Let m_{ij} indicate the relation (fosterer or adversary) between objects i and j. Each object is considered with its rank as weights, and so the following relation needs to be solved for values r_i: $r_i = \sum_j m_{ij} \cdot r_j$ for all $i \in U$.

These equations combine to $R = MR$ in matrix notation.

So that different values m_{ij} must not be used from task to task–depending on its size– a constant c is introduced to normalise the values m_{ij}. This changes the previous relation as follows $r_i = \sum_j (c \cdot m_{ij}) \cdot r_j$, but it can be transformed to $r_i = c \cdot \sum_j m_{ij} \cdot r_j$, which gives us $R = cMR$. If one takes the reciprocal of c to be denoted by λ, then a known relation: $MR = \lambda R$ appears, i.e. R is an eigenvector of the matrix M.

Taking into account the ideas from the previous chapter, the normalized value m_{ij} should be 1 for fosterer objects, -1 for adversary objects, and 0 for all other cases (where the tolerance relation is partial).

According to the much-cited example, a bald man does not resemble a hairy man, although an uprooted hairline does not change a person. As a person can have up to two million hairlines, the linear model of this example is reduced to only four states. Here 1 (bald) and 4 (hairy) correspond to the two end states, while 2 and 3 are two intermediate states. 1 and 2 are in a common cluster, and so are 3 and 4. Figure 22 shows two graphs demonstrating this problem, where clusters are denoted by assigning different shades to the vertices. The similarities (denoted by solid lines) are the same in both cases, but the dissimilarities (denoted by dashed lines) have changed: dissimilarity appears between the second neighbours in the latter case. Hence, the first graph/relation is partial, and the second is total.

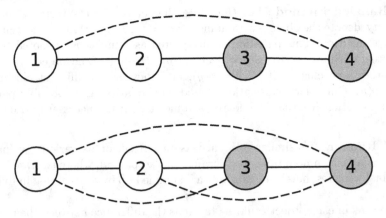

Fig. 22. Two simple symmetric linear model

Two matrices based on the relations and partitions for each graph are:

$$M_1 = \begin{pmatrix} 1 & 1 & 0 & 1 \\ 1 & 1 & -1 & 0 \\ 0 & -1 & 1 & 1 \\ 1 & 0 & 1 & 1 \end{pmatrix} \text{ and } M_2 = \begin{pmatrix} 1 & 1 & 1 & 1 \\ 1 & 1 & -1 & 1 \\ 1 & -1 & 1 & 1 \\ 1 & 1 & 1 & 1 \end{pmatrix}$$

The only adversary relation is between 2 and 3, because they are similar, but are in different clusters. In the first case, 1–3 and 2–4 are not comparable, so the corresponding values of the matrix are 0. In the second case, these pairs of objects are dissimilar, but they are in different clusters, so these are fosterer relations.

Calculating the eigenvalues and eigenvectors for the first graph, the following is obtained:

r_1	r_2	r_3	r_4	λ
0.7071	0.7071	0.3344	0.1368	2.4142
0.5000	−0.5000	0.6770	−0.5873	−0.4142
0.0000	0.0000	−0.6230	−0.6938	2.4132
0.5000	−0.5000	−0.2041	0.3939	−0.4142

The graph is symmetric, so it is expected, that $r_1 = r_4$ and $r_2 = r_3$. Unfortunately, none of the eigenvectors satisfy this. Therefore, another method is needed.

Power-Method. The algorithm of von Mises [34] for a diagonal matrix M results in the biggest eigenvalue (with the highest absolute value), and the corresponding eigenvector. The method starts with an arbitrary vector R_0 that in our case should be $1 = (1, \ldots, 1)^T$. Then R_{k+1} is determined as follows: the rank vector R_k is multiplied by the matrix M and normalized as shown by (5).

$$R_{k+1} \leftarrow \frac{M R_k}{||M R_k||} \tag{5}$$

Unfortunately, this iteration is converges slowly, but it is easy to use even for large sparse matrices. This is why it is used in PageRank implementation. If $R_i \approx R_{i+1}$, the method is stopped and the values in the vector R_i are considered the rank of the objects. If matrix M has an eigenvalue that is strictly greater in magnitude than its other eigenvalues, then R_i converges.

If this method is applied to the matrices shown in Fig. 22, then the result are: $R = (1, 0.414, 0.414, 1)^T$ and $R = (1, 0.618, 0.618, 1)^T$, where $r_1 = r_4$ and $r_2 = r_3$ as it was expected. These values also fit to the naive ideas: for the first graph, for object 1 both objects 2 and 4 are fosterers, while object 1 is a fosterer and object 2 is an adversary for object 2. So the expected relation $r_1 > r_2$ is fulfilled. In the second case, when there are more fosterer relations, the rank gained is also higher for objects 2 and 3.

Summarising it gives:

$$R_{k+1} \leftarrow \frac{M R_k}{||M R_k||} \approx \frac{M^{k+1} 1}{||M^{k+1} 1||}$$

If k is a power of 2, then the same values can be calculated by repeated squaring using the following recurrence relation:

$$B_1 \leftarrow M \text{ and } N_{i+1} \leftarrow \frac{N_i B_i}{||N_i B_i||} \tag{6}$$

Algorithm 2. Python implementation of the ranking method

```python
def power_method(M, eps = 1e-9):
    N = M
    R = np.ones((len(M),))
    N2 = N@N
    N2 /= m = np.max(N2@R)
    while np.linalg.norm(N@R - N2@R) > eps:
        N, N2 = N2, N2@N2
        N2 /= np.max(N2@R)
    return N2
```

Table 10. Ranking numbers...

A) 1,...,12 in a common cluster.

1	2	3	4	5	6	7	8	9	10	11	12
−0.62	0.97	0.06	0.97	−0.32	1.00	−0.62	0.97	0.06	0.89	−0.62	1.00

B) 1,...,100 in a common cluster.

1	2	3	4	5	6	7	8	9	10	11	12	...
−0.55	0.99	−0.03	0.99	−0.28	1.00	−0.37	0.99	−0.03	0.95	−0.44	1.00	...

C) 1,...,12 using optimal partition.

[1]	[2]						[3]		[5]	[7]	[11]
1	2	4	6	8	10	12	3	9	5	7	11
1.00	1.00	1.00	0.73	1.00	0.84	0.73	0.73	0.73	0.84	1.00	1.00

D) 1,...,15 using optimal partition.

[1]	[2]							[3]			[5]	[7]	[11]	[13]
[1]	[2]							[3]			[5]	[7]	[11]	[13]
1	2	4	6	8	10	12	14	3	9	15	5	7	11	13
1.00	1.00	1.00	0.67	1	0.79	0.67	0.86	0.79	0.79	0.54	0.79	0.86	1.00	1.00

E) 1,...,12 using common cluster and the weakened relation.

1	2	3	4	5	6	7	8	9	10	11	12
−0.53	1.00	0.06	0.83	−0.34	0.68	−0.53	0.70	−0.17	0.42	−0.53	0.81

F) 1,...,12 using optimal partition and the weakened relation.

[1]	[2]						[3]		[5]	[7]	[13]
1	2	4	6	8	10	12	3	9	5	7	11
1.00	1.00	0.89	0.53	0.83	0.53	0.62	0.77	0.88	0.89	1.00	1.00

If $N_i 1 \approx N_{i+1} 1$, then let $R \leftarrow N_i 1$. Not surprisingly, the two calculations give the same result.

Based on these, it is not difficult to write the ranking program in Python using the services of the Numpy package (Algorithm 2). Remark, that for Numpy, the operator @ is the matrix multiplication operation.

Figure 23 shows how each value r_i changes for a random matrix M. Here, the axis x represents the number of applications of (6), while the axis y represents

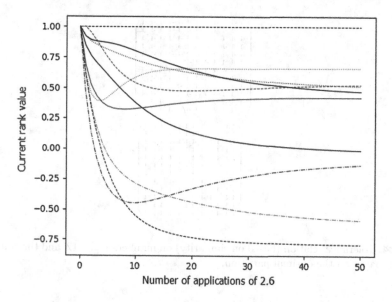

Fig. 23. Changes in the rank of the objects in a random tolerance relation

the current values of the ranks. Due to normalisation, the highest rank is always 1, but as it can be seen from the chart, the rank of the objects changes from time to time. The algorithm is terminated when the ranks cease changing.

Ranks of Numbers. As there is no standard tolerance relation (based on similarity) for larger databases, the various clustering/classification methods are usually tested on random graphs [39]. Here the following relation is used: let two numbers be similar if they have a non-trivial common divisor, i.e. $\gcd(x, y) > 1$, where $x, y \in \mathbb{N}^+$. 1 is considered to be similar to itself. Then $4\mathcal{R}6$ and $6\mathcal{R}9$ are fulfilled (the common divisors are 2 and 3), but $4\mathcal{R}9$ is not, so this relation is not transitive. If someone is interested in the correlation clustering of numbers $1, \ldots, n$, it can be easily formulated if $n < 111\,546\,435$, otherwise the situation becomes complicated [4].

In the following, the rank of each element is determined by using the optimal clustering of the set of numbers, except in the first case, where the numbers $1, \ldots, 12$ are placed in a common cluster.

As there is a single cluster in the Eq. (6), the matrix M can be replaced in the calculation with the matrix of the relation $\mathcal{R}$ in Fig. 24. In this matrix $\mathcal{R}$, the relation between the numbers i and j is given by the jth number from the ith row: similar numbers are denoted by 1, dissimilar numbers by -1. 12 and 6 are similar to the multiples of 2 and 3, these are eight numbers including themselves and different from four of them (1, 5, 7, and 11). The powers of 2 are similar to every even number (six numbers) and different from all odd numbers (six numbers). The powers of 3 (itself and 9) are similar to four numbers and different from

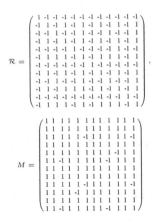

Fig. 24. Tolerance relation (based on similarity) on numbers $1, \ldots, 12$, and the suitable matrix based on the optimal partition.

eight numbers. 5 is similar to its duplicate, but there is no such number for 7 and 11 and for 1. Because in each case the numbers mentioned together are similar to the same numbers, so—according to the symmetry—their rank is the same (Table 10/A). In this table, the ranks are rounded to the closest hundreds, to fit on the page.

If one does the same calculation for numbers $1, \ldots, 100$ (Table 10/B), then the numbers will typically increase. There are many more even numbers which will increase the rank of even numbers, but they have the opposite effect on the rank of powers of 3. In this set, there are numbers similar to 7 or 11, so their rank increases; and there are more such numbers for 7, so its rank grows more.

Consider the partition of natural numbers obtained by correlation clustering [4]. The largest such cluster is the set of even numbers. This is followed by a set of odd numbers divisible by 3; next, the set of numbers which are divisible by 5, but not by 3 or 2, and so on. This can be formulated as $[2] = \{x \in U : 2|x\}$, $[3] = \{x \in U : 3|x\} \setminus [2]$, $[5] = \{x \in U : 5|x\} \setminus [2] \setminus [3]$, ... Of course, each prime number may have one cluster.

If the fosterer and adversary objects are considered based on the optimal partition and the tolerance relation (based on similarity) $\mathcal{R}$ in Fig. 24, then the matrix M presented in Fig. 24 is obtained. The partition mentioned above is optimal because it minimises the number of negative numbers in the matrix M. This, of course, also has an impact on rankings. While keeping the numbers in a common cluster, multiple ranks were negative due to dissimilarities, by using the optimal partition each cluster as a cluster by applying similarity (Table 10/C). For singletons (containing big primes and 1) the rank 1.0 is a proper value, as there are no similar numbers. In the set of even numbers, the rank of the powers of two will also be 1.0, as they are similar to all even numbers, and cluster [2] has only even numbers, and all even numbers are here. Numbers with other divisors will have similar numbers in other clusters. The more prime divisors a number

Table 11. Rank of numbers $1, \ldots, 1000$ using the optimal partition

x	$r(x)$	$r'(x)$
1	1.0	1.0
2	1.0	1.0
3	0.71587172	0.79465596
5	0.78447065	0.81182099
7	0.82770503	0.84160798
11	0.88038405	0.88841938
13	0.89607481	0.90279018
17	0.91498232	0.91946078
4	1.0	0.8102978
6	0.75571551	0.55292589
8	1.0	0.71548803
10	0.85121315	0.5761841

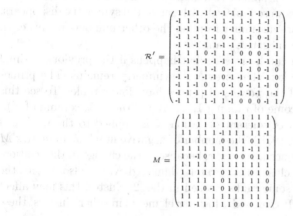

Fig. 25. Weakened tolerance relation (based on similarity) on numbers $1, \ldots, 12$, and the suitable matrix based on the optimal partition.

has, the more clusters contain similar numbers, so its rank will be reduced. In cluster [3], the rank of the powers of three will be the highest, but it will not reach level 1.0, because there are nearly as many numbers that are dividable by 3—that is, similar—in cluster [2], as in cluster [3].

If the method on numbers $1, \ldots, 15$ is applied instead of on numbers $1, \ldots, 12$, then the ranks change (Table 10/D). The rank of 7 fell, as its adversary number (14) appeared. As number 15 is an adversary for both 5 and 10, it reduces the rank of both of them. The reader may be wondering how these ranks look when there are more numbers, e.g. $1 \ldots, 1000$ (Table 11, column r). Perhaps it is clear from above that in the case of an optimal clustering, the representatives of the individual clusters come from the powers of primes. Each power of a prime is given the same rank because it is completely symmetrical in terms of similarity.

Weakening the Tolerance Relation. The method was also executed on a different relation. Here two numbers are similar if one of the numbers is a divisor of the other number instead of the existence of a real common divisor. The partial relation here is made from a complete relation. If two numbers were dissimilar at the original tolerance relation, they will be dissimilar at the weakened relation, too. Moreover, numbers 4 and 6 were similar before, but not anymore. Therefore, the relation holds less often.

At first glance in Table 10/E the ranks of 2, 4, 8 are different, as well as ranks of 3 and 9. Once there is only one cluster, it is sufficient to count how many positive and negative values are in each row of the first matrix in Fig. 25. The number 2 is similar to every even number, that is, to every object in its cluster [2], and there are no other similar numbers anywhere else. The number 4 is similar to half of the numbers of its cluster, but not dissimilar to any numbers in this cluster. The number 8 is similar to a quarter of the numbers of its cluster, etc. The fact that some ones were replaced with zeros, the symmetry disappears. The numbers 1, 7 and 11 are dissimilar from all the other numbers in this case, so now they have the same (negative) rank.

An interesting question can be to see what happens if the previous optimal partition (Table 10/F) is taken. The previous asymmetry remains. The primes have the highest rank, and the powers of primes have lower ranks. To see this tendency, let's see the outcome of ranking $1, \ldots, 1000$ (Table 11, column of r').

If the optimal partition of tolerance relation $\mathcal{R}$ is applied to the weakened tolerance relation $\mathcal{R}'$ (Table 10/F), the number of negative numbers in matrix M will also be significantly reduced. However, because of the change in the relation, the symmetry within the clusters is severely damaged which also affects the rankings. Therefore, when some ranks weaken in the [2] cluster, this may affect elements in other clusters. If there exist fosterer elements in other clusters, these will in turn increase in rank.

Which partition could be optimal for this weakened relation? The number of similar and dissimilar numbers in the clusters should be taken into consideration according to some given number. If the difference between these numbers not in their cluster is maximal, by moving the actual number to the maximal cluster, a better partition will be obtained. Let $U = \{1, \ldots, 1000\}$, and $n = 75$, which is $3 \cdot 5^2$. With the original relation $\mathcal{R}$, [2] contains 267 numbers that are similar to n (which are divisible by 3 or 5), and 233 which are dissimilar to it. [3] contains 167 similar numbers (all of them divisible by 3) and [5] contains 67 similar numbers. At the weakened relation $\mathcal{R}'$, the previously dissimilar numbers remain dissimilar, and in [2] there are no divisors of 75, just its even multiples. These, by definition, are similar to it, so [2] contains 6 members, similar to 75. In [3] there are three divisors of 75 (including itself), as well as 6 of its odd multiples, so [3] contains 9 similar numbers. In [5] there are two divisors of 75 and it does not contain any of its multiples (because any multiple needs to have

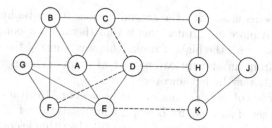

Fig. 26. Multiple representatives are needed

a prime divisor 3, so it would either be in [3], or in [2] if the number is even). If the difference between similar and dissimilar numbers is taken for each cluster, this will be maximal in the number's cluster, as here for [3]. This means that this kind of partition is stable for $\mathcal{R}'$, and with a very high probability that it is the optimal partition, but this needs to be proven.

8.5   Selecting the Representatives[10]

In many applications, however, it might not be enough to have only one representative in each set of objects. Figure 26 shows a very simple example of this problem. Clearly object A has the highest r value so it is the most representative object of the set. However, it is only similar to objects B, C, D, E, F and G and does not have any kind of connections with the rest of the objects. So the aforementioned property for samples is not satisfied as object A alone cannot represent the entire set.

Three possible ways are offered to generate more than one representative from which only the third option proves to be appropriate for real-world applications. A representative member x is said to cover the member y if x is similar to y.

1. The user gives a threshold value k. Then k percent of random objects are
   treated as representatives.
2. The user gives an interval for the rank values. If the rank of an object is in this interval, then it is considered as a representative.
3. Use an algorithm based on similarity to generate the necessary number of representatives.

The main issue with the first option is that the user must have some knowledge about the given set to choose an optimal k value. Due to randomness, critical information may get disregarded. Another problem is that sometimes one object can be enough to represent the whole set, but the user forces the system to choose additional representatives.

---

[10] The work described in this section was based on [35].

Similar problems arise with the second option. It can be hard to choose a proper interval. A more important issue is that the first few points selected will always be the ones with the highest rank. This way, some of the representatives may be picked from an already covered set of points. This is redundant, and some of the points may be left uncovered.

The pseudo-code of the third option can be seen in Algorithm 3. The input of the function is a set of data points D and the output is the set of representatives REP. It is an iterative method, in each step the algorithm keeps a record of the covered objects (i.e. the objects that are similar to one of the representatives) which is empty in the first iteration (line 6). In every iteration, the object with the highest rank is selected from the uncovered objects (line 10–15). In line 16-20, the data points, that are covered by the currently selected representative, are inserted into the set C. At the end of each step, the chosen representative is moved into the set REP. The algorithm stops when there are no uncovered members left.

The strength of the method is that it uses the similarity between objects, and so it generates the optimal number of representatives. The other two methods can create too few or too many representatives. Another advantage is that it does not need any user-defined parameters. The algorithm can be treated as a directed sampling method which can be a very powerful tool in many applications.

A political party contains members that share a common political ideology. However, in some parties, it can happen that even though the members follow the same vision, there are some disagreements. So the group can be divided into smaller groups. In this case, one politician is not enough to represent the entire party. The aforementioned algorithm could be a solution as it takes into account the variety of the members.

Figure 8.5 presents the steps of the algorithm for the data set shown in Fig. 26. The grey ellipses contain the covered objects by a chosen representative member. In the first step, object A is chosen. In the second step, objects B, C, D, E, F and G are not considered as they are covered by A. The second method, mentioned at the beginning of this section, could have chosen B or G as possible representatives because they have the second-highest r value. Naturally, it is pointless to select them because the object A makes it redundant (both of them are similar to A). After four steps, the algorithm finishes and the four representatives are objects A, K, H, I. It can be easily seen that these 4 members share the diversity of the original data set.

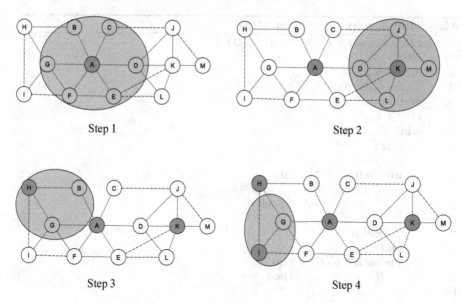

Fig. 27. The execution of the algorithm

9 Approximation Pairs Based on Representatives[11]

As mentioned before, samples and representatives play a very important role in data mining. In Sect. 8.2 some possible ways were shown to select representatives for a group of objects. The importance of the representatives lies in reducing the execution time of the algorithms. So a natural question can be: can the representatives be used in the set-approximation process?

In the next two sections, two new approaches are introduced which try to answer the previous question.

1. Approximation Pairs Based on Similarity-based Rough Sets
2. Set-based Approximation Pairs

---

[11] The work described in this chapter was based on [6,35].

Algorithm 3. Selecting representatives

1: **function** SELECT REPRESENTATIVES(D)
2:     $REP \leftarrow \emptyset$
3:     **for each** $p \in \mathcal{D}$ **do**
4:         calculate the rank of point p
5:     **end for**
6:     $C \leftarrow \emptyset$
7:     **while** $C \neq D$ **do**
8:         $max \leftarrow -\inf$
9:         $max_p \leftarrow None$
10:        **for each** $p \in (D \setminus C)$ **do**
11:            **if** rank of point $p > max$ **then**
12:                $max \leftarrow$ rank of point p
13:                $max_p \leftarrow p$
14:            **end if**
15:        **end for**
16:        **for each** $p \in (D \setminus C)$ **do**
17:            **if** max_p covers p **then**
18:                $C \leftarrow C \cup \{p\}$
19:            **end if**
20:        **end for**
21:        $REP \leftarrow REP \cup \{max_p\}$
22:    **end while**
23:    **return** REP
24: **end function**

---

10   Approximation Pairs Based on Similarity-based Rough Sets[12]

Similarity-based rough sets is an approximation space which is based on the partition generated by correlation clustering. The lower approximation of a set S is the union of those base sets that are subsets of S. To get these base sets, every point in each base set must be considered. It can be a time-consuming task if the number of points is high. The effectiveness of the representatives lies in situations when the number of objects is very large. It can be practical to use the power of representatives in the approximation process. For each base set, let us consider only its representatives. Let $B \in \mathfrak{B}$ be a base set, and $REP(B)$ be the set of its representatives. The approximation functions are defined as the following:

- $l_r(S) = \bigcup \{B \mid B \in \mathfrak{B} \text{ and } \forall x \in REP(B) : x \in S\}$;
- $u_r(S) = \bigcup \{B \mid B \in \mathfrak{B} \text{ and } \exists x \in REP(B) : x \in S\}$.

This way, the lower approximation of a set S becomes the union of those base sets for which every representative is a member of S. A base set belongs

---

[12] The work described in this section was based on [35].

to the upper approximation if at least one of its representatives is in the set
S. Naturally, the certainty of the lower approximation might be lost, but as
the number of points is increasing a lot of resources can be saved. This a very
important feature for which it can be worth giving up the certainty property.

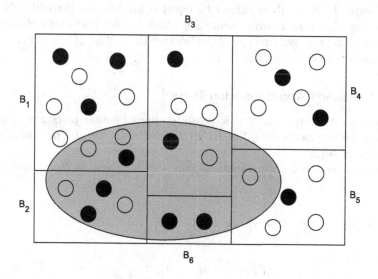

Fig. 28. Approximation based on representatives

In Fig. 28 a simple example is provided for the method. The base sets are
denoted by solid-line rectangles, and the set to be approximated (S) is denoted by
a grey ellipse. For each base set, the black circles symbolise the representatives.

The approximation of the set S is the following based on the representatives:

– $l_r(S) = B_2 \cup B_6$
– $u_r(S) = B_1 \cup B_2 \cup B_3 \cup B_6$

The approximation of the set S is the following based on the classical approximation pair:

– $l(S) = B_2 \cup B_6$
– $u(S) = B_1 \cup B_2 \cup B_3 \cup B_5 \cup B_6$

The lower approximation is the same in both cases which is, of course, not
necessary. The upper approximation differs in one base set (B_5). When there is a
huge number of points and there are several sets to be approximated, approximation using representatives is recommended. In this case, the method can reduce
the run-time of the approximation significantly. Determining the approximation
with the classical functions 32 objects needed to be considered. Using the proposed method, only 13 of them had to be tested, so almost 60% of the original

points were left-out. Of course, with 32 to 13 points is not a significant change, but in the case of millions of objects, it can be very useful. Working with only the representatives, time and resources can always be saved, because it is sure that the number of representatives is less than that of U. Proving this is very straightforward. Naturally, there cannot be more representatives than objects in the universe. Their numbers cannot be equal either because it could only happen if every object were a representative which implies that every cluster was singleton. Using this system is pointless because the system of base sets is empty (every singleton cluster is discarded).

10.1   Set-Based Approximation Pairs[13]

In this subsection, two other new possible approximation pairs are proposed based on the representatives. Let $REP(S)$ denote the set of representatives of any arbitrary set S.

The two proposed approximation pairs can be given as $\langle \mathsf{l}, \mathsf{u}_1 \rangle$ and $\langle \mathsf{l}, \mathsf{u}_2 \rangle$, where

$$\mathsf{l}(S) = \bigcup_{\substack{x \in REP(S) \\ [x] \subseteq S}} \{[x]\}$$

$$\mathsf{u}_1(S) = \bigcup_{x \in REP(S)} \{[x]\}$$

$$\mathsf{u}_2(S) = \bigcup_{\substack{x \in REP(S) \\ \|[x] \cap S\| > \|[x] \setminus S\|}} \{[x]\}$$

In this space, a base set contains objects that are similar to a given representative, so each of them generates a base set. The representatives depend on the set to be approximated and so does the system of base sets. As a result, the base sets change as the set changes (they are relative to the set).

Figure 29 shows how the lower and upper approximation of a set S looks like based on $\langle \mathsf{l}, \mathsf{u}_1 \rangle$ and $\langle \mathsf{l}, \mathsf{u}_2 \rangle$. In the middle, the lower approximation can be seen. On the left side, the upper approximation based on u_1 is shown and the right figure illustrates the upper approximation based on u_2 500 points were generated randomly in the unit square. The set to be approximated forms a triangle and its points are marked with larger squares. The base of the tolerance relation (based on similarity) was the Manhattan distance of these objects (d) this time. A similarity ($SIMM$) and a dissimilarity threshold ($DIFF$) were defined. $SIMM$ was set to 0.1 and $DIFF$ was set to 0.2. The objects that are similar to an object x are denoted by circles in Fig. 29. The colours of the circles refer to the process of approximation. The environments of the first representatives (objects that are similar to it) are represented by darker circles while the environments of the later representatives are represented by brighter ones.

---

[13] The work described in this section was based on [6].

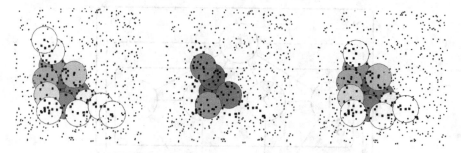

Fig. 29. Three different set approximations of a triangle

The topmost circle—in the left picture—contains 4 objects from the set S, but 3 of them are similar to a former representative. The remaining one becomes a representative. However, this object is very similar to many non-set objects, so this circle is not included in the right picture, as this object is not a representative in this case.

10.2 Properties of Approximation

Proposition 1. *None of the aforementioned approximation pairs are monotone.*

Proof. The first counterexample is presented in Fig. 31. In the figures, the solid/-dashed edge between nodes x and y denotes that the relation $x\mathcal{R}y$ holds/does not hold, respectively. If there is no edge between some nodes, then the given relation is partial [32, 49], and the two nodes that aren't connected are neither similar nor dissimilar, i.e. they are incomparable or not have been compared yet. The ranks of the objects are for the two given sets:

S: $\langle 0.126, 0.626, 0.626, 0.126, -1.000, 0.547, 0.547 \rangle$,
S': $\langle 0.126, 0.626, 0.626, 0.126, 1.000, 0.547, 0.547 \rangle$.

In the case of the set S $r_1 = r_4 < r_2 = r_3$, henceforth one possible $REP(S)$ can be $\{2, 4\}$, where $[2] = \{1, 2, 3, 5\}$ and $[4] = \{3, 4, 5, 7\}$. Object 4 is similar to 5 and 7, and 2 is similar to 5, hence $\mathsf{I}(S) = \emptyset$.

In the case of the set S' $r_1 = r_4 < r_2 = r_3 < r_5$, and object 5 which covers every member of S', so object 5 becomes the only representative. Hence $\mathsf{I}(S') = \{1, 2, 3, 4, 5\}$, thus $\mathsf{I}(S) \subseteq \mathsf{I}(S')$ holds.

However, although $S \subseteq S'$ $\mathsf{u}_1(S) \not\subseteq \mathsf{u}_1(S')$, where $\mathsf{u}_1(S) = \{1, 2, 3, 4, 5, 7\}$ and $\mathsf{u}_1(S') = \{1, 2, 3, 4, 5\}$. This proves that the approximation pair $\langle \mathsf{I}, \mathsf{u}_1 \rangle$ is not monotone.

To disprove the monotonicity of function u_2 a different counterexample is needed. Figure 30 show this tolerance relation and the sets S and S'. The ranks of the objects in this case are:

S: $\langle 0.585, 0.759, 0.698, 0.668, -1.000, -0.009, 0.318 \rangle$,
S': $\langle 0.585, 0.759, 0.698, 0.668, 1.000, -0.009, 0.318 \rangle$.

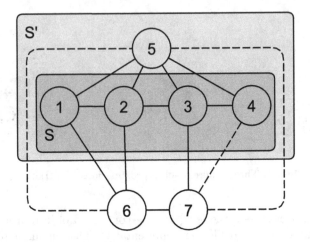

Fig. 30. Monotonicity does not hold for u_2.

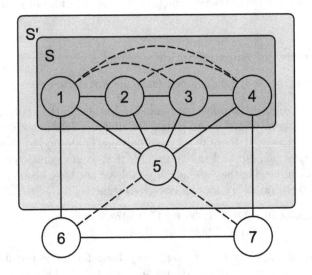

Fig. 31. Monotonicity does not hold for u_1.

For ranks in S $r_2 > r_3 > r_4 > r_1$, so at first the sizes of sets $[2] \cap S = \{1, 2, 3\}$ and $[2] \setminus S = \{5, 6\}$ need to be checked . The first of the two sets is bigger, so $2 \in REP(S)$. The set $S \setminus [2]$ only contains object 4, so the sizes of the sets $[4] \cap S = \{3, 4\}$ and $[4] \setminus S = \{5\}$ need to be checked. Again, the first is the bigger set, so $4 \in REP(S)$. Therefore $u_2(S) = [2] \cup [4] = \{1, 2, 3, 4, 5, 6\}$. Summarising this: $S \subseteq S'$ but $u_2(S) \not\subseteq u_2(S')$. Hence, the approximation pair $\langle l, u_2 \rangle$ is not monotone, too.

Proposition 2. *The weak approximation property holds for both approximation pairs.*

Proof. In both cases, the functions use the same representatives based on the same order of ranks. Therefore the lower approximation cannot be a larger set than the upper approximation.

Proposition 3. *For* $\langle l, u_1 \rangle$ *the strong approximation property holds.*

Proof. By definition $l(S) \subseteq S$ is always true. In the case of u_1, every member of the set S is covered by at least one of the representatives. Thus $S \subseteq u_1(S)$ is also true.

Proposition 4. *For* $\langle l, u_2 \rangle$ *the strong approximation property does not hold.*

Proof. In Fig. 32 an example can be seen that contradicts this property. The ranks of the objects are the following:

$$r = \langle 1.00, 0.85, 0.36, 0.36, 0.65, 0.55, -0.32, -0.80, -0.11, -0.11, -0.40 \rangle.$$

Hence the possible representatives of the set S are the objects 1 and 7, where $[1] = \{1, 2, 3, 4, 5, 6\}$ and $[7] = \{6, 7, 8, 9, 10, 11\}$. The upper approximation of the set S is $u_2(S) = \{1, 2, 3, 4, 5, 6\}$, because $[7] \cap S = \{6, 7, 8\}$ and $[7] \setminus S = \{9, 10, 11\}$, therefore object 7 cannot be a representative of S. So $S \not\subseteq u_2(S)$, meaning that the strong approximation is not necessarily true.

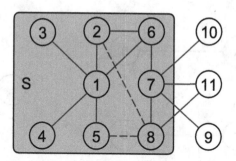

Fig. 32. Strong approximation property does not hold for the second approximation pair

Proposition 5. *The normality of the lower and upper approximation holds.*

Proof. The representatives are selected from the members of the set. If $S = \emptyset$ then there are no representatives, so $l(S) = u_1(S) = u_2(S) = \emptyset$. So the normality of the lower approximation holds because an empty set does not have any representative members. The same holds for the normality of the upper approximation.

11   A Novel Area of Application – Graph Approximation on Similarity-based Rough Sets

Rough set theory is a possible way to handle uncertainty using set-approximations. Similarity-based rough sets is a new way for the same problem and it is based on the similarity of objects. A tolerance relation (based on similarity) can be represented by a signed graph. This graph is complete if the relation is total, and it is not complete if the relation is partial. Because of the symmetry, it must be an undirected graph. Due to the reflexivity, every vertex in the graph has a self-loop edge. If two objects are similar, then a positive edge runs between them, and if they are different, then the edge is negative. Every graph can be represented by a set that contains ordered pairs. In this case, it can be represented by a set of 3-tuples. If a similarity graph can be treated as a set, then a natural question arises: can this graph be also approximated? In this chapter, a possible way is shown to define the lower and upper approximation of a graph and also how this method can be applied in data pre-processing.

The graph in Fig. 33 can be represented by the following set:

$$\left\{ \begin{array}{l} \langle A,A,+\rangle\,,\ \langle B,B,+\rangle\,,\ \langle C,C,+\rangle\,,\ \langle D,D,+\rangle\,,\ \langle A,B,+\rangle\,,\ \langle A,C,+\rangle\,, \\ \langle A,D,-\rangle\,,\ \langle B,A,+\rangle\,,\ \langle B,D,-\rangle\,,\ \langle B,E,-\rangle\,,\ \langle C,A,+\rangle\,,\ \langle C,D,+\rangle\,, \\ \langle D,A,-\rangle\,,\ \langle D,B,-\rangle\,,\ \langle D,C,+\rangle\,,\ \langle E,B,-\rangle \end{array} \right\}$$

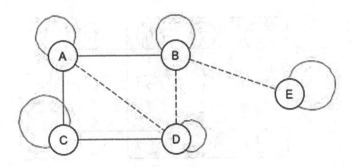

Fig. 33. A tolerance relation (based on similarity) represented by a signed graph

The main goal of this part of the dissertation is to extend the principles of similarity-based rough sets to graphs that represent tolerance relation (based on similarity). As mentioned before, from the theoretical point of view, a Pawlakian approximation space can be characterized by an ordered pair $\langle U, \mathcal{R} \rangle$, where U denotes the universe (a nonempty set of objects) and $\mathcal{R}$ denotes an equivalence relation based on indiscernibility. In the similarity-based rough sets, $\mathcal{R}$ is a tolerance relation (based on similarity). In the case of graph approximation, the universe is a complete undirected signed graph in which every node is connected to every node by 2 edges (one positive and one negative) and every node has also 2 self-loop edges. Formally $G = U \times U \times \{+, -\}$.

Let $G_1 \subseteq G$ $(G_1 \neq G)$ be a graph representing an arbitrary tolerance relation (based on similarity). This graph defines the background knowledge. The relation $\mathcal{R}$ is also defined by G_1. For all objects $x, y \in U$ if $(x, y, +)$, then the objects are similar and if $(x, y, -)$, then they are different. In rough set theory (and also in its similarity-based version), the base sets provide the background knowledge. Here, there are base graphs, and they represent the same. The system of base graphs is also determined by the correlation clustering, and it can be given by the following formula:

Definition 12.

$$\mathfrak{B}_G = \{g \mid g \subseteq G_1, \ and \ \langle x, y, s \rangle \in g \ if \ p(x) = p(y) \ and \ s \in \{+, -\}\}$$

where p is the partition gained from the correlation clustering.

Definition 13. Let $G_2 \subseteq G_1$ be an arbitrary subgraph of G_1. The lower and upper approximation of G_2 can be given by the following way:

$$\mathsf{l}(G_2) = \bigcup \{g \mid g \in \mathfrak{B}_G \ and \ g \subseteq G_2\}$$

$$\mathsf{u}(G_2) = \bigcup \{g \mid g \in \mathfrak{B}_G \ and \ g \cap G_2 \neq \emptyset\}$$

The lower approximation is the disjoint union of those base graphs that are subgraphs of G_2. The upper approximation is the disjoint union of those base graphs for which there exists a graph which is a subgraph of both G_1 and G_2.

Definition 14. The accuracy of the approximation can be calculated by the following fraction, where $N(G)$ denotes the number of self-loop edges in an arbitrary graph G:

$$\alpha_{G_2} = \frac{|\mathsf{l}(G_2)/2| - N(\mathsf{l}(G_2))}{|\mathsf{u}(G_2)/2| - N(\mathsf{u}(G_2))}$$

Graph approximation uses the same concepts as the set approximation. However, it is a stricter method, as it takes into consideration not only the objects but the edges too.

12   Attribute Reduction with Graph Approximation

The three main steps of data mining are pre-processing, knowledge discovery and post-processing. Data pre-processing is a very important step in the data mining process which involves transforming raw data into an understandable format. Real-world data is often incomplete, inconsistent, noisy, and is likely to contain many errors. Data pre-processing is a method of resolving such issues. One of the main issues with raw data is that there can be too many attributes. So a natural question can be if some of these attributes can be removed from the system preserving its basic properties (whether some attributes can be considered as superfluous). In this subsection, a method is proposed to measure the dependency between two attributes (or set of attributes). If this dependency value is above a threshold, then one of the attributes can be removed. Let $IS = (U, A)$ an information system and $A', A'' \subseteq A$ two sets of attributes. Let G_1 be the graph representing the tolerance relation (based on similarity), which is based on the attribute set A'. Let G_2 be the graph representing the tolerance relation (based on similarity) which is based on the attribute set A''. To measure the dependency between A' and A'' the following method is proposed:

1. Determine the system of base graphs based on G_1;
2. Approximate G_2 using the base graphs defined in the first step;
3. Calculate the accuracy of approximation;
4. If the accuracy is higher than a threshold, then A'' can be treated as superfluous.

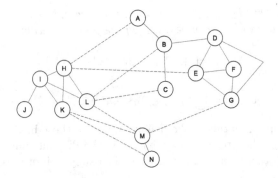

Fig. 34. G_1 graph representing a tolerance relation (based on similarity) based on a set of attributes

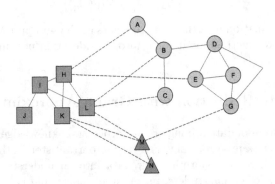

Fig. 35. The base graphs generated by the correlation clustering

In the following figures, a very simple example is shown with 14 objects. In Fig. 34 a graph can be seen, which denotes a similarity-based on a set of attributes. In the figure, the solid lines denote the similarity, while the dashed ones denote the difference between objects. In Fig. 35 the base graphs can be observed which were generated by the correlation clustering. In this example, there are three base graphs.

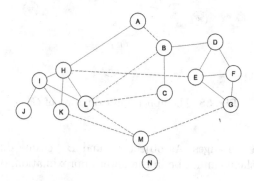

Fig. 36. G_2 graph representing a tolerance relation (based on similarity) based on a set of attributes with the same objects

Figure 36 shows another graph, which also illustrates another set of attributes. It is important that the similarity is based on the same objects as before. The difference between G_1 and G_2 is negligible. It can be observed only in 4 edges. Between objects A and H, the negative edge was replaced by a positive one. Between objects K and N, the edge was deleted as well as between D and G. Between A and B the positive edge was changed to negative.

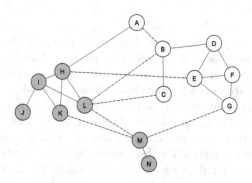

Fig. 37. The lower approximation of G_2

In Fig. 37 the lower approximation and in Fig. 38 the upper approximation can be seen. It is interesting that even though there are only slight differences between the two graphs the lower approximation contains only the two smallest base graphs. The reason is that graph approximation is stricter because it takes

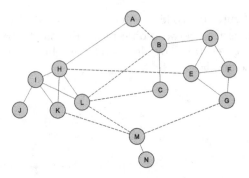

Fig. 38. The upper approximation of G_2

into consideration the edges. As objects A and B became different, the base graph containing them cannot be in the lower approximation, only in the upper approximation.

The accuracy of the approximation is: $\alpha_{G_2} = \frac{14}{29} = 0.48$

This means that based on G_1 the percentage of the available information about G_2 is only 48% even though they are almost equivalent. This method takes into account the real similarity among objects; therefore it can give appropriate results in situations, where other algorithms proved to be a dead end. For example, in mathematical statistics a common method to measure the dependency between attributes is correlation. Although, it works only if there is a linear relationship between the attributes. Our proposed method can work in various applications for any type of relationship.

13   Summary

The main result of my research was developing a completely new approximation space which was based on a tolerance relation (based on similarity). In this space, the system of base sets is generated by the correlation clustering. Correlation is a clustering technique that is based on a tolerance relation. The proposed space has many good qualities and it is different from the covering spaces induced by a tolerance relation (based on similarity). The main difference between our space and these covering spaces is that ours considers the real similarity among the objects while the covering spaces generated by a tolerance relation only consider similarity to a distinguished member. An algorithm was successfully developed, based on physics, by which tolerance relations (based on similarity) can easily be visualized. In data mining, to reduce the execution time of an algorithm, it is common to use samples. A sample contains points from the original data set. There are numerous ways to choose a part of the input data set which can be treated as a sample. However, in every method, it is crucial that the chosen objects must represent the entire population. In this case, representativeness means that the specific properties are as similar in the sample as in the entire set. Without this property, important information might be disregarded. In this dissertation, a method was provided to generate the representatives of a given set. One of the good properties of this algorithm is that it chooses the necessary number (not too small or big) of representatives. The importance of the representatives lies in reducing the execution time of the algorithms. In our research, it was applied to set approximation. The lower approximation of a set is the union of those base sets that are subsets of the given set. In order to get these base sets, every object in each base set must be considered. It can be a time-consuming task if the number of points is high. In this situation, it could be helpful to consider only the representatives for each base set. New approximation pairs have also been proposed based on the representatives. Feature selection is an important part of machine learning. Feature selection refers to the process of reducing the number of attributes in a dataset, or of finding the most meaningful attributes. This process improves the quality of the model and it also makes the model more efficient. In the last section of the dissertation, our graph-approximation method is described. It is based on the fact that any tolerance relation (based on similarity) can be represented by a signed graph and any graph can be represented by a set. Therefore the proposed method (similarity-based rough sets) can be used for graphs as well. In the last part of this work, a possible way is shown to use graph approximation in feature selection.

Acknowledgement. This work was supported by the construction EFOP-3.6.3-VEKOP-16-2017-00002. The project was supported by the European Union, co-financed by the European Social Fund.

Appendix A Algorithms

Between 2006 and 2016, "Advanced Search Methods" was a compulsory subject for some Computer Science master students. Initially, students learned about the well known NP-hard problems (SAT, NLP, TSP, etc.) and various popular optimization methods. Later, to help some physicists from University Babes-Bolyai in Cluj, one co-author began to research the problem of correlation clustering. This problem can easily be formulated (which equivalence relation is the closest to a given tolerance relation?), can be quickly understood, is freely scalable, but NP-hard, and if there are more than 15 objects in a general case an approximate solution can only be provided. That is why this problem got a central role from 2010. In this year, the students with the co-author's lead implemented the learned algorithms, and used correlation clustering to test and compare them. There were several didactic goals of this development: they worked as a team, where the leader changed from algorithm to algorithm, whose duty was to distribute the subtasks among the others and compose/finalize their work. The implementations together gave thousand of LOCs, so the students got experience with a real-life size problem. When designing this system as a framework, the OOP principles were of principal importance, to be able to apply it for other optimization problems. The system was completed with several refactorisations and extended with methods developed directly for correlation clustering, and several special data structures which allowed to run programs several magnitudes faster. Finally, the full source of the whole system with detailed explanations was published at the Hungarian Digital Textbook Repository [3]. According to the students' requests, the system was written in Java. Jason Brownlee published a similar book using Ruby [11].

In the next subsections, there is some brief information about the used algorithms and their parameters. The whole descriptions can also be seen in [3].

A.1 Hill Climbing Algorithm

This method is very well-known. Each state in the search plane represents a partition. A state is considered better than another state if its number of conflicts is less than that of the other state. In each step, it is checked, whether there is a better state in the neighborhood of the actual state. If there is not, then the algorithm stops. If there is, then the next step goes from this point.

A.2 Stochastic Hill Climbing Algorithm

The original hill climbing search is greedy, it always moves to the best neighbor. Stochastic hill climbing is a variant, where the algorithm chooses from the neighbors in proportion to their goodness, allowing the algorithm to move in a worse direction as well.

A.3 Tabu Search

Each state in the search plane represents a partition like in the previous algorithms. The tabu search defines a list of banned states or directions to where it cannot move at a time. This is called a tabu list or memory. There are many types of memories. In the experiments, a short-term memory was used with a size of 50.

The neighborhood of the actual state consists of banned and permitted states. In each step, there are two possibilities:

- If one of the neighbors is so good that it is better than the best state so far, then it should go there even if it is banned. This is called the aspirant condition.
- The algorithm moves to the best-permitted neighbor of the actual state.

If the new state is better than the best state so far, then this state will be the new best. The previous state will also be added to the tabu list, so the algorithm could not go backwards immediately. If the list is full, then the last state will be deleted. The algorithm stops if it reaches the 1000th step and it returns the best state.

A.4 Simulated Annealing

Each state in the search plane represents a partition. In each step, the algorithm chooses a neighbor of the actual state. Let f denote the number of conflicts in the actual state and f' denote the number of conflicts in the neighbor. If $f' < f$, then the algorithm moves to the neighbor. If not, then it chooses this state with the probability of $\frac{e^{f-f'}}{T}$. The value T is the temperature value which is a crucial parameter. It should decrease in each step. Determining the starting temperature value is a hard task. The common method for the issue is heating. In each temperature (starting from 1), 500 attempts are made to move to a neighbor, and the number of successful movements are counted. If the ratio of the number of successful movements and the number of attempts reaches 0.99, then the heating procedure stops, otherwise the temperature value is increased. After the heating, the annealing (search) step comes. It is important that how much time the algorithm spends in each temperature value. A minimal step count was defined and it increases each time the temperature is decreased until it reaches a maximal value when the algorithm stops and returns the best state. The minimal step count was set to 100 and the maximal was set to 1000. The temperature values are always decreased by 97%.

A.5 Parallel Tempering

The simulated annealing runs on a single thread. This is a parallelized version, where threads can cooperate. In this method, 3 threads were used.

A.6 Genetic Algorithm

In this algorithm, each partition is represented by an entity. In each search step, there is a population of entities with a size of 100. In the beginning, each entity represents a random partition. This population contains the actual generation of entities and the best entities from the old generation. In each step, the best 25 entities stay in the population. The rest of the spaces are filled with the descendants of the entities of the old generation. In step 1, the algorithm defines the new generation. Step 2 is the reproduction step. A descendant is created in this step with the crossover of 2 parent entities. In the experiments, one-point crossover was used. For the crossover, not a random or the best element is chosen but an element from a set defined by a parameter. The size of this set was 4. After step 2, each child entity goes through a mutation phase (step 3) whose probability was 2/3. After each child entity is created, the actual generation is overwritten by the new generation, and the algorithm goes back to step 1. The algorithm stops when it reaches the 2000th generation and returns the best entity of the population.

A.7 Bees Algorithm

This algorithm is based on the society of honey bees. Each partition is represented by a "bee". There are two types of bees: scout bees and recruit bees. Scout bees scout the area and they report back to the hive about their findings. Then the necessary number (in proportion to the goodness of the finding) of recruit bees go to the area to forage. In this case, the scouts are scattered across the search plane and recruit bees were assigned only to the best of them. These bees are called elites. The rest of the scout bees wander in the plane. It changes dynamically which scout bees are considered as elite and how many recruits are assigned to them. The recruits search around the elite bee to which they were assigned, and if they find a better state, then the scout bees move to that position. In the experiments, the number of scout bees was set to 50 and the number of elites was 5 and 1000 recruit bees follow the elites. In the beginning, the scout bees start from a random position. The algorithm stops when it reaches the 2000th step and it returns the partition represented by the best bee.

A.8 Particle Swarm Optimization

In this algorithm, each partition is represented by an insect (particle). Each insect knows its best position and the best position of the swarm. The size of the swarm was set to 50. In each step, each insect moves in the search plane. In the beginning, the insects start from a random position. There are 3 possibilities for them to move:

– Randomly move
– Move towards its best position
– Move toward the best position of the swarm

The possibilities of the moves was set to 0.2, 0.3, 0.5 respectively. After reaching the 6000th step, the algorithm stops and returns the insect with the best position.

A.9 Firefly Algorithm

In this algorithm, each partition is represented by a firefly. The fireflies are unisex and their brightnesses are proportionate to the goodness of the partition they represent. In the beginning, the fireflies start from a random position, and in each step, each firefly moves to its brightest neighbor. If the brightest neighbor of a firefly is itself, then it moves randomly. Brightness is dependent on the distance of the insects. The intensity of a firefly is defined by the following formula: $I_d = \frac{I_0}{1+\gamma d^2}$, where I_0 denotes the starting intensity, γ is the absorption coefficient (was set to 0.03) and d is the distance between the two fireflies. After 10000 steps the algorithm stops, and the result is the partition that is represented by the brightest firefly. The number of fireflies was set to 50.

Appendix B Software

I developed a program that helps us approximate different types of sets using the proposed new approximation space (similarity-based rough sets). The software can be downloaded from: https://github.com/Nagy-David/Similarity-Based-Rough-Sets. For giving the input datasets, the user has two options:

1. Generating random coordinate points
2. Reading a dataset from a file

1. *Random Points*

The user gives the number of points, and then the points are generated in a 2-dimensional interval which is also given by the user In this option, the base of the tolerance relation (representing the similarity) is the Euclidean distance of the objects (d). A similarity $(SIMM)$ and a dissimilarity threshold $(DIFF)$ were defined. The tolerance relation $\mathcal{R}$ can be given this way for any objects x, y:

$$x\mathcal{R}y = \begin{cases} +1 & d(x,y) \le SIMM \\ -1 & d(x,y) > DIFF \\ 0 & \text{otherwise} \end{cases} \tag{7}$$

2. *Continuous Data*

Each row represents a single entity. In the software, there is an option to normalize the data in the way described below. Let A be an attribute and v the value to be normalized. After the normalization:

$$v = \frac{v - min(A)}{max(A) - min(A)} \tag{8}$$

The similarity is defined in two steps.

(a) step: Let $A_1, A_2 \ldots A_n$ be some attributes, $t_1, t_2 \ldots t_n$ be threshold values and x, y be two objects. Let $z(A_i)$ denote the attribute value of A_i for any object $(i = 1 \ldots n)$. If $\exists i \in \{1 \ldots n\} : |x(A_i) - y(A_i)| \ge t_i$, then the objects x and y are treated as different.

(b) step: If the condition in the first step does not hold, then the tolerance relation $\mathcal{R}$ can be defined in the following way for any objects x, y using a similarity threshold $SIMM$ and a dissimilarity threshold $DIFF$:

$$x\mathcal{R}y = \begin{cases} +1 & d(x,y) \le SIMM \\ -1 & d(x,y) > DIFF \\ 0 & \text{otherwise} \end{cases} \tag{9}$$

The d "distance" value is calculated for any objects x, y by the following method:

$$d(x,y) = \sqrt{\sum_{i=1}^{n} (x(A_i) - y(A_i))^2} \tag{10}$$

The necessity of the first step can be explained by the following simple example. Let us assume that the objects are patients. Two patients may differ only in the blood pressure level and the other attribute values are relatively close to one another. So the distance between these two entities can be a small value. However, the patients cannot be treated as similar because a high blood pressure level can indicate an illness. This fact remains hidden without the first step, because the similarity value can be small for the two patients. The same holds for normalized data.

After getting the input points the software runs a search/optimization algorithm that finds a quasi-optimal partition. As mentioned earlier, the singleton clusters mean little information, so the software leaves them out and creates the system of base sets. After defining the base sets, the user can select a set of points for approximation.

In the software. the user has the option to insert the members of the left-out singleton clusters to any base set. Two singleton clusters cannot be merged together due to the tolerance relation based on similarity (their members are different). It was mentioned earlier that there are two types of singletons:

- Its member is different from most of the objects so it forms a cluster alone.
- Due to the background knowledge the system decided that this object cannot be a member of any other group.

The software does not examine for a singleton to which type it belongs, so there is no mandatory annotation for any singleton. It is up to the user to decide.

References

1. Aigner, M.: Enumeration via ballot numbers. Discrete Math. **308**(12), 2544–2563 (2008). https://doi.org/10.1016/j.disc.2007.06.012, http://www.sciencedirect.com/science/article/pii/S0012365X07004542
2. Altman, A., Tennenholtz, M.: Ranking systems: the PageRank axioms. In: Proceedings of the 6th ACM Conference on Electronic Commerce, pp. 1–8. ACM (2005)
3. Aszalós, L., Bakó, M.: Advanced search methods (in Hungarian) (2012). http://morse.inf.unideb.hu/~aszalos/diak/fka
4. Aszalós, L., Hajdu, L., Pethő, A.: On a correlational clustering of integers. Indagationes Mathematicae **27**(1), 173–191 (2016)
5. Aszalós, L., Nagy, D.: Visualization of tolerance relations. In: Gábor, Kusper; Roland, K. (ed.) Proceedings of the 10th International Conference on Applied Informatics, pp. 15–22 (2018)
6. Aszalós, L., Nagy, D.: Iterative set approximations based on tolerance relation. In: Mihálydeák, T., et al. (eds.) Rough Sets, pp. 78–90. Springer, Cham (2019). https://doi.org/10.1007/978-3-030-22815-6_7
7. Aszalós, L., Nagy, D.: Selecting representatives. In: Ganzha, M., Maciaszek, L., Paprzycki, M. (eds.) Communication Papers of the 2019 Federated Conference on Computer Science and Information Systems. Annals of Computer Science and Information Systems, vol. 20, pp. 13–19. PTI (2019). https://doi.org/10.15439/2019F95
8. Bansal, N., Blum, A., Chawla, S.: Correlation clustering. Mach. Learn. **56**(1–3), 89–113 (2004)
9. Becker, H.: A survey of correlation clustering. Advanced Topics in Computational Learning Theory, pp. 1–10 (2005)
10. Bello, R., Falcon, R.: Rough Sets in Machine Learning: A Review, pp. 87–118. Springer, Cham (2017). https://doi.org/10.1007/978-3-319-54966-8_5
11. Brownlee, J.: Clever Algorithms: Nature-Inspired Programming Recipes. 1st edn. Lulu.com (2011)

12. Caballero, Y., Bello, R., Alvarez, D., Gareia, M.M., Pizano, Y.: Improving the K-NN method: rough set in edit training set. In: Debenham, J. (ed.) Professional Practice in Artificial Intelligence, pp. 21–30. Springer, Boston (2006)
13. Ciucci, D., Mihálydeák, T., Csajbók, Z.E.: On Definability and Approximations in Partial Approximation Spaces, pp. 15–26. Springer, Cham (2014). https://doi.org/10.1007/978-3-319-11740-9_2
14. Csajbók, Z., Mihálydeák, T.: From vagueness to rough sets in partial approximation spaces. In: JRS2014 (submitted, 2014)
15. Csajbók, Z., Mihálydeák, T.: A general set theoretic approximation framework. In: Greco, S., Bouchon-Meunier, B., Coletti, G., Fedrizzi, M., Matarazzo, B., Yager, R. (eds.) Advances on Computational Intelligence, Communications in Computer and Information Science, vol. 297, pp. 604–612. Springer, Heidelberg (2012). https://doi.org/10.1007/978-3-642-31709-5_61
16. Csajbók, Z., Mihálydeák, T.: Partial approximative set theory: a generalization of the rough set theory. Int. J. Comput. Inf. Syst. Ind. Manag. Appl. **4**, 437–444 (2012)
17. Delic, D., Lenz, H.J., Neiling, M.: Improving the quality of association rule mining by means of rough sets. In: Grzegorzewski, P., Hryniewicz, O., Gil, M.Á. (eds.) Soft Methods in Probability, Statistics and Data Analysis, pp. 281–288. Physica-Verlag HD, Heidelberg (2002)
18. Dubois, D., Prade, H.: Twofold fuzzy sets and rough sets–some issues in knowledge representation. Fuzzy Sets Syst. **23**(1), 3–18 (1987). https://doi.org/10.1016/0165-0114(87)90096-0, http://www.sciencedirect.com/science/article/pii/0165011487900960. fuzzy Information Processing in Artificial Intelligence and Operations Research
19. Dubois, D., Prade, H.: Rough fuzzy sets and fuzzy rough sets*. Int. J. Gener. Syst. **17**(2–3), 191–209 (1990). https://doi.org/10.1080/03081079008935107
20. Düntsch, I., Gediga, G.: Approximation Operators in Qualitative Data Analysis, pp. 214–230. Springer, Heidelberg (2003). https://doi.org/10.1007/978-3-540-24615-2_10
21. Dávid, N., László Aszalós, T.M.: Finding the representative in a cluster using correlation clustering. Pollack Periodica **14**(1), 15–24 (2019). https://doi.org/10.1556/606.2019.14.1.2
22. Erdös, P., Rényi, A.: On random graphs I. Publ. Math. Debrecen **6**, 290 (1959)
23. Erdős, P., Rényi, A.: On the evolution of random graphs. In: Publication of the Mathematical Institute of the Hungarian Academy of Sciences, pp. 17–61 (1960)
24. Filiberto, Y., Caballero, Y., Larrua, R., Bello, R.: A method to build similarity relations into extended rough set theory. In: 2010 10th International Conference on Intelligent Systems Design and Applications, pp. 1314–1319, November 2010. https://doi.org/10.1109/ISDA.2010.5687091
25. Gogoi, P., Bhattacharyya, D.K., Kalita, J.K.: A rough set-based effective rule generation method for classification with an application in intrusion detection. Int. J. Secur. Netw. **8**(2), 61–71 (2013). https://doi.org/10.1504/IJSN.2013.055939
26. Greco, S., Matarazzo, B., Słowiński, R.: Parameterized rough set model using rough membership and bayesian confirmation measures. Int. J. Approx. Reason. **49**(2), 285–300 (2008). https://doi.org/10.1016/j.ijar.2007.05.018, http://www.sciencedirect.com/science/article/pii/S0888613X0700151X. special Section on Probabilistic Rough Sets and Special Section on PGM'06
27. Grzymala-Busse, J.W.: LERS-A System for Learning from Examples Based on Rough Sets, pp. 3–18. Springer, Dordrecht (1992). https://doi.org/10.1007/978-94-015-7975-9_1

28. Guan, J.W., Bell, D.A., Liu, D.Y.: The rough set approach to association rule mining. In: Third IEEE International Conference on Data Mining, pp. 529–532, November 2003. https://doi.org/10.1109/ICDM.2003.1250969

29. Hu, Q., Yu, D., Liu, J., Wu, C.: Neighborhood rough set based heterogeneous feature subset selection. Inf. Sci. **178**(18), 3577–3594 (2008). https://doi.org/10.1016/j.ins.2008.05.024, http://www.sciencedirect.com/science/article/pii/S0020025508001643

30. Janusz, A.: Algorithms for similarity relation learning from high dimensional data. Trans. Rough Sets **17**, 174–292 (2014)

31. Lenarcik, A., Piasta, Z.: Discretization of Condition Attributes Space, pp. 373–389. Springer, Dordrecht (1992). https://doi.org/10.1007/978-94-015-7975-9_23

32. Mani, A.: Choice inclusive general rough semantics. Inf. Sci. **181**(6), 1097–1115 (2011)

33. Mihálydeák, T.: Logic on similarity based rough sets. In: Nguyen, H.S., Ha, Q.T., Li, T., Przybyła-Kasperek, M. (eds.) Rough Sets, pp. 270–283. Springer, Cham (2018). https://doi.org/10.1007/978-3-319-99368-3_21

34. Mises, R.V., Pollaczek-Geiringer, H.: Praktische verfahren der gleichungsauflösung. ZAMM - Journal of Applied Mathematics and Mechanics / Zeitschrift für Angewandte Mathematik und Mechanik **9**(2), 152–164 (1929). https://doi.org/10.1002/zamm.19290090206, https://onlinelibrary.wiley.com/doi/abs/10.1002/zamm.19290090206

35. Nagy, D., Aszalós, L.: Approximation based on representatives. In: Mihálydeák, T., et al. (eds.) Rough Sets, pp. 91–101. Springer, Cham (2019). https://doi.org/10.1007/978-3-642-31900-6_48

36. Nagy, D., Mihálydeák, T., Aszalós, L.: Similarity Based Rough Sets, pp. 94–107. Springer, Cham (2017). https://doi.org/10.1007/978-3-319-60840-2_7

37. Nagy, D., Mihálydeák, T., Aszalós, L.: Similarity based rough sets with annotation. In: Nguyen, H.S., Ha, Q.T., Li, T., Przybyła-Kasperek, M. (eds.) Rough Sets, pp. 88–100. Springer, Cham (2018). https://doi.org/10.1007/978-3-319-99368-3_7

38. Nagy, D., Mihalydeak, T., Aszalos, L.: Different types of search algorithms for rough sets. Acta Cybern. **24**(1), 105–120 (2019). https://doi.org/10.14232/actacyb.24.1.2019.8, http://cyber.bibl.u-szeged.hu/index.php/actcybern/article/view/3999

39. Néda, Z., Sumi, R., Ercsey-Ravasz, M., Varga, M., Molnár, B., Cseh, G.: Correlation clustering on networks. J. Phys. A: Math. Theor. **42**(34), 345003 (2009). https://doi.org/10.1088/1751-8113/42/34/345003, http://www.journalogy.net/Publication/18892707/correlation-clustering-on-networks

40. Néda, Z., Sumi, R., Ercsey-Ravasz, M., Varga, M., Molnár, B., Cseh, G.: Correlation clustering on networks. J. Phys. A: Math. Theor. **42**(34), 345003 (2009)

41. Nguyen, H.S.: Discretization problem for rough sets methods. In: Polkowski, L., Skowron, A. (eds.) Rough Sets Curr. Trends Comput., pp. 545–552. Springer, Heidelberg (1998). https://doi.org/10.1007/3-540-69115-4_75

42. Pawlak, Z.: Rough sets. Int. J. Parallel Prog. **11**(5), 341–356 (1982)

43. Pawlak, Z., Skowron, A.: Rudiments of rough sets. Inf. Sci. **177**(1), 3–27 (2007)

44. Pawlak, Z., Wong, S., Ziarko, W.: Rough sets: probabilistic versus deterministic approach. Int. J. Man-Mach. Stud. **29**(1), 81–95 (1988). https://doi.org/10.1016/S0020-7373(88)80032-4

45. Pawlak, Z., et al.: Rough sets: theoretical aspects of reasoning about data. In: System Theory, Knowledge Engineering and Problem Solving, Kluwer Academic Publishers, Dordrecht, 1991, September 1991

46. Pawlak, Z., Skowron, A.: Rough sets: some extensions. Inf. Sci. **177**(1), 28–40 (2007). https://doi.org/10.1016/j.ins.2006.06.006, http://www.sciencedirect.com/science/article/pii/S0020025506001496, zdzisław Pawlak life and work (1926–2006)
47. Polkowski, L., Skowron, A., Zytkow, J.: Rough foundations for rough sets soft computing: rough sets, fuzzy logic, neural networks, uncertainty management, knowledge discovery, pp. 55–58, May 2020
48. do Prado, H.A., Engel, P.M., Filho, H.C.: Rough clustering: an alternative to find meaningful clusters by using the reducts from a dataset. In: Alpigini, J.J., Peters, J.F., Skowron, A., Zhong, N. (eds.) RSCTC 2002. LNCS (LNAI), vol. 2475, pp. 234–238. Springer, Heidelberg (2002). https://doi.org/10.1007/3-540-45813-1_30
49. Skowron, A., Stepaniuk, J.: Tolerance approximation spaces. Fundamenta Informaticae **27**(2), 245–253 (1996)
50. Ślęzak, D., Ziarko, W.: The investigation of the bayesian rough set model. Int. J. Approx. Reason. **40**(1), 81–91 (2005). https://doi.org/10.1016/j.ijar.2004.11.004, http://www.sciencedirect.com/science/article/pii/S0888613X04001410. data Mining and Granular Computing
51. Słowiński, R., Vanderpooten, D.: A generalized definition of rough approximations based on similarity. IEEE Trans. Knowl. Data Eng. **12**(2), 331–336 (2000)
52. Słowiński, R., Vanderpooten, D.: Similarity relation as a basis for rough approximations. Adv. Mach. Intell. Soft-Comput. **4**, 17–33 (1997)
53. Swiniarski, R.W., Skowron, A.: Rough set methods in feature selection and recognition. Pattern Recogn. Lett. **24**(6), 833–849 (2003). https://doi.org/10.1016/S0167-8655(02)00196-4, http://www.sciencedirect.com/science/article/pii/S0167865502001964
54. Yang, Y., Chen, D., Dong, Z.: Novel algorithms of attribute reduction with variable precision rough set model. Neurocomputing **139**, 336–344 (2014). https://doi.org/10.1016/j.neucom.2014.02.023
55. Yao, Y.Y.: Combination of Rough and Fuzzy Sets Based on α-Level Sets, pp. 301–321. Springer, Boston (1997). https://doi.org/10.1007/978-1-4613-1461-5_15
56. Yao, Y.: Decision-theoretic rough set models. In: Yao, J., Lingras, P., Wu, W.Z., Szczuka, M., Cercone, N.J., Ślęzak, D. (eds.) Rough Sets and Knowledge Technology, pp. 1–12. Springer, Heidelberg (2007). https://doi.org/10.1007/978-3-540-72458-2_1
57. Yao, Y., Greco, S., Słowiński, R.: Probabilistic Rough Sets, pp. 387–411. Springer, Heidelberg (2015). https://doi.org/10.1007/978-3-662-43505-2_24
58. Yao, Y., Yao, B.: Covering based rough set approximations. Information Sciences **200**, 91–107 (2012). https://doi.org/10.1016/j.ins.2012.02.065, http://www.sciencedirect.com/science/article/pii/S0020025512001934
59. Zakowski, W.: Approximations in the space (u,π). Demonstratio Math. **16**(3), 761–770 (1983). https://doi.org/10.1515/dema-1983-0319
60. Zimek, A.: Correlation clustering. ACM SIGKDD Explor. Newslett. **11**(1), 53–54 (2009)

Author Index

Printed in the United States
By Bookmasters